Dynamics and Transport in Macromolecular Networks

Dynamics and Transport in Macromolecular Networks

Theory, Modeling, and Experiments

Edited by Li-Tang Yan

WILEY-VCH

Editor

Li-Tang Yan
State Key Laboratory of Chemical Engineering
Department of Chemical Engineering
Tsinghua University
China

Cover Image: © cybrain/Shutterstock

Library of Congress Card No.: applied for

British Library Cataloguing-in-Publication Data
A catalogue record for this book is available from the British Library.

Bibliographic information published by the Deutsche Nationalbibliothek
The Deutsche Nationalbibliothek lists this publication in the Deutsche Nationalbibliografie; detailed bibliographic data are available on the Internet at <http://dnb.d-nb.de>.

Print ISBN: 978-3-527-35098-8
ePDF ISBN: 978-3-527-83954-4
ePub ISBN: 978-3-527-83955-1
oBook ISBN: 978-3-527-83956-8

Typesetting: Straive, Chennai, India

Contents

Preface

Macromolecular networks form the structural basis of many important classes of materials. For example, macromolecular polymer networks are among the most widely used polymeric materials, with applications in rubbers, cosmetics, adhesives, medical devices, soft actuators, membranes, gas storage, catalysis, and electronic materials. The networks of biomacromolecules also form principal structural components throughout biology, from the intracellular scaffold known as the cytoskeleton to extracellular matrices of collagen. Hence, understanding the dynamics of nanoscale particles or objects in the confined environments of networks is important in many fields, ranging from polymer network-based nanocomposites with advantageous properties, drug delivery through mucosal tissues, cellular manipulation by designed DNA hydrogels, to controlled exchange of molecules in biological gels such as extracellular matrices. However, in order to successfully exploit them in technological applications and to ensure efficient scale-up, an in-depth understanding of the structure organization, dynamic mechanism, and structure–property relationship is not only required but also essential for their development. In particular, experimental studies and theoretical modeling offer diverse approaches to investigate the network structures as well as to determine their effects on the dynamic behaviors of particles in such networks. In light of the growing interest in the transport behaviors of various objects in macromolecular networks, the aim of this book is to provide a unique introduction to the currently emerging, highly interdisciplinary field of those transport processes that exhibit various dynamic patterns and even anomalous behaviors of dynamics. In so doing, it extracts and emphasizes common principles and research advancement from various disciplines while providing up-to-date coverage of this new field of research.

This book begins with discussions on the structures and dynamics of some typical types of macromolecular or biomacromolecular networks. In Chapter 1, Meng introduces various theoretical models based on statistical descriptions of polymer networks and the phenomenological ones well matching with experimental measurements. The development and application of these models significantly prompt the understanding of the elasticity/viscoelasticity of permanently/transiently cross-linked macromolecular networks. In addition to the theoretical models, computer simulations play crucial roles in understanding the structures and dynamics of macromolecular networks. In Chapter 2, Palkar and Kuksenok develop molecular simulation methods based on dissipative particle dynamics (DPD) to model reactive hydrogels including characterization of erosion and reverse gelation in degrading networks, where they also survey the continuum modeling of the reactive hydrogels with the focus on modeling of degradation of hydrogels on the continuum length scales.

Recently, new experimental approaches have also been developed to understand the structures and dynamics of macromolecular or biomacromolecular networks. In Chapter 3, Cao and coworkers present the recent progress achieved in experimentally studying the dynamic bonds in associating polymer networks, including fundamental understanding of the relaxation and morphology of these bonds as well as the emergent design of dynamic polymer materials with the superior performance and energy-related applications. The contents conceived in this chapter are expected to inspire readers/researchers to develop new dynamic polymer systems with advanced dynamic features as well as to synergize with the presented dynamic networks. Seeing is believing – this simple intuition promotes researchers to directly visualize polymer dynamics in entangled solutions or cross-linked networks, and the developments in optical imaging with enough spatial and temporal resolutions have recently enabled us to do so in real space and on real time. In Chapter 4, Zhou and Jiang give a nice summary regarding this important aspect.

The advanced applications based on the tailored structures and dynamics of macromolecular or biomacromolecular networks are particularly concerned in this book. For example, in Chapter 5, Liu and coworkers introduce the microstructures and mechanical properties of the meniscus and summarize the biomaterial requirements for constructing meniscal scaffolds, inspiring new ideas and facilitating the design of new-generation meniscal scaffolds. Load-bearing biological tissues, such as muscles and cartilages, exhibit a unique combination of properties that include high elasticity, exceptional toughness, and swift recovery. In Chapter 6, Xue and Cao delve into novel methods for constructing hydrogels that exhibit a desirable combination of strength, toughness, and quick recovery, emulating the attributes in synthetic biomaterials of load-bearing biological tissues.

The content then moves on to computational and theoretical aspects of transport and dynamics in the macromolecular or biomacromolecular networks. Three topics are reviewed and consolidated. The first topic is the applications of theoretical and simulation to the transport of small-molecule solvents in macromolecular networks, which are described in Chapter 7. In this chapter, Man and coworkers review the history of the diffusion–mechanical coupling theory of the swelling of gel—one of the typical systems of macromolecular networks. The general dynamical models for a spherical gel, a cylindrical gel, and a disk-shaped gel are given based on Onsager variational principle calculations, and then the behaviors of the three types of gels for small deformation are discussed, separately.

The second topic in this section is the application and development of theoretical approaches to the particle transport in macromolecular polymer networks, which are described in two chapters. In Chapter 8, Ge presents a comprehensive description of theoretical and computational perspective on hopping diffusion of nanoparticles in cross-linked polymer networks, where the theories of nonsticky spherical nanoparticles in cross-linked flexible polymers as well as entropic origin of the hopping energy barrier are particularly emphasized. In Chapter 9, contributed by Chen and Liu, the particle diffusion in one of the important macromolecular polymer networks, that is, the polymer elastomer, is concerned. It provides useful knowledge on the structures and dynamics of some typical elastomer systems and inspires the development of next-generation high-performance elastomer-based materials.

The third topic in this section turns to the theory and simulation studies of the nanoparticle transport in biomacromolecular networks. In Chapter 10, Tian and Shi discuss the impact of two parameters, shape and stiffness of nanoparticle on their diffusivities in mucus—one of the typical biomacromolecular networks, elucidate the underlying mechanisms, and establish theoretical regulatory models. The findings conceived in this chapter can provide guidance and theoretical support for the rational design of novel carriers. In the last chapters, the theoretical and simulation results regarding this topic in our group are also delineated.

Written by specialists in various disciplines, such as polymer, soft matter, mechanics, and biomedicine, this book provides comprehensive knowledge in this emerging and important field, although covering all aspects of the field is impossible. From the editor's part, I am very pleased and honored with the theoretical and experimental approaches, principles, and perspectives conceived in each chapter and would also like to express my most heartfelt thanks to all contributors who delivered truly excellent topics of research and took the time to write up detailed and pedagogical chapters. We are also very grateful to Lifen Yang, Katherine Wong, and Manoj Kumar Mohanasundaram from the Wiley-VCH for helpful suggestions and excellent cooperation.

June 2023

Li-Tang Yan
Tsinghua University

1

Modeling (Visco)elasticity of Macromolecular and Biomacromolecular Networks

Fanlong Meng[1,2,3]

[1] *Chinese Academy of Sciences, CAS Key Laboratory for Theoretical Physics, Institute of Theoretical Physics, Beijing 100190, China*
[2] *University of Chinese Academy of Sciences, School of Physical Sciences, Beijing 100049, China*
[3] *University of Chinese Academy of Sciences, Wenzhou Institute, Wenzhou 325000, China*

When polymers are permanently crosslinked into a network, e.g., by covalent bonds which possess a very high energy barrier to break, then the network is elastic and can sustain mechanical loadings, resulting from the conformational entropy of the polymers if not considering other specific intramolecular/intermolecular interactions. By replacing the permanent crosslinks with the transient ones (low energy barriers to break and/or re-form), which can be either physical crosslinks formed by physical interactions (hydrogen bonds, guest-host interactions, hydrophobicity, etc.) or chemical ones whose breakage energy barrier is low (e.g., by adding catalysts), then the crosslinked network can dynamically re-organize its topological structure by the crosslink breakage/re-formation; the network can be treated as complex fluids, exhibiting interesting viscoelastic properties. Due to their characteristic rheological properties, the products made of the macromolecular networks (both permanently and transiently crosslinked ones) are widely utilised in various applications such as rubber bands, tires, self-healing materials, etc. As a special class of the macromolecular networks, the biomacromolecular networks are ubiquitous in nature such as cytoskeletons and extracellular matrices, which are relevant with various bio-functions including shape maintenance of cells, cell division, and movements. In the following, we will introduce the well-known theories of permanent macromolecular networks consisting of flexible polymers in Section 1.1 and those of permanent biomacromolecular networks consisting of semiflexible polymers in Section 1.2, discuss the viscoelastic responses of transient macromolecular networks in Section 1.3, and then finish this chapter with a brief discussion about some possible developments in the future.

Dynamics and Transport in Macromolecular Networks: Theory, Modeling, and Experiments,
First Edition. Edited by Li-Tang Yan.

1.1 Permanent Macromolecular Networks

As polymers are the main entity of the macromolecular network, the physical properties of *a single polymer* are very important in determining the overall responses of the network.

1.1.1 Mechanic Properties of a Single Polymer Chain

For a single polymer, one can define a *persistence length* [1, 2] to quantify the bending rigidity of the polymer ($l_p = \kappa/k_BT$, with κ as the bending rigidity, k_B as the Boltzmann constant, and T as the temperature), over which the correlations in the tangent direction along the polymer are lost. A relevant physical quantity is *Kuhn length* [1], which is $l_k = 2l_p$, over which the polymer can be treated as freely joint. By comparing the contour length l_c and persistence length l_p (as shown in Figure 1.1), polymers can be categorized into flexible ($l_c \gg l_p$), semiflexible ($l_c \simeq l_p$) and rigid ($l_c \ll l_p$) ones. Usually, synthetic polymers such as polyethylene can be taken as flexible and bio-polymers of interested lengths such as microtubules are semiflexible. There exist many famous models for describing the mechanic properties of a single polymer [1, 3–6], and three of them are listed here, two for flexible polymers and one for semiflexible polymers.

- **Gaussian chain**. By assuming a flexible polymer with a bead-spring structure (shown in Figure 1.1a) where the lengths of the springs can fluctuate obeying a Gaussian distribution and the orientations of the springs are independent of the others, one can get the probability of finding the polymer which consists of N monomers (or springs in the bead-spring description) with the end-to-end vector, $\boldsymbol{R}' - \boldsymbol{R}$, as shown in Figure 1.1 [3], as: $P(\boldsymbol{R}' - \boldsymbol{R}; N) = \left(\frac{3}{2\pi Nb^2}\right)^{3/2} \exp\left[-\frac{3(\boldsymbol{R}'-\boldsymbol{R})^2}{2Nb^2}\right]$, where b is the monomer size. Then the free energy of the polymer as a function of the end-to-end vector, $\boldsymbol{R}' - \boldsymbol{R}$, is: $F_{gc} = \frac{3k_BT}{2Nb^2}(\boldsymbol{R}' - \boldsymbol{R})^2$, which describes the Gaussian chain as a Hookean spring with the elastic constant: $k = 3k_BT/Nb^2$. Straightforwardly, one can obtain the force-extension relation as:

$$f = \frac{3k_BT}{Nb^2}|\boldsymbol{R}' - \boldsymbol{R}|. \tag{1.1}$$

  As noted, the Gaussian chain can be stretched infinitely with $|\boldsymbol{R}' - \boldsymbol{R}| \to \infty$ if the tensile force is large enough, which is not correct for realistic polymers. Actually, this model can only describe the cases where the polymer undergoes small deformations rather than finite ones, i.e., $|\boldsymbol{R}' - \boldsymbol{R}| \ll l_c$.

Figure 1.1 Illustration of a (a) flexible polymer ($l_c \gg l_p$) (the inset shows the bead-spring structure), (b) semiflexible polymer ($l_c \simeq l_p$), and (c) rigid polymer ($l_c \ll l_p$).

- **Langevin chain**. For a flexible polymer chain which is stretched finitely, $|\boldsymbol{R}' - \boldsymbol{R}| \sim l_c$, one can adopt the freely joint chain model, where the length of the springs in Figure 1.1a is a constant b (infinitely rigid and not deformable) and the orientation of each spring is independent of others. Then the force-extension relation can be obtained as [6]:

$$|\boldsymbol{R}' - \boldsymbol{R}| = Nb\,\mathcal{L}(fb/k_\mathrm{B}T), \tag{1.2}$$

where $\mathcal{L}(x) = \coth(x) - 1/x$ is the Langevin function. One can easily show that Equation (1.2) reduces to Equation (1.1) if $|\boldsymbol{R}' - \boldsymbol{R}| \ll l_c$.
- **Semiflexible chain**. For semiflexible polymer chains, one needs to consider the bending rigidity (the orientations of the springs are dependent on those of their neighbors if still taking the bead-spring picture for an intuitive understanding). A semiflexible polymer of contour length l_c can be coordinated as $\boldsymbol{r}(s)$ with $0 \leq s \leq l_c$ the arc length coordinate along the polymer, and the bending energy of the semiflexible polymer can be written as: $E_B = \frac{\kappa}{2}\int_0^{l_c} ds \left[\frac{d^2\boldsymbol{r}(s)}{ds^2}\right]^2$ where κ is the bending modulus, and $d^2\boldsymbol{r}(s)/ds^2$ is the local curvature at s. There are different models reviewed in Ref. [7, 8] for describing the elasticity of a single semiflexible polymer, such as *Marko-Siggia model* [9], *Ha-Thirumalai model* [10], *MacKintosh-Käs-Janmey model* [11], and *Blundell-Terentjev model* (shown below as an example) [12]. By taking the mean inextensibility assumption, one can obtain the free energy of a semiflexible polymer as a function of the end-to-end factor of the polymer $x = |\boldsymbol{R}' - \boldsymbol{R}|/l_c$, approximated as (details in Ref. [12]): $F_{sc}(x) = k_\mathrm{B}T\pi^2 c\left(1 - x^2\right) + \frac{k_\mathrm{B}T}{\pi c(1-x^2)}$, where $c = l_p/2l_c = \kappa/2k_\mathrm{B}Tl_c$ describes the bending rigidity of the polymer. From this, one can obtain the force-extension relation as:

$$f = \frac{2k_\mathrm{B}T}{l_c}\left[-\pi^2 c + \frac{1}{\pi c\left(1 - x^2\right)^2}\right] x. \tag{1.3}$$

Obviously, the force diverges at large $x \to 1$ when the polymer is stretched to its length limit.

With the mechanical models of a single polymer as exemplified above, one can try to construct theoretical models for macromolecular networks by taking account of the crosslinked structure of the network. The theories for a permanently crosslinked network have been developed for a relatively long time, and there are many successful models as reviewed in Ref. [13, 14]. The theoretical models of permanently crosslinked macromolecular networks can be roughly categorised into two types: statistical ones based on assumed network structures (Section 1.1.2) and phenomenological ones with fitting parameters well matching experimental observations (Section 1.1.3).

1.1.2 Statistical Models

With the mechanic models of a single polymer as shown above, one can try to construct the constitutive models of a macromolecular network with proper assumptions of the network structure. Here, we introduce four models with different network architectures, as shown in Figure 1.2.

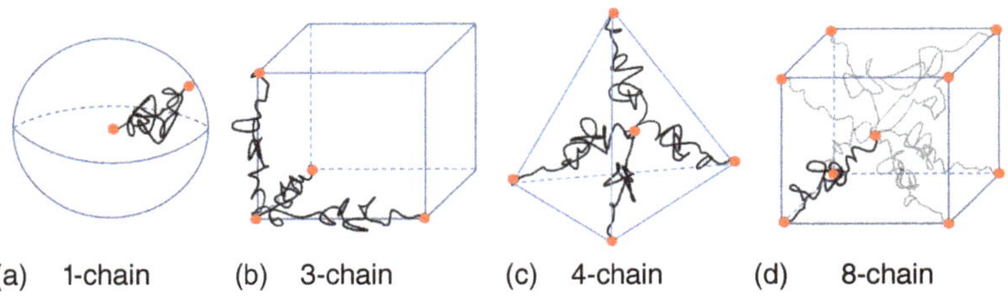

Figure 1.2 Four different architectures of macromolecular networks: (a) 1-chain model (full network model), (b) 3-chain model, (c) 4-chain model, (d) 8-chain model.

- **1-chain model (also called as full network model).** 1-chain model may be the most straightforward assumption of the network architecture, where the orientation of the polymer connecting two neighboring crosslinks (one located at the center of the sphere and the other at the sphere surface) is randomly distributed [15]. Then, the free energy density of the macromolecular network can be obtained by averaging the energy contributions of all polymers. Suppose the network is uniformly deformed, and the material point located at $\boldsymbol{r}$ is displaced to a new location $\boldsymbol{r}' = \mathbf{E} \cdot \boldsymbol{r}$ where $\mathbf{E}$ is the deformation gradient tensor. Note that $\mathbf{E}$ can be written as $\mathbf{E} = \mathbf{Q} \cdot \mathbf{S}$ with $\mathbf{Q}$ as an orthogonal tensor and $\mathbf{S}$ as a diagonal tensor; in other words, the deformation can be decomposed into a stretch denoted by $\mathbf{S}$ and a rotation denoted by $\mathbf{Q}$. The diagonal components ($\lambda_{1,2,3}$) of $\mathbf{S}$, satisfy the relation: $I_1 = \lambda_1^2 + \lambda_2^2 + \lambda_3^2 = E_{\alpha\beta}^2$ where I_1 is the first invariant which will be used later. Upon deformations, the sphere of radius ξ (denoting the mesh size of the network) will be deformed into an ellipsoid, where the lengths of the three semi-axes become $\lambda_1\xi$, $\lambda_2\xi$, and $\lambda_3\xi$, respectively. The energy density of the network can be calculated as:

$$F_{1c} = n_c \int \sin\theta d\theta\, d\varphi F_{\text{chain}}[\lambda(\theta,\varphi)\,\xi]/4\pi, \tag{1.4}$$

where n_c is the density of the subchains (polymers connecting two neighboring crosslinks), $\lambda(\theta,\varphi) = \sqrt{\sin^2\theta(\cos^2\varphi\lambda_1^2 + \sin^2\varphi\lambda_2^2) + \cos^2\theta\lambda_3^2}$ is the deformation ratio at orientation (θ,φ), and F_{chain} is the energy density of a single polymer, which can be that of Gaussian chain model, Langevin chain model, semiflexible chain model, etc. Then, by averaging the contributions of all polymers, one can obtain the free energy density of the macromolecular network. However, it is usually not possible to obtain an analytic form of the free energy of the network consisting of the non-Gaussian polymers due to the complexity lying in the integration of the polar and azimuthal angles in the spherical coordinate system. The exception is the macromolecular networks of Gaussian polymers ($F_{\text{chain}} \propto \frac{3}{2}k_B T\lambda^2$), of which the free energy density can be explicitly obtained as:

$$F_{1c} = \frac{1}{2}G\left(\lambda_1^2 + \lambda_2^2 + \lambda_3^2 - 3\right), \tag{1.5}$$

where G is the shear modulus of the material as a function of n_c; this energy form is also called as neo-Hookean model.

- **3-chain model.** In the 3-chain model [16], a non-deformed macromolecular network is composed of repeated cubic cells; in each cell, the cell edges of length l_0 denote polymer chains, and the cell vertices denote crosslinks. Upon deformations, the cubic becomes a cuboid, where the directions of the edges are along the principle directions of the deformation and the lengths of three orthogonal edges become $\lambda_1 l_0$, $\lambda_2 l_0$, $\lambda_3 l_0$, respectively. Then, by averaging the energy contributions of these three orthogonal chains, one can obtain an analytical form of the free energy density of the macromolecular networks as:

$$F_{3c} = \frac{1}{3} n_c \sum_i F_{\text{chain}}(\lambda_i). \tag{1.6}$$

  Note that the three orthogonal chains can be deformed with different stretch ratios. The advantage of the 3-chain model (also of 8-chain model) is that: one can write down the analytic form of the free energy density of the macromolecular network as long as the explicit form of the free energy of a single chain exists.
- **4-chain model.** In the 4-chain model, the macromolecular network is composed of repeated regular tetrahedrons before deformation, which was first proposed by Flory and Rehner [17]. In each tetrahedron, there are five crosslinks (four at the vertices and one in the tetrahedron body whose position can fluctuate) and four chains (connecting the center crosslink and each vertex crosslink). Upon deformation, the four vertices are deformed to new positions affinely with the applied deformation. The free energy density of the network can be obtained by averaging the contributions of the four deformed chains. However, due to several factors such as the non-affine displacement of the center crosslink, fluctuation of the center crosslink relying on the strains, etc., one can not write down an analytical form of the free energy density of the macromolecular networks, which needs to be calculated numerically. We shall not introduce the calculations in detail, and interested readers can refer to Ref. [13, 18].
- **8-chain model.** Similarly as the 3-chain model, the macromolecular network is also treated as repeated cubes before deformation and the cubes becomes cuboid after deformation in the 8-chain model [19]. The difference between the 8-chain model and 3-chain model lies in the chain and crosslink arrangements in the cube: there are nine crosslinks in each cube for the 8-chain model (eight at the cube vertices and one at cube center) and eight polymer chains connecting the center crosslink and each vertex crosslink (no chain along the edge anymore). Note that all polymers are deformed identically in this model and their energy is a function of the stretch ratios along the three orthogonal directions. The free energy density of the macromolecular network in this model is simply:

$$F_{8c} = n_c F_{\text{chain}}(\lambda_{1,2,3}), \tag{1.7}$$

  where F_{chain} is chosen according to the properties of a single polymer as in other models.

One can also adapt the network architectures to meet realistic demands in theoretical modeling, for example, one can show the microscopic physical picture in Mullins effect by introducing changes in the chain contour lengths with deformations [20]. Also, by introducing a nonuniform size distribution of the unit cell or proposing other more realistic network architectures, one can try to study the non-affine deformation in real macromolecular networks.

1.1.3 Phenomenological Models

There are many phenomenological models, which are constructed with fitting parameters to match the experimental observations and are usually portable to use. Here, we shall briefly introduce a very useful class: invariants-based continuum models. For describing the deformation of the material, we have introduced the deformation gradient tensor in the above section 1.1.2, based on which one can define the left Cauchy-Green deformation tensor as: $\mathbf{B} = \mathbf{E}\mathbf{E}^{\mathrm{T}}$. For isotropic materials, we can obtain the principle invariants [scalar valued function $I(\mathbf{B})$ with $I(\mathbf{B}) = I(\mathbf{Q} \cdot \mathbf{B} \cdot \mathbf{Q}^T)$ for all orthogonal tensor $\mathbf{Q}$] of the left Cauchy-Green deformation tensor:

$$I_1 = \mathrm{tr}\mathbf{B} = \lambda_1^2 + \lambda_2^2 + \lambda_3^2, \tag{1.8}$$

$$I_2 = \frac{1}{2}\left[(\mathrm{tr}\mathbf{B})^2 - \mathrm{tr}\mathbf{B}^2\right] = \lambda_1^2\lambda_2^2 + \lambda_2^2\lambda_3^2 + \lambda_3^2\lambda_1^2, \tag{1.9}$$

$$I_3 = \det\mathbf{B} = \lambda_1^2\lambda_2^2\lambda_3^2. \tag{1.10}$$

I_3 is equal to 1 for incompressible materials, which is usually the case for rubbers. Thus, the free energy of incompressible materials can be constructed in terms of I_1 and I_2. Rivlin proposed a general form of the free energy density based on the invariants, I_1 and I_2 [21]:

$$F_{\mathrm{R}} = \sum_{ij} C_{ij}(I_1 - 3)^i(I_2 - 3)^j, \tag{1.11}$$

where the coefficients C_{ij} are usually phenomenological parameters to be fitted with experimental data. It can act as a framework guiding how to construct continuum models based on invariants.

- **Neo-Hookean model.** From the above expression (Eq. (1.11)), one can immediately recognize that the case of $C_{10} \neq 0$ and other $C_{ij} = 0$ corresponds to the neo-Hookean model obtained above from statistical arguments,

$$F_{\mathrm{nH}} = C_{10}(I_1 - 3), \tag{1.12}$$

  and the phenomenological parameter $C_{10} = n_c k_B T/2$ has its microscopic origin as described above.
- **Mooney–Rivlin model.** Another simple example is Mooney–Rivlin model, which reads as [22]:

$$F_{\mathrm{MR}} = C_{10}(I_1 - 3) + C_{01}(I_2 - 3), \tag{1.13}$$

  and it is also widely used for mechanics of rubber materials at small deformations.

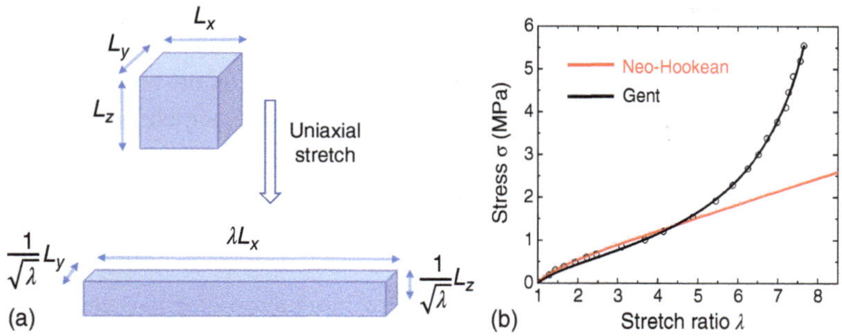

Figure 1.3 (a) Uniaxial stretch test, where the stretch ratio along the stretch direction is λ and the stretch ratios along the other two orthogonal directions are identical, which is equal to $1/\sqrt{\lambda}$ due to incompressibility of the material. (b) Stress–strain relation of the uniaxial stretch test, where the symbols denote the data extracted from the experiment. Source: Adapted from Treloar [13]. The red curve is fitted with the neo-Hookean model (fitting parameter $G \approx 0.30$ MPa), and the black curve is fitted with Gent model (fitting parameters $G \approx 0.25$ MPa and $J_m \approx 85$).

- **Gent model.** When macromolecular networks undergo large deformations, there is a very good phenomenological model for describing the mechanics of them–Gent model [23], with the free energy density defined as:

$$F_G = -\frac{G}{2} J_m \ln\left(1 - \frac{I_1 - 3}{J_m}\right), \tag{1.14}$$

  where J_m characterizes the finite stretchability of the macromolecular networks, and one can show that the Gent model reduces to neo-Hookean model for small deformations, i.e., $I_1 - 3 \to 0$. Figure 1.3 shows the comparison of the experiment with the neo-Hookean/Gent model. By performing Taylor expansion of Eq. (1.14), one can notice that Gent model is still a specific form of Rivlin framework as shown in Eq. (1.11). Such models incorporating the finite stretchability of the materials can be applied to understand instabilities in rubber systems [24, 25].

There are other models based on tensor invariants, which we shall not show them here for saving space, and interested readers can refer to Ref. [14].

Apart from the above models based on invariants of the left Cauchy–Green tensor, there are many other types of phenomenological models, e.g., stretch-based models which include the widely used Ogden model [26, 27]. For practical needs, one can choose the most convenient one to use.

1.2 Permanent Biomacromolecular Networks

Macromolecular networks widely exist in biological systems, such as cytoskeleton and extracellular matrices, and these networks are simply called as biomacromolecular networks here. In these networks, the consisting polymers are usually semiflexible, such as microtubules, actin filaments and vimentin filaments, and the crosslinks can be motor proteins. In reality, constructing elastic models of biomacromolecular networks is similar to that of flexible macromolecular networks, as discussed in Section 1.1. We shall first show several useful elastic models of biomacromolecular networks in Section 1.2.1 and then focus on how to apply them to analyze the characteristic properties of these networks which make them special from flexible ones in Section 1.2.2.

1.2.1 Elastic Models

Here we shall introduce several elastic models of biomacromolecular networks which are widely used for understanding their mechanical properties, which are reviewed in Ref. [7, 8].

- **Storm et al. model.** Suppose that $\mathbf{E}$ is the deformation gradient tensor and $\rho \sim 1/\xi^2$ is the polymer length per unit volume (ξ denotes mesh size), a measure of network density. After deformation, the length density of the semiflexible polymers per unit volume which cross the plane perpendicular to j axis would change to $\rho E_{jk} n_k / \det \mathbf{E}$, where $\mathbf{n}$ is the orientation of the end-to-end vector of the polymer in an initially undeformed network, and $\det \mathbf{E}$ can measure the relative volume change of the deformed network. The tension acting on the polymer which connects two neighboring crosslinks is: $f(|\mathbf{E} \cdot \mathbf{n}| - 1)$, where $|\mathbf{E} \cdot \mathbf{n}| - 1$ is the axial strain of the filament, and $f(|\mathbf{E} \cdot \mathbf{n}| - 1)$ denoting the force-extension relationship of the polymer can be chosen from the models discussed in the above section. The stress tensor can then be obtained as a function of the strain tensor [3, 28]:

$$\sigma_{ij} = \frac{\rho}{\det \mathbf{E}} \left\langle f(|\mathbf{E} \cdot \mathbf{n}| - 1) \frac{\mathrm{E}_{il} n_l \mathrm{E}_{jk} n_k}{|\mathbf{E} \cdot \mathbf{n}|} \right\rangle. \tag{1.15}$$

- **8-chain model.** By assuming the 8-chain network architecture, similarly as that of flexible macromolecular networks, one can immediately know that all eight chains in the cube are stretched in the same way, with the length changing from $\sqrt{3}\xi/2$ (ξ denotes the edge length of the cube before deformation) to $\sqrt{I_1}\xi/2$. Palmer and Boyce provided the free energy density of the deformed macromolecular network as [29]:

$$F_{8\mathrm{c}}(I_1) = n_\mathrm{c} F_\mathrm{chain}(\sqrt{I_1/3}\xi). \tag{1.16}$$

  where n_c denotes the chain density and F_chain denotes the free energy of a single semiflexible polymer. Based on the free energy, one can obtain the constitutive relation describing the elasticity of the networks.
- **3-chain model.** In 2016, the free energy density of the biomacromolecular network is obtained based on the 3-chain architecture of the network [30]. The analytic form of the free energy density is:

$$F_{3\mathrm{c}}(\{\lambda_{i=1,2,3}\}) = \frac{n_\mathrm{c}}{3} \sum_{i=1,2,3} F_\mathrm{chain}(\lambda_i \xi) = \frac{n_\mathrm{c} k_\mathrm{B} T}{3} \times \left[\pi^2 c \left(3 - x^2 I_1\right) + \frac{3 - 2 I_1 x^2 + I_2 x^4}{\pi c \left(1 - I_1 x^2 + I_2 x^4 - I_3 x^6\right)} \right], \tag{1.17}$$

where $c = l_p/2l_c$ and $x = \xi/l_c$ are defined in Eq. (1.3). From the free energy of the biomacromolecular network, one can obtain the constitutive relation as [31]:

$$\sigma_{ij} = 2\left[\left(\frac{\partial F_{3c}}{\partial I_1} + I_1\frac{\partial F_{3c}}{\partial I_2}\right)B_{ij} - \left(I_1\frac{\partial F_{3c}}{\partial I_1} + 2I_2\frac{\partial F_{3c}}{\partial I_2}\right)\frac{\delta_{ij}}{3} - \frac{\partial F_{3c}}{\partial I_2}B_{ik}B_{kj}\right] - P\delta_{ij}, \tag{1.18}$$

where $\mathbf{B} = \mathbf{E}\mathbf{E}^{\mathrm{T}}$ is the left Cauchy-Green deformation tensor, and P is the Lagrangian multiplier in charge of the assumed incompressibility, the value of which can be determined by the boundary conditions.

In the next section, the 3-chain model is taken as an example to analyze the elastic properties of biomacromolecular networks.

1.2.2 Nonlinear Elasticity, Stability, and Normal Stress

When a biomacromolecular network undergoes deformations, e.g., a simple shear, the networks can exhibit very different mechanic responses compared with flexible macromolecular networks. Here we shall show how to analyze the characteristic properties of the biomacromolecular networks by taking the 3-chain model as an example for describing the free energy of the networks.

- **Nonlinear elasticity.** For a biomacromolecular network undergoing a simple shear deformation denoted by the deformation gradient tensor as:

$$\mathbf{E} = \begin{pmatrix} 1 & \gamma & 0 \\ 0 & 1 & 0 \\ 0 & 0 & 1 \end{pmatrix}, \tag{1.19}$$

where γ is the shear strain, the stress–strain relation in the 3-chain model for the network of given parameters (x, c) can be explicitly written as [30]:

$$[\sigma_{xy}]_{3c}(\gamma; x, c) = \frac{2}{3}nk_B T\gamma x^2\left[\frac{(1-x^4)}{c\pi[1-(2+\gamma^2)x^2+x^4]^2} - c\pi^2\right], \tag{1.20}$$

from which one can know that the stress is a linear function of the shear strain for small strains ($\sigma_{xy} \propto \gamma$ at $\gamma \to 0$), and diverges at a finite strain $\gamma_c = 1/x - x$, as shown in Figure 1.4a. There are two definitions of the shear modulus: nominal shear modulus $G(\gamma) = \sigma_{xy}(\gamma)/\gamma$ and differential shear modulus $K(\gamma) = \partial\sigma_{xy}/\partial\gamma$, which are identical at $\gamma \to 0$. An interesting mechanic property of biomacromolecular networks is about the relationship between the differential shear modulus K and the shear stress σ_{xy}, which is found to show a universal scaling at large strains $K \propto \sigma_{xy}^{3/2}$ (shown in Figure 1.4b). When the strain is approaching the divergence point $\gamma \to \gamma_c = 1/x - x$, then the stress and the differential shear modulus can be approximated as:

$$\sigma_{xy}(\gamma) \simeq nk_B T\frac{2\gamma x^2(1-x^4)}{3\pi c[1-(2+\gamma^2)x^2+x^4]^2}, \tag{1.21}$$

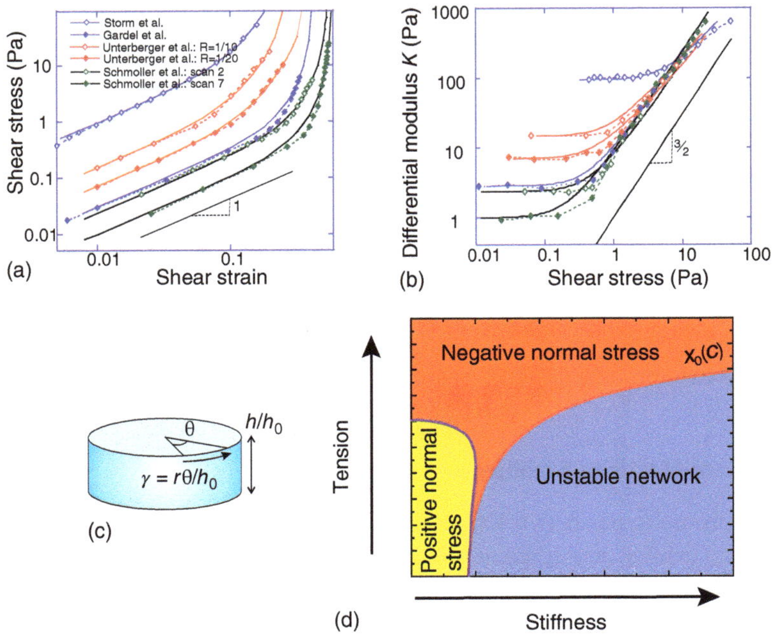

Figure 1.4 (a) Stress–strain relation of sheared biomacromolecular networks, with dots from different experiments. Source: Adapted from [28, 32–34]. And lines obtained by fitted Eq. (1.20). (b) Relationship between differential shear modulus and shear stress, which exhibit 3/2 scaling at large stresses [same data source as (a)]. (c) Twisting deformation of the biomacromolecular network and the phase diagram showing positive/negative normal stress regions in the plane of (biomacromolecule stiffness and pre-tension) and the unstable region. Source: The figure is adapted from Ref. [30].

$$K(\gamma) \simeq nk_BT\frac{2(1-x^4)\left[x^2-(2-3\gamma^2)x^4+x^6\right]}{3\pi c\left[1-(2+\gamma^2)x^2+x^4\right]^3}, \tag{1.22}$$

from which one can easily notice $K \propto \sigma_{xy}^{3/2}$. This 3/2 scaling actually originates from the mechanic properties of a single biopolymer (semiflexible), since there is the relation as: $d\sigma/d\gamma \sim df/dx \sim 1/(1-x^2)^3 \sim f^{3/2}$ at $x \to 1$ (f denotes force) from the elastic model of a single semiflexible polymer.

- **Network stability.** For a single biomacromolecule (semiflexible polymer), the relaxed state is given by x_0 satisfying zero-force condition $f(x_0) = 0$, and the value of x_0 depends on the rigidity of the polymer described by c. However, for a biomacromolecular network, the consisting polymers are not necessarily at the force-free states, in other words, the end-to-end factor x can be $>, <, = x_0$, meaning the network is in a prestretched, precompressed, and relaxed state, respectively. The criterion to judge whether a network of a given parameter set (x, c) is stable is given by the linear shear modulus (the constant modulus at small strains): $G_0 \geq 0$. In the 3-chain model, this condition can be expressed as [30]:

$$c \leq \frac{1}{\pi^{3/2}}\frac{\sqrt{1+x^2}}{(1-x^2)^{3/2}}. \tag{1.23}$$

Note that the right-hand side in Eq. (1.23) is always larger than $c^* = 1/\pi^{\frac{3}{2}}$, meaning that for sufficiently flexible chains with $c < c^*$, the network is always stable. On the other hand, for filament networks composed of the stiffer filaments with $c \geq c^*$, the network strands have to be crosslinked with x exceeding the pre-tension threshold given by Eq. (1.23), in order for the network to have a finite linear shear modulus. In other words, there are usually forces acting on the crosslinked filaments in the network in order to be mechanically stable with a non-zero shear modulus.

- **Normal stress.** Consider a cylindrical macromolecular network which is subjected to a twist deformation, as shown in Figure 1.4c. For typical rubbers, the height of the cylinder (length along the longitudinal direction) increases together with a decrease in its radius, and this phenomenon is called as "positive Poynting effect" corresponding to a positive normal stress [35, 36]. However, a biomacromolecular network usually behaves oppositely, i.e., its height decreases and radius increases, showing the "negative Poynting effect" which corresponds to a negative normal stress [37]. Suppose the height and the radius of the undeformed cylinder are h_0 and R_0, respectively, which change to $h = \lambda_h h_0$ and $R = \lambda_R R_0$ after being twisted with $\lambda_{h,R}$ as the stretch ratio along the longitudinal and the radial directions, respectively ($\lambda_h \lambda_R^2 = 1$ due to the incompressibility of the material). In the cylindrical coordinate system, the material point at (r, θ, z) changes its position to $(r/\sqrt{\lambda_h}, \theta + \Theta z/h_0, \lambda_h z)$ after rotating the top plate with angle Θ, and the deformation gradient tensor of the macromolecular network can be expressed as:

$$\mathbf{E} = \begin{pmatrix} 1/\sqrt{\lambda_h} & 0 & 0 \\ 0 & 1/\sqrt{\lambda_h} & \gamma(r)/\sqrt{\lambda_h} \\ 0 & 0 & \lambda_h \end{pmatrix}, \tag{1.24}$$

where $\gamma(r) = r\Theta/h_0$ is the shear strain depending on the radial position. Given the shear strain at the outermost surface, $\gamma_0 = \gamma(R_0)$, the total free energy of the deformed network can be expressed as a function of the stretching ratio λ_h along the z-axis after integration over the radius [30],

$$F_{\text{cylinder}}(\lambda_h; \gamma_0) = \int_0^{R_0} dr F_{3c}(\lambda_h, r; \gamma_0). \tag{1.25}$$

By minimizing the above free energy, we can obtain the optimal λ_h, which can tell us whether the material shows positive or negative Poynting effect. The phase diagram in the space of (c, x) is shown in Figure 1.4d, where the boundary line separating the positive/negative normal stress regions is given. It is clear from the phase diagram that a sheared flexible network with $c < c^* = 1/\pi^{\frac{3}{2}}$ usually shows a positive normal stress, while a semiflexible filament network exhibits a negative normal stress.

In this section, we have introduced useful theoretical models of the *permanent* macromolecular/biomacromolecular networks, which can be utilized for analyzing the characteristic elastic properties.

1.3 Transient Macromolecular/Biomacromolecular Networks

Transient macromolecular/biomacromolecular networks are important in both industrial applications and biological systems, as reviewed in Ref. [38, 39]. In transient networks, the crosslinks usually can break and re-form dynamically at high temperatures, so the network shape (and microscopic structure) can be altered; meanwhile, at low temperatures, the crosslinks are stable so that the network just act as a permanent one as described in the above section. Such dynamic crosslinks can be either physical which is formed by physical interactions (hydrogen bonds, guest–host interactions, hydrophobicity, protein–protein interactions, etc.), or chemical covalent bonds where polymer chains exchange by transesterification reaction, transamination of vinylogous urethanes, etc. In industry, such transient feature of crosslinks allows the networks to behave as thermoplastics at high temperatures which are remoldable and recyclable, and then as thermosets at low temperatures showing excellent mechanic properties. In bio-systems, the transient crosslinks such as motor proteins enable the biomacromolecular networks to change their structures accordingly for accommodating to different mechanical or chemical environments.

The theoretical studies of transient macromolecular networks have a long history, which can be roughly categorized into statistical and macroscopic ones, as recently reviewed in Ref. [40]. For statistical works based on the dynamic evolution of polymer chains (with dynamic breakage from and re-crosslinked to the crosslinks), one can analyze the viscoelasticity of the networks. Such studies can date back to the work by Green and Tobolsky [41], where they introduced the breakage and the re-formation of the crosslinks for understanding the relaxation of the polymer networks. Following this, a consistent framework for treating the crosslink dynamics subjected to external forces was constructed with the representative works by Tanaka and Edwards in 1990s and then further developed recently [42–49]. In such models, the crosslink dynamics depend on the loadings on the polymer chains, which can be explicitly expressed in terms of the deformations. Separately, the concepts of Rouse dynamics and reptation were used for studying the dynamics of a transient network by Leibler et al. [50], and also by Rubinstein and Semenov [51]. For macroscopic models, Drozdov et al. developed continuum models for various systems involving transient networks, where the crosslink dynamics can be expressed in terms of the energy density with fitting parameters [52–54]. Similar ideas were further developed and successfully applied to deal with dual networks by Long and Hui et al. [55–57]. Apart from these, there are also theoretical works on modeling how the viscoelasticity of the transient macromolecular networks depend on the nonaffinity of the network [46, 53], network defects [58], crosslinker mobility [59], multifunctional crosslink dynamics [60], etc.

Following Tanaka–Edwards model, we shall introduce a microscopic-macroscopic hybrid model in this section, where the microscopic crosslink dynamics depending on the mechanic forces acting on the polymer chains is incorporated into a macroscopic continuum theory (Figure 1.5).

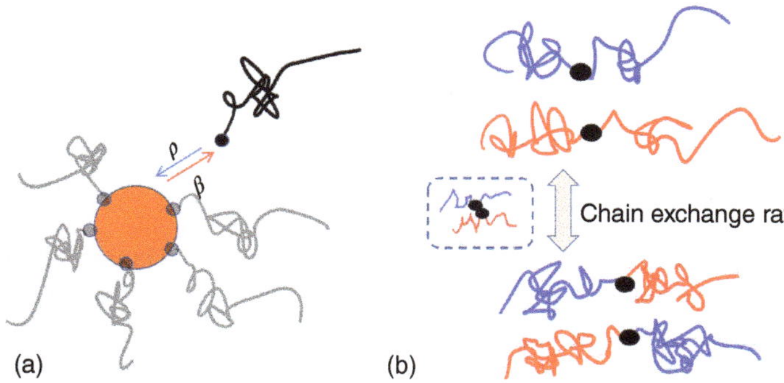

Figure 1.5 Schematics of (a) a physical crosslink where chains can dynamically break from and re-attach to, and (b) a chemical crosslink where chains can dynamically exchange.

1.3.1 Theoretical Framework

We here introduce the modeling of physically crosslinked transient macromolecular networks, and it can be easily generalized to chemical ones, for which we shall talk about vitrimers [a new type of associative covalent adaptive networks (CAN)] as an example. Suppose the energy barrier for a *flexible* polymer to break from a crosslink is W_b, and then the rate for a relaxed chain to break from its bonded crosslink is: $\beta_0 = \omega_0 \exp(-W_b/k_B T)$, which increases to $\beta = \omega_0 \exp[(-W_b + fb)/k_B T]$ when the chain experiences a tensile force f lowering the energy barrier [45]. Note that the force is tensile for *flexible* polymer, i.e., $f > 0$; for *semiflexible* polymers, one may need to use the absolute value of f, i.e., $|f|$, since it can be stretched or compressed. Correspondingly, one can define the re-crosslinking rate for a dangling polymer to be crosslinked: $\rho_0 = \omega_0 \exp(-W_c/k_B T)$ with W_c as the re-crosslinking barrier; note that in cases of physical crosslinks, the practical re-crosslinking rate is usually much smaller than ρ_0 as the dangling polymer needs to diffuse to "find" the crosslink and the effective re-crosslinking rate is $\rho = 1/(\tau + 1/\rho_0)$ where τ is the diffusion time. The newly crosslinked chains are assumed to be relaxed, i.e., their reference state needs to be defined according to when they are crosslinked.

Suppose there are N_0 chains in the polymer network, where at time t the number of the crosslinked chains is $N_c(t)$ and the number of the dangling chains is $N_b(t) = N_0 - N_c(t)$. After an infinitesimal time interval Δt, the number of the initially crosslinked chains decreased to $N_c(0)\exp(-\beta\Delta t)$, and meanwhile, there are chains which are newly crosslinked to the network, of which the number is $N_b(0)\rho\Delta t$; the total number of crosslinked chains is: $N_c(0)\exp[-\beta(\Delta t; 0)\Delta t] + N_b(0)\rho\Delta t$. Note that the breakage rate $\beta(t; t')$ depends both on the initial time t' when the chain is crosslinked and the current time t, since the forces acting on the chains depend on the deformation referenced to when the chains get crosslinked. After another time interval, the number of the chains remaining crosslinked at $t = \Delta t$ decreases to $N_c(0)\exp[-\beta(\Delta t; 0)\Delta t]\exp[-\beta(2\Delta t; 0)\Delta t] + N_b(0)\rho\Delta t \exp[-\beta(2\Delta t; \Delta t)\Delta t]$, and meanwhile, there are chains newly crosslinked to the network, of which the number is $N_b(\Delta t)\rho\Delta t$; the total number of crosslinked chains becomes: $N_c(0)\exp[-\beta(\Delta t; 0)\Delta t]\exp[-\beta(2\Delta t; 0)\Delta t] + N_b(0)\rho\Delta t \exp[-\beta(2\Delta t; \Delta t)\Delta t] + N_b(\Delta t)$

$\rho\Delta t$. By repeating the above calculations, the total number of the crosslinked chains after N time intervals becomes: $\prod_{l=1}^{N} N_c(0)\exp[-\beta(l\Delta t;0)\Delta t] + \sum_{m=1}^{N} N_b((m-1)\Delta t)\Delta t \prod_{l=m+1}^{N} \exp[-\beta(l\Delta t; m\Delta t)\Delta t]$, which can be written in a continuous form:

$$N_c(t) = N_c(0)\, e^{-\int_0^t dt'\, \beta(t';0)} + \int_0^t dt'\, N_b(t)\, \rho\, e^{-\int_{t'}^t dt''\, \beta(t'';t')}. \tag{1.26}$$

Consider the fact that the relaxed chains are almost isotropically distributed, then the free energy of the network can be regarded as the summation of that of all polymers, as what Tanaka and Edwards did in their papers [44, 45]. As mentioned in the previous sections on permanent macromolecular networks, it is usually difficult (or impossible) to obtain an analytical expression of the free energy of a macromolecular network by statistically averaging the contributions of the polymers which distributed uniformly in all possible directions, and this is also the case for transient macromolecular networks. So, one needs to make simplifications in order to obtain a compact form of the free energy of a transient macromolecular network; the network can be treated as an assembly of subnetworks which form at different time according to when the chains get crosslinked. Then the free energy density of the transient polymer network can be expressed as [61]:

$$F_{tr}(t) = e^{-\int_0^t dt'\, \beta(t';0)} F_{per}(t;0) + \int_0^t dt' \frac{N_b(t)}{N_c(0)} \rho\, e^{-\int_{t'}^t dt''\, \beta(t'';t')} F_{per}(t;t'), \tag{1.27}$$

where $F_{per}(t;t')$ denotes the free energy density contributed by the subnetwork which forms at time t' (reference time), and its form can be taken from those of permanent macromolecular networks depending on the network properties.

1.3.2 Applications

In the following part, we shall show how to utilize the above theoretical framework to study the viscoelasticity of a transient macromolecular network by taking the uniaxial stretch test as an example, where the global deformation gradient tensor referenced at time $t = 0$ can be expressed as:

$$\mathbf{E}(t;0) = \begin{pmatrix} \lambda(t;0) & 0 & 0 \\ 0 & 1/\sqrt{\lambda(t;0)} & 0 \\ 0 & 0 & 1/\sqrt{\lambda(t;0)} \end{pmatrix}, \tag{1.28}$$

where λ denotes the stretch ratio along x direction and incompressibility of the network is assumed.

- **Small deformations.** When the transient macromolecular network consisting of flexible polymers undergoes a small deformation, i.e., $\lambda \to 1$, the consisting polymers can be treated as Gaussian chains and the form of the free energy density of the permanent network can be taken as that of neo-Hookean model, $F_{per}(t;t') = \frac{1}{2}G[\lambda(t;t')^2 + 2/\lambda(t;t') - 3] = \frac{1}{2}G[\lambda(t)^2/\lambda(t')^2 + 2\lambda(t')/\lambda(t) - 3]$, where the relation for the deformation gradient tensor $\mathbf{E}(t;t') = \mathbf{E}(t;0)\mathbf{E}^{-1}(t';0)$ is utilized. Meanwhile, the breakage rate depends on the force acting on the polymer, where the force is a function of the end-to-end distance of the polymer (the distance r depends on when the polymer is crosslinked to the network and the deformation of the network), $f = \frac{3k_BT}{Nb^2}r$. Here, we can use the averaged end-to-end distance of the polymers distributed uniformly in all possible directions to obtain an approximate value of the breakage rate $\beta(t;t') = \omega_0 \exp[(-W_b + 3k_BT\langle r\rangle/Nb)/k_BT] = \beta_0 \exp[3\langle r\rangle/Nb]$, and it increases with the stretch ratio λ as shown in Ref. [61, 62]. The stress–strain relation in this case can be explicitly expressed as [61]:

$$\sigma_{xx}(t) = Ge^{-\int_0^t dt'\,\beta(t';0)}\left[\lambda(t) - \frac{1}{\lambda(t)^2}\right] + G\int_0^t dt'\,\frac{N_b(t)}{N_0}\,\rho\, e^{-\int_{t'}^t dt''\,\beta(t'';t')} \times \left[\frac{\lambda(t)}{\lambda(t')^2} - \frac{\lambda(t')}{\lambda(t)^2}\right], \tag{1.29}$$

where the first term on the right-hand-side shows contributions of the polymers which are crosslinked at time $t = 0$ and remain crosslinked until time t, and the second term denotes the contributions of the polymers which are crosslinked at different time t' and remain crosslinked until time t during the deformation. Other stress components are *zero*. Here, we consider two cases. *Case (1)*: the network is instantaneously stretched and then kept with a constant stretch ratio λ_0. Then, the stress relaxes exponentially with time and the relaxation rate is equal to breakage rate; this is obvious from Eq. (1.29), where the first term on the right-hand-side shows the exponential decay of the stress, and the second term has no contribution to the stress because of $\lambda(t) = \lambda(t') = \lambda_0$. Usually in realistic transient macromolecular networks, the stress relaxation does not follow a simple exponential function of time, but a stretched exponential function, $e^{-(\beta t)^\alpha}$ with the stretching factor $\alpha < 1$ depending on system properties, for example, a wide distribution of the energy barriers for crosslink breakage [61]. *Case (2)*: another common rheology test is strain ramp deformation, where the network is stretched with a constant strain rate, $\lambda(t) = 1 + \dot{\gamma}t$. The relationship between the stress and the strain/time can be obtained from Eq. (1.29), which is plotted in Figure 1.6 for different strain rates. For small strain rates, the stress first increases with the strain and then decreases, and the turnover is named as yielding point located at a small

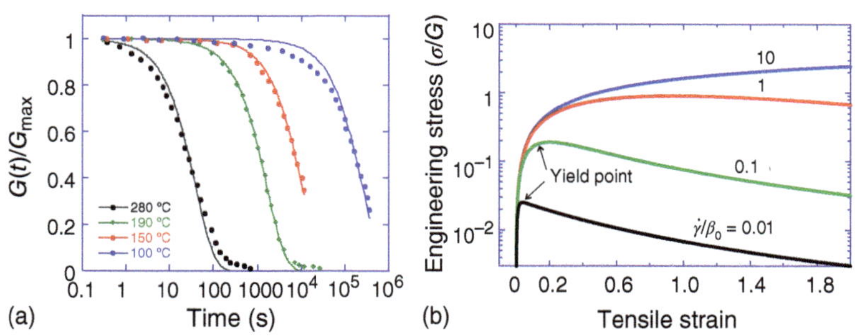

Figure 1.6 (a) Stress relaxation. Dots are from the experiment [63] and lines are obtained by fitting the exponential decay equation. (b) Stress–strain relation in the ramp deformation tests with different strain rates. Source: Ref. [61] American Chemical Society/CC BY 4.0.

strain $\gamma_y \simeq \dot{\gamma}/\beta$ (yielding time $t_y \simeq 1/\beta$). When the strain rate is large, then the stress–strain relation is similar to that of a permanent macromolecular network, since most of the initially crosslinked chains have not broken from the crosslinks at small strains.

- **Large deformations.** When the transient macromolecular networks consisting of flexible polymers undergo a large deformation, Gent model is usually taken for describing the energy density of the permanent macromolecular networks, which is $F_{\text{per}}(t;t') = -\frac{1}{2}GJ_{\text{m}}\ln[1 - \frac{J(t;t')}{J_m}]$ with $J(t;t') = \lambda^2(t;t') + 2/\lambda(t;t') - 3$, and the stress–strain relationship is [64]:

$$\sigma_{xx} = \frac{GJ_{\text{m}}}{J_{\text{m}} - J(t;0)}\left[\lambda(t) - \frac{1}{\lambda(t)^2}\right]e^{-\int_0^t \beta(t';0)dt'} + \int_0^t \rho\frac{GJ_{\text{m}}}{J_{\text{m}} - J(t;t')}\left[\frac{\lambda(t)}{\lambda(t')^2} - \frac{\lambda(t')}{\lambda(t)^2}\right]\frac{N_{\text{b}}(t')}{N_0}e^{-\int_{t'}^t \beta(t'',t')dt''}dt'. \tag{1.30}$$

Note that there is a divergence point, $J(t;0) \to J_{\text{m}}$, where the stress diverges upon deformation. In the *stress relaxation* test, the stress of the network decreases exponentially with time in the same way as in the case of small deformations. In the *strain ramp deformation* test, the stress–strain relationship of the transient macromolecular networks becomes more interesting as shown in Figure 1.7a. When

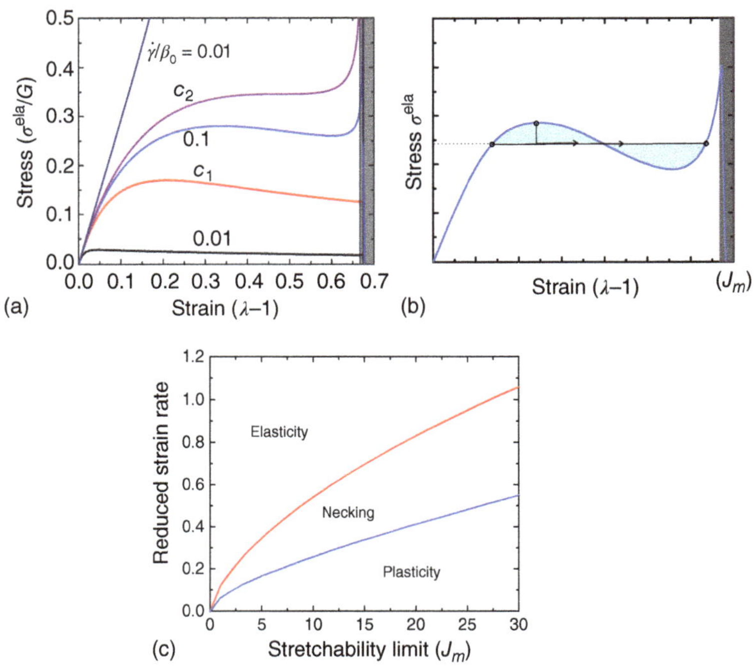

Figure 1.7 (a) Stress–strain relation in the ramp deformation tests with different strain rates. (b) A schematic of "S-shaped" stress–strain relation. (c) Phase diagram showing different viscoelastic responses of the transient macromolecular networks (elastic, plastic, or necking) in the plane of (stretchability limit, strain rate). Source: The figure is adapted from Ref. [64].

the strain rate is large, the stress increases with the strain until the divergence point, resembling that of a permanent macromolecular network. When the strain rate is small, the stress first increases and then decreases, similarly as that in the small deformation case. When the strain rate is moderate, the stress first increases, then decreases (turnover at the yielding point), and then increases again until the divergence point; such "S-shaped" stress–strain relation indicates the possibility to have two co-existing phases (one with large λ and the other with small λ, as necking instability) during the deformation if the energy barrier to create the interface between the two phases is not high (Figure 1.7b). The different responses of the transient macromolecular networks in the ramp deformation test are shown in the phase diagram (Figure 1.7c).

- **A special example: vitrimers.** Since Leibler et al. discovered a new type of associative CAN, which is named as vitrimers [63], such materials quickly attracted increasing attentions (reviewed in [65]); these materials exhibit characteristic properties including (i) the number of the crosslinks remains constant as other associative CAN and (ii) thermal viscosity changes with temperature in the form of Arrhenius law as inorganic silica materials. The covalently crosslinked polymers in vitrimers can dynamically exchange when adding catalysts or increasing temperature, and such bond exchange reactions can be implemented by transesterification reaction [63, 66], transamination of vinylogous urethanes [67], etc. Differently from the crosslink breakage and re-formation dynamics in physical transient networks, the exchangeable chains in vitrimers first associate and then dissociate, and one can treat that the crosslink breakage and re-formation occur simultaneously. The number of crosslinked chains can then be expressed as [68]:

$$N_c(t) = N_c(0)\,e^{-\beta t} + N_c(0)\int_0^t dt'\,\beta\,e^{-\beta(t-t')} \equiv N_c(0), \tag{1.31}$$

where the chain exchange rate β is taken as a constant for simplicity. As an obvious extension, the free energy of a dual network (two subnetworks, a transient one of fraction ν and a permanent one of fraction $1-\nu$) can be expressed as,

$$F_{tr}(t) = \left[\nu e^{-\beta t} + (1-\nu)\right] F_{per}(t;0) + \nu\int_0^t dt'\beta e^{-\beta(t-t')}F_{per}(t;t'), \tag{1.32}$$

where the reference state of the transient part changes with time. A detailed study about the rheological properties of the vitrimers in tests such as stress relaxation, strain ramp and creep, can be found in Ref. [68, 69].

- **Transient biomacromolecular networks.** In a biomacromolecular network, e.g., these existing in the form of cytoskeleton, the crosslinks such as bio-motors can dynamically attach to and detach from the biomacromolecules in the network, which endows the network with dynamic properties; in other words, the transient biomacromolecular network is a special example of macromolecular networks. Then by substituting the free energy density of the permanent biomacromolecular networks (examples given in Section 1.2.1) into Eq. (1.27), one can study how a transient biomacromolecular network responds to external mechanic stimuli [70]. Suppose that a transient biomacromolecular network is instantly stretched with the stretch ratio λ_0 and then kept stretched for a time interval τ, after which the tensile force is released and the network relaxes to a new state characterized by a new stretch ratio λ_r, which can be calculated by solving $\sigma_{ela}(\tau) = \sigma_0 e^{-\beta\tau} = \sigma\left(\lambda_0/\lambda_r(\tau)\right)$ [β is the breakage rate of the crosslinks as a function of λ_0] (Figure 1.8a). Then one can show that the recovery ratio (defined as $S(\tau) = \frac{\lambda_0 - \lambda_r(\tau)}{\lambda_0 - 1}$) decreases with increasing τ due to the fact that the stress relaxes during this time period (shown in Figure 1.8b). In the strain ramp deformation, one finds that the "S-shaped" stress–strain relation of the material stretched with a moderate strain rate resembles that of flexible macromolecular networks undergoing large deformations, as shown in Figure 1.8c. Here, re-crosslinking is neglected, as it was shown in Ref. [61, 64] that re-crosslinking process usually alters the viscoelastic response of the materials in the ramp deformation test in a quantitative rather than qualitative way. Also, one can study how polydispersity in the mesh size of the network influences the viscoelasticity of the transient macromolecular networks, e.g., in the strain ramp deformation process, the network is softer at small strains and stiffer at large strains if the distribution of the mesh size is narrower as shown in Figure 1.8d, where one can assume the length of the polymer connecting two crosslinks l_c obeys a normal distribution $p(l_c) = (1/\sqrt{2\pi}\Delta)\exp[-(l_c - l_c^0)/2\Delta^2]$, with the expectation value l_c^0 and standard variance Δ as shown in (l_c has a correspondence with the mesh size $\xi(l_c)$ if the network resides in a stress-free state).

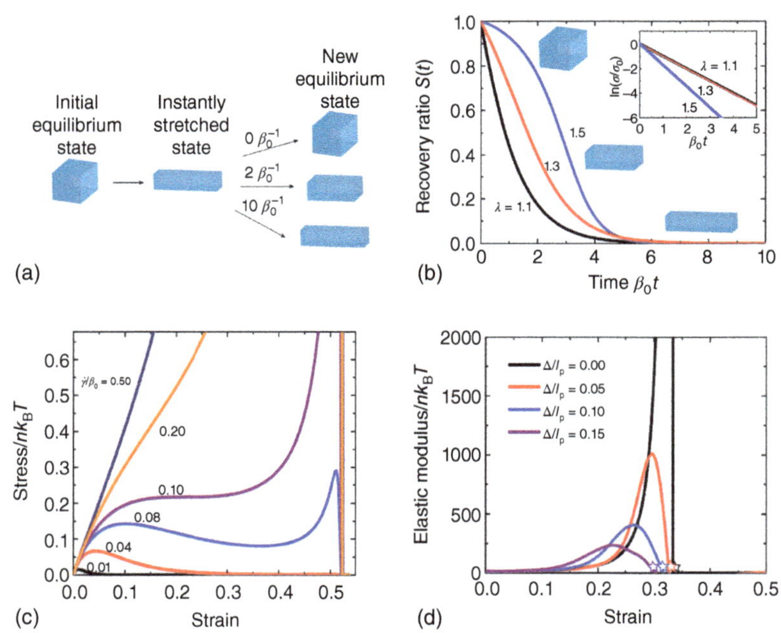

Figure 1.8 (a) Schematic showing the process of shape recovery and (b) the recovery ratio as a function of keeping time, where β_0 is the breakage rate of the crosslinks under the force-free condition (inset denotes the stress relaxation). (c) Stress–strain relationship in the ramp deformation test. (d) The evolution of modulus with applied strain in the ramp deformation test for networks of different distributions of mesh sizes. Source: Ref. [70]/ American Chemical Society.

In this section, we show how to model a transient macromolecular network including biological ones by incorporating the crosslink dynamics into a continuum theory, which can be applied to investigate the viscoelasticity of such interesting materials.

1.4 Outlooks

In this chapter, we have introduced several basic theories on macromolecular and biomacromolecular networks, which can be utilized to analyze the (visco)elasticity of different macromolecular network-based materials. These standard models can also be generalized to study other macromolecular systems, e.g., micro/nanocomposites [71, 72], ferrogels [73–75], electro-gels [76, 77], active gels [78], phase separation in polymer networks [79, 80], particle motion in a transient polymer network [81], exchangeable liquid crystal elastomers [82], etc. Interesting readers can refer to the literatures on these specific topics.

References

1 Flory, P.J. and Volkenstein, M. (1969). *Statistical Mechanics of Chain Molecules*. Wiley Online Library.
2 IUPAC (1997). *Compendium of Chemical Terminology*, 2e. Oxford: Blackwell Scientific Publications.
3 Doi, M. and Edwards, S.F. (1988). *The Theory of Polymer Dynamics*, vol. 73. Oxford University Press.
4 Rubinstein, M. and Colby, R.H. (2003). *Polymer Physics*. Oxford University Press.
5 Kawakatsu, T. (2004). *Statistical Physics of Polymers: An Introduction*. Springer Science & Business Media.
6 Doi, M. (2013). *Soft Matter Physics*. Oxford University Press.
7 Broedersz, C.P. and MacKintosh, F.C. (2014). Modeling semiflexible polymer networks. *Reviews of Modern Physics* 86: 995.
8 Meng, F. and Terentjev, E.M. (2017). Theory of semiflexible filaments and networks. *Polymers* 9: 52.
9 Marko, J.F. and Siggia, E.D. (1995). Stretching DNA. *Macromolecules* 28: 8759–8770.
10 Ha, B.-Y. and Thirumalai, D. (1997). Semiflexible chains under tension. *The Journal of Chemical Physics* 106: 4243–4247.
11 MacKintosh, F.C., Käs, J., and Janmey, P.A. (1995). Elasticity of semiflexible biopolymer networks. *Physical Review Letters* 75: 4425.
12 Blundell, J.R. and Terentjev, E.M. (2009). Stretching semiflexible filaments and their networks. *Macromolecules* 42: 5388–5394.
13 Treloar, LRG (1975). *The Physics of Rubber Elasticity*. Clarendon Press, Oxford.
14 Boyce, M.C. and Arruda, E.M. (2000). Constitutive models of rubber elasticity: a review. *Rubber Chemistry and Technology* 73: 504–523.

15 Wu, P.D. and Van Der Giessen, E. (1993). On improved network models for rubber elasticity and their applications to orientation hardening in glassy polymers. *Journal of the Mechanics and Physics of Solids* 41: 427–456.
16 Wang, M.C. and Guth, E. (1952). Statistical theory of networks of non-Gaussian flexible chains. *The Journal of Chemical Physics* 20: 1144–1157.
17 Flory, P.J. and Rehner, J. Jr. (1943). Statistical mechanics of cross-linked polymer networks I. Rubberlike elasticity. *The Journal of Chemical Physics* 11: 512–520.
18 Treloar, L.R.G. (1954). The photoelastic properties of short-chain molecular networks. *Transactions of the Faraday Society* 50: 881–896.
19 Arruda, E.M. and Boyce, M.C. (1993). A three-dimensional constitutive model for the large stretch behavior of rubber elastic materials. *Journal of the Mechanics and Physics of Solids* 41: 389–412.
20 Qian, D. and Meng, F. (2021). Modelling Mullins effect induced by chain delamination and reattachment. *Polymer* 222: 123608.
21 Rivlin, R.S. (1948). Large elastic deformations of isotropic materials IV. Further developments of the general theory. *Philosophical Transactions of the Royal Society of London. Series A* 241: 379–397.
22 Mooney, M. (1948). The thermodynamics of a strained elastomer. I. General analysis. *Journal of Applied Physics* 19: 434–444.
23 Gent, A.N. (1996). A new constitutive relation for rubber. *Rubber Chemistry and Technology* 69: 59–61.
24 Meng, F., Chen, J.Z.Y., Doi, M., and Ouyang, Z. (2014). Phase diagrams and interface in inflating balloon. *AIChE Journal* 60: 1393–1399.
25 Meng, F., Chen, J.Z.Y., Doi, M., and Ouyang, Z. (2015). The phase diagram and radial collapse of an inflated soft tube under twist. *Soft Matter* 11: 7046–7052.
26 Ogden, R.W. (1972). Large deformation isotropic elasticity–on the correlation of theory and experiment for incompressible rubberlike solids. *Proceedings of the Royal Society of London. Series A* 326: 565–584.
27 Ogden, R.W. (1997). *Non-Linear Elastic Deformations*. Courier Corporation.
28 Storm, C., Pastore, J.J., MacKintosh, F.C. et al. (2005). Nonlinear elasticity in biological gels. *Nature* 435: 191–194.
29 Palmer, J.S. and Boyce, M.C. (2008). Constitutive modeling of the stress–strain behavior of f-actin filament networks. *Acta Biomaterialia* 4: 597–612.
30 Meng, F. and Terentjev, E.M. (2016). Nonlinear elasticity of semiflexible filament networks. *Soft Matter* 12: 6749–6756.
31 Bower, A.F. (2009). *Applied Mechanics of Solids*. CRC Press.
32 Gardel, M.L., Shin, J.H., MacKintosh, F.C. et al. (2004). Elastic behavior of cross-linked and bundled actin networks. *Science* 304: 1301–1305.
33 Unterberger, M.J., Schmoller, K.M., Bausch, A.R., and Holzapfel, G.A. (2013). A new approach to model cross-linked actin networks: multi-scale continuum formulation and computational analysis. *Journal of the Mechanical Behavior of Biomedical Materials* 22: 95–114.
34 Schmoller, K.M., Fernandez, P., Arevalo, R.C. et al. (2010). Cyclic hardening in bundled actin networks. *Nature Communications* 1: 1–8.

35 Poynting, J.H. (1909). On pressure perpendicular to the shear planes in finite pure shears, and on the lengthening of loaded wires when twisted. *Proceedings of the Royal Society of London. Series A* 82: 546–559.
36 Mihai, L.A. and Goriely, A. (2011). Positive or negative poynting effect? The role of adscititious inequalities in hyperelastic materials. 467: Proceedings of the Royal Society A: Mathematical, Physical and Engineering Sciences, 3633–3646.
37 Janmey, P.A., McCormick, M.E., Rammensee, S. et al. (2007). Negative normal stress in semiflexible biopolymer gels. *Nature Materials* 6: 48–51.
38 Yang, Y. and Urban, M.W. (2013). Self-healing polymeric materials. *Chemical Society Reviews* 42: 7446.
39 Wei, Z., Yang, J.H., Zhou, J. et al. (2014). Self-healing gels based on constitutional dynamic chemistry and their potential applications. *Chemical Society Reviews* 43: 8114–8131.
40 Guo, Q. and Long, R. (2020). Mechanics of polymer networks with dynamic bonds. In: *Self-Healing and Self-Recovering Hydrogels*, Advances in Polymer Science, vol. 285 (ed. C. Creton and O. Okay), 127–164. Cham: Springer.
41 Green, M.S. and Tobolsky, A.V. (1946). A new approach to the theory of relaxing polymeric media. *The Journal of Chemical Physics* 14 (2): 80–92.
42 Fricker, H.S. (1973). On the theory of stress relaxation by cross-link reorganization. *Proceedings of the Royal Society of London. A. Mathematical and Physical Sciences* 335 (1602): 289–300.
43 Baxandall, L.G. and Edwards, S.F. (1988). Deformation-dependent properties of polymer networks constructed by addition of crosslinks under strain. *Macromolecules* 21 (6): 1763–1772.
44 Tanaka, F. and Edwards, S.F. (1992). Viscoelastic properties of physically crosslinked networks. 1. Transient network theory. *Macromolecules* 25 (5): 1516–1523.
45 Tanaka, F. and Edwards, S.F. (1992). Viscoelastic properties of physically crosslinked networks: Part 1. Non-linear stationary viscoelasticity. *Journal of Non-Newtonian Fluid Mechanics* 43 (2–3): 247–271.
46 Koga, T., Tanaka, F., Kaneda, I., and Winnik, F.M. (2009). Stress buildup under start-up shear flows in self-assembled transient networks of telechelic associating polymers. *Langmuir* 25: 8626–8638.
47 Vernerey, F.J., Long, R., and Brighenti, R. (2017). A statistically-based continuum theory for polymers with transient networks. *Journal of the Mechanics and Physics of Solids* 107: 1–20.
48 Brighenti, R. and Vernerey, F.J. (2017). A simple statistical approach to model the time-dependent response of polymers with reversible cross-links. *Composites Part B: Engineering* 115: 257–265.
49 Lamont, S.C., Mulderrig, J., Bouklas, N., and Vernerey, F.J. (2021). Rate-dependent damage mechanics of polymer networks with reversible bonds. *Macromolecules* 54 (23): 10801–10813.
50 Leibler, L., Rubinstein, M., and Colby, R.H. (1991). Dynamics of reversible networks. *Macromolecules* 24 (16): 4701–4707.
51 Rubinstein, M. and Semenov, A.N. (2001). Dynamics of entangled solutions of associating polymers. *Macromolecules* 34 (4): 1058–1068.

52 Drozdov, A.D. (1999). A constitutive model in finite thermoviscoelasticity based on the concept of transient networks. *Acta Mechanica* 133: 13–37.

53 Drozdov, A.D. and Christiansen, J.deC. (2006). Constitutive equations for the nonlinear viscoelastic and viscoplastic behavior of thermoplastic elastomers. *International Journal of Engineering Science* 44: 205–226.

54 Drozdov, A.D. and Christiansen, J.C. (2022). Tuning the viscoelastic response of hydrogel scaffolds with covalent and dynamic bonds. *Journal of the Mechanical Behavior of Biomedical Materials* 130: 105179.

55 Hui, C.-Y. and Long, R. (2012). A constitutive model for the large deformation of a self-healing gel. *Soft Matter* 8 (31): 8209–8216.

56 Long, R., Mayumi, K., Creton, C. et al. (2014). Time dependent behavior of a dual cross-link self-healing gel: theory and experiments. *Macromolecules* 47 (20): 7243–7250.

57 Guo, J., Liu, M., Zehnder, A.T. et al. (2018). Fracture mechanics of a self-healing hydrogel with covalent and physical crosslinks: a numerical study. *Journal of the Mechanics and Physics of Solids* 120: 79–95.

58 Ciarella, S., Sciortino, F., and Ellenbroek, W.G. (2018). Dynamics of vitrimers: defects as a highway to stress relaxation. *Physical Review Letters* 121: 058003.

59 Mulla, Y. and Koenderink, G.H. (2018). Crosslinker mobility weakens transient polymer networks. *Physical Review E* 98: 062503.

60 Ozaki, H. and Koga, T. (2020). Theory of transient networks with a well-defined junction structure. *The Journal of Chemical Physics* 152 (18): 184902.

61 Meng, F., Pritchard, R.H., and Terentjev, E.M. (2016). Stress relaxation, dynamics, and plasticity of transient polymer networks. *Macromolecules* 49: 2843–2852.

62 Serero, Y., Jacobsen, V., Berret, J.-F., and May, R. (2000). Evidence of nonlinear chain stretching in the rheology of transient networks. *Macromolecules* 33 (5): 1841–1847.

63 Montarnal, D., Capelot, M., Tournilhac, F., and Leibler, L. (2011). Silica-like malleable materials from permanent organic networks. *Science* 334: 965–968.

64 Meng, F. and Terentjev, E.M. (2016). Transient network at large deformations: elastic–plastic transition and necking instability. *Polymers* 8 (4): 108.

65 Denissen, W., Winne, J.M., and Du Prez, F.E. (2016). Vitrimers: permanent organic networks with glass-like fluidity. *Chemical Science* 7: 30–38.

66 Brutman, J.P., Delgado, P.A., and Hillmyer, M.A. (2014). Polylactide vitrimers. *ACS Macro Letters* 3: 607–610.

67 Denissen, W., Rivero, G., Nicolaÿ, R. et al. (2015). Vinylogous urethane vitrimers. *Advanced Functional Materials* 25: 2451–2457.

68 Meng, F., Saed, M.O., and Terentjev, E.M. (2019). Elasticity and relaxation in full and partial vitrimer networks. *Macromolecules* 52 (19): 7423–7429.

69 Meng, F., Saed, M.O., and Terentjev, E.M. (2022). Rheology of vitrimers. *Nature Communications* 13: 5753.

70 Meng, F. and Terentjev, E.M. (2018). Fluidization of transient filament networks. *Macromolecules* 51 (12): 4660–4669.

71 Mai, Y.-W. and Yu, Z.-Z. (2006). *Polymer Nanocomposites*. Woodhead Publishing.

72 Hubbe, M.A., Rojas, O.J., Lucia, L.A., and Sain, M. (2008). Cellulosic nanocomposites: a review. *BioResources* 3: 929–980.

73 Zrinyi, M., Barsi, L., and Büki, A. (1997). Ferrogel: a new magneto-controlled elastic medium. *Polymer Gels and Networks* 5: 415–427.

74 Awasthi, S. (2021). A review on hydrogels and ferrogels for biomedical applications. *JOM* 73: 2440–2451.

75 Junot, G., Wei, X., Ortín, J. et al. (2022). Elastically-mediated collective organisation of magnetic microparticles. *Soft Matter* 18: 5171–5176.

76 Shahinpoor, M., Bar-Cohen, Y., Simpson, J.O., and Smith, J. (1998). Ionic polymer-metal composites (IPMCs) as biomimetic sensors, actuators and artificial muscles-a review. *Smart Materials and Structures* 7: R15.

77 Bahramzadeh, Y. and Shahinpoor, M. (2014). A review of ionic polymeric soft actuators and sensors. *Soft Robotics* 1: 38–52.

78 Prost, J., Jülicher, F., and Joanny, J.-F. (2015). Active gel physics. *Nature Physics* 11: 111–117.

79 Style, R.W., Sai, T., Fanelli, N. et al. (2018). Liquid-liquid phase separation in an elastic network. *Physical Review X* 8: 011028.

80 Wei, X., Zhou, J., Wang, Y., and Meng, F. (2020). Modeling elastically mediated liquid-liquid phase separation. *Physical Review Letters* 125: 268001.

81 Sprakel, J., van der Gucht, J., Cohen Stuart, M.A., and Besseling, N.A.M. (2008). Brownian particles in transient polymer networks. *Physical Review E* 77: 061502.

82 Zhao, J. and Meng, F. (2023). Modelling viscoelasticity and dynamic nematic order of exchangeable liquid crystal elastomers. *Physical Review Letters* 131: 068101.

2

Modeling Reactive Hydrogels: Focus on Controlled Degradation

Vaibhav Palkar and Olga Kuksenok

Clemson University, Department of Materials Science and Engineering, 161 Sirrine Hall, 515 Calhoun Drive, Clemson, SC 29634, USA

2.1 Introduction

Hydrogels are cross-linked polymers typically highly swollen in aqueous environments; these swollen polymer networks can be further functionalized to be able to respond to the specific stimuli in a controlled and programmable manner. Significant efforts have been recently directed toward designing functional hydrogels with properties tailored via a range of reversible and/or irreversible chemical reactions. Recent advances in fabrication of various functionalized hydrogels, from shear-thinning hydrogels incorporating reversible bonds to self-healing networks, are surveyed by Zhang and Khademhosseini [1]. Synthetic approaches to design hydrogels which can effectively mimic a dynamic environment of extracellular matrix (ECM) had been reviewed by Rosales and Anseth [2]. In particular, dynamic chemistries for such reversible hydrogels can incorporate reversible covalent bonds, ionic bonds, hydrogen bonds, host–guest bonds, and protein assemblies [2]. Tissue engineering and drug delivery are among the most important examples of hydrogel applications. Li and Mooney [3] reviewed recent advances in synthesis and fabrication of hydrogels with tunable properties and controlled degradability suitable for a range of drug delivery applications.

Reactivity and tunable properties of hydrogels can be controlled via a broad range of external stimuli, including illumination with external light of a selected wavelength, variations in pH, addition of catalysts, or application of mechanical forcing to the systems incorporating chromophores. For example, Wang et al. [4] recently synthesized polymer networks that toughen under the external forcing via force-triggered chemical reactions. Further, using light to control reactivity poses well-known advantages including an ability to control reactions remotely and noninvasively so that the properties of the polymer network can be altered dynamically. In particular, photo-controlled degradation of polymer networks often permits spatially resolved dynamic control of physical and chemical properties of the materials [5–11]. In a number of application of degradable hydrogels, either the

Dynamics and Transport in Macromolecular Networks: Theory, Modeling, and Experiments, First Edition. Edited by Li-Tang Yan.

characteristic size of the degradable gel particles [12] or the characteristic features of these polymer networks [5, 7, 13] are on nanometers to microns scales, length scales referred to as mesoscopic. To understand the dynamics of degradation of polymer network at the mesoscale, diffusion of all the network fragments with chemical kinetics during controllable degradation and hydrodynamic interactions need to be taken into account. Dissipative particle dynamics (DPD) approach [14–16], a mesoscale approach utilizing soft repulsive interactions between beads representing groups of atoms, allows relatively low computational cost of simulations and has been widely used to model a variety of complex systems [17–28], including modeling dynamics of hydrogels in various environments [29–43].

In this review, we will primarily focus on some of the most recent work on modeling reactions in hydrogels, with particular emphasis on degradation in swollen polymer networks. We will also introduce selected examples of catalyst-initiated reactions in polymer networks, such as industrially relevant hydrosilylation reaction and enzymatic degradation of hydrogels. In what follows, we begin with modeling reactive hydrogels on the mesoscale using DPD approach. While it is critically important to understand dynamics of reactive hydrogels on the mesoscale, the characteristic length scales of the hydrogel matrix in many relevant applications significantly exceed the length scales accessible via mesoscale modeling. Thereby, in this review, we also briefly review select approaches to continuum modeling of the reactive hydrogels with the primary focus on modeling of degradation of various hydrogels on the continuum length scales.

In majority of the examples of reactive hydrogels considered in this review, polyethylene glycol (PEG)-based hydrogels are chosen as model systems. Notably, PEG-based hydrogels are often utilized in various biomedical applications due to their lack of cytotoxicity, their hydration properties, and a relative ease of end group modification [44–46].

2.2 Mesoscale Modeling of Reactive Polymer Networks

2.2.1 Introducing Dissipative Particle Dynamics Approach for Reactive Polymer Networks

We begin with the brief outline of the main features of the DPD approach. DPD was originally proposed by Hoogerbrugge and Koelman [14] and refined by Español and Warren [15] to ensure that the Gibbs distribution is recovered as the equilibrium solution of the proposed algorithm. Shortly after, Groot and Warren [16] derived the relationships between the intrinsic parameters of the DPD model and physical parameters of the systems of interest, such as Flory–Huggins interaction parameter describing affinity between various moieties in the system and interfacial tension between the different liquids. The latest methodological developments and applications of the DPD model for a range of systems are surveyed in recent comprehensive reviews by Español and Warren [17] and Santo and Neimark [47]. The beads in DPD represent groups of atoms; the motion of these beads is governed by the Newton's equations of motion [16]. For the nonbonded beads, the pairwise additive force consists of purely repulsive conservative, dissipative, and random contributions; the contributions of all forces vanish beyond a cutoff radius, r_c, which introduces an intrinsic length scale of the approach [16]. A typical choice of the conservative force in DPD is the force corresponding to the soft repulsion potential [16, 17]:

$$\boldsymbol{F}_{ij}^{C} = \begin{cases} a_{ij}\left(1 - \frac{r_{ij}}{r_c}\right)\boldsymbol{e}_{ij} & (r_{ij} < r_c) \\ 0 & (r_{ij} \geq r_c) \end{cases}, \tag{2.1}$$

where a_{ij} is the repulsion coefficient between the beads i and j separated by a distance $r_{ij} = |\boldsymbol{r}_{ij}|$, $\boldsymbol{r}_{ij} = \boldsymbol{r}_i - \boldsymbol{r}_j$, and $\boldsymbol{e}_{ij} = \frac{\boldsymbol{r}_{ij}}{r_{ij}}$. Two remaining contributions to the total force, dissipative and random force, are [16] $\boldsymbol{F}_{ij}^{D} = -\gamma\omega^{D}(r_{ij})(\boldsymbol{e}_{ij} \cdot \boldsymbol{v}_{ij})\boldsymbol{e}_{ij}$, and $\boldsymbol{F}_{ij}^{R} = \sigma\omega^{R}(r_{ij})\zeta_{ij}\Delta t^{-1/2}\boldsymbol{e}_{ij}$; here γ and σ are the strengths of the respective contributions, $\boldsymbol{v}_{ij} = \boldsymbol{v}_i - \boldsymbol{v}_j$ is the relative velocity, Δt is the simulation time step, and ζ_{ij} is a symmetric Gaussian distributed random variable with unit variance and zero mean. The dissipative and random forces are coupled through the fluctuation–dissipation theorem [15, 16] so that the following two conditions are imposed [16]: $\omega^{D}(r_{ij}) = (\omega^{R}(r_{ij}))^2$ and $\sigma^2 = 2\gamma k_B T/m$, where k_B is the Boltzmann constant, T is temperature, and m is a mass of the bead. While the typical and most convenient choice of the weight function is [16] $\omega^{R}(r_{ij}) = (1 - r_{ij}/r_c)$ for $r_{ij} < r_c$ and zero otherwise, other choices are also permitted as long as the fluctuation–dissipation theorem is satisfied.

The repulsion coefficient between the dissimilar beads in DPD is often calculated based on the Flory–Huggins interaction parameter, χ_{ij}, as [16] $a_{ij} = a_{ii} + 3.27\chi_{ij}$, where a_{ii} is the repulsion coefficient between the same type of beads; herein, the bead number density of three is chosen. Most commonly, the value of a_{ii} is derived based on the degree of coarse-graining as proposed by Groot and Rabone [48]. However, a number of alternative parametrizations, which in some instances allow one to access even larger length scales, have been recently proposed [49–51]. As one example, Anderson et al. [51] proposed a top-down parametrization based on the known partition coefficients between the different phases. It should also be noted that for a broad range of multicomponent systems, the bottom-up parametrization of the DPD parameters based on molecular dynamics (MD) simulations can be performed [47, 52].

For the bonded beads, in addition to the conservative, dissipative, and random forces acting between these beads as discussed earlier, the bonded interactions are introduced. Most commonly, the bonded beads are assumed to interact via the harmonic potential:

$$U_b = \frac{K_b}{2}(r_{ij} - r_0)^2, \tag{2.2}$$

where r_0 is an equilibrium bond length and K_b is a spring constant. In some instances, angle potential, $U_\theta = K_\theta(\theta_{ijk} - \theta_0)^2$, and dihedral potential, $U_\phi = K_\phi(1 + \cos(n\phi - \phi_0))$, are also introduced to capture interactions between the bonded beads; herein, θ_{ijk} is the angle between the two consecutive bonds between the ij and jk pairs of beads, and ϕ and n are the dihedral angle and phase, respectively, while θ_0, and ϕ_0 are the equilibrium values of the respective angles. The value of K_θ defines the rigidity of the polymer chain; the angle potential is needed to capture relatively rigid chains or chain segments [53, 54]. While the dihedral potential is typically used in various coarse-grained MD (CGMD) approaches [55], it is not commonly implemented in DPD unless required to capture some specific structures or properties, such as formation of α-helical structures [56]. Finally, Morse potential had also been implemented in DPD studies of a number of systems with hydrogen bonds [57–59] .

Chemical reactions are often incorporated within DPD as a stochastic process [36–39, 60, 61], similar to the approaches used in CGMD simulations [62, 63]. Typically, the three main parameters controlling the reaction processes are the probability of bond breaking, P, the reaction time step τ_r, and a reaction cutoff distance r_{react}. The common choice of the reaction time step in DPD simulations is $\tau_r = 10\Delta t$ [36–39, 60]. The reaction cutoff distance is used to filter possible reaction partners. Bond formation reactions are only considered for pairs of atoms within r_{react} distance from each other [36, 37]. Then at each reaction time, a random number is generated for possible reactions; if the generated number is lower than P, the reaction is allowed. DPD approaches had been utilized to model gelation via atom transfer radical polymerization [36], free radical polymerization [37], iniferter-mediated photo-growth of hydrogels [38, 39], and complexation and decomplexation reactions within hydrogels [41]. It is worth noting that in some reactive systems, the DPD approach can be integrated with another computational technique. Among the most recent examples, DPD approach integrated with quantum-chemical reaction path calculation was recently utilized to model the process of curing of thermoset resin [64]. DPD had also been recently used to quantify the effect of cross-linking reaction on drug diffusion in hyaluronic acid microneedles; in this work, atomistic MD simulations were performed to derive DPD parameters [65].

2.2.2 Addressing Unphysical Crossing of Polymer Bonds in DPD Along with Reactions

The soft conservative force in DPD provided in Eq. (2.1) does not prevent polymer chains to cross through each other. This is a known limitation of the standard DPD approach and significantly affects simulations of polymers, since such topological violations imply that entanglements are not captured. Among the few different efforts to effectively minimize topological violations, Kumar and Larson [66] first developed a segmental repulsive potential (SRP) effectively adding extra repulsion between bonds. The additional SRP force is given as:

$$\boldsymbol{F}_{ij}^{\mathrm{SRP}} = b\left(1 - \frac{d_{ij}}{d_{\mathrm{c}}}\right)\boldsymbol{e}_{ij}^{\mathrm{S}}, \tag{2.3}$$

where d_c is the SRP cutoff distance, b is the strength of the SRP repulsion, $d_{ij} = |\boldsymbol{d}_{ij}|$ is the distance between centers of bonds, and $\boldsymbol{e}_{ij}^{\mathrm{S}} = \boldsymbol{d}_{ij}/d_{ij}$ is the unit vector in the direction from one bond to another. The original SRP approach defines d_{ij} as the minimum distance between the centers of the bonds. Sirk et al. [67] later modified the SRP approach by redefining d_{ij} as the distance between bonds. Choosing the parameters $b = 80$ and $d_c = 0.8$ for this modified SRP (mSRP) approach leads to minimization of topological violations [67] which we confirmed with additional simulations in our recent work [68].

The mSRP approach described earlier was developed for polymers with a fixed topology, i.e. polymeric systems without chemical reactions. For modeling chemical reactions, the additional repulsion between bonds needs to be switched on/off as bond formation/breaking occurs. We recently incorporated this ability into the mSRP framework and implemented it as part of the LAMMPS simulation software [69–71]. Within LAMMPS, the mSRP repulsion between bonds is introduced via adding pseudo beads into the system at the locations of the bonds. The pseudo beads only interact with other pseudo beads and have no interactions with any other beads. Hence, in order to switch the extra repulsion on and off, we introduced the ability to insert and delete pseudo beads via the pair style srp/react command in LAMMPS [72]. The kinetics of bond breaking and formation occur via the stochastic process described earlier. The corresponding pseudo bead is either inserted or deleted during the reaction time step. A schematic representing the process discussed earlier is shown in Figure 2.1.

2.2.3 Modeling Cross-linking Due to Hydrosilylation Reaction

The mSRP DPD approach introduced earlier ensures that the unphysical bonds crossing is minimized for all the bonds in the system including the new bonds formed. This approach was recently utilized by Xiong et al. [73] to model cross-linking via hydrosilylation reaction in polymer blends containing linear preceramic polyhydromethylsiloxane (PHMS) precursors and sacrificial polymer

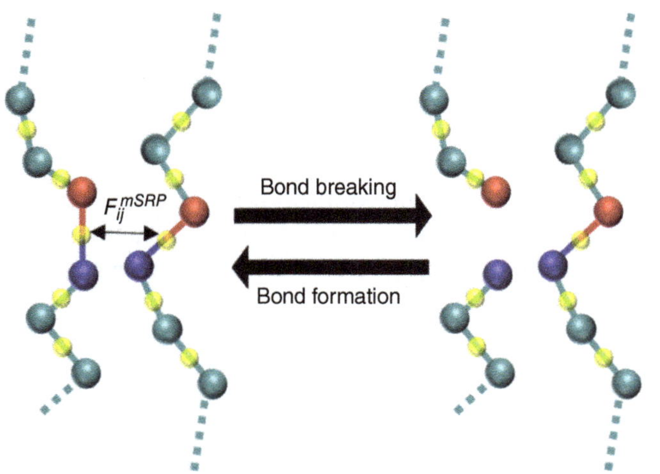

Figure 2.1 Schematic of the bond breaking and formation mechanism highlighting the role of mSRP pseudo beads (shown in yellow). When a bond is broken, the corresponding pseudo bead is deleted while a new bead is inserted upon bond formation.

chains. As reactive sacrificial component, vinyl-terminated polydimethylsiloxane (ν-PDMS) linear chains were considered in this work. The hydrosilylation reaction between the silicon-hydrogen group of PHMS and the vinyl groups of ν-PDMS results in formation of a ladder-type network structure [74, 75]. PDMS is often selected as the sacrificial component to fabricate porous polymer-derived ceramics (PDCs) due to the complete decomposition of PDMS during pyrolysis [76–78]. Hydrosilylation reaction takes place in a presence of a catalyst [79]; furthermore, hydrosilylation is one of the most important industrial applications of homogeneous catalysis [80].

Xiong et al. [73] introduced a suitable DPD framework to model hydrosilylation reaction along with the phase separation in the binary polymer blends incorporating PHMS and reactive sacrificial component (ν-PDMS) and in the ternary blends, which in addition to PHMS and ν-PDMS linear chains also incorporate second sacrificial nonreactive component. The following linear chains were considered as an additional nonreactive sacrificial component: methyl-terminated PDMS (m-PDMS), poly(methyl methacrylate) (PMMA), and polyacrylonitrile (PAN). To quantitatively characterize evolution in the binary and ternary reactive blends, the time evolution of the characteristic length scale within the system representing the size of the sacrificial domains, as well as fraction of vinyl groups of ν-PDMS that remain unreacted at a given time instant, were characterized as a function of time. It is instructive to calculate partial radial distribution functions (RDFs) $g_{\alpha\beta}(r, t)$ to characterize the structural properties as a function of time and upon reaching an equilibrium. The $g_{\alpha\beta}(r, t)$ for the beads of types α and β is computed via the beads' coordinates as [81]:

$$c_\alpha c_\beta g_{\alpha\beta}(r,t) = \frac{1}{\rho}\left\langle \frac{1}{N}\sum_{i=1}^{N_\alpha}\sum_{j\neq i}^{N_\beta}\delta(\boldsymbol{r}(t) + \boldsymbol{r}_i(t) - \boldsymbol{r}_j(t))\right\rangle, \tag{2.4}$$

where ρ is the total number density of the system of N beads, $c_{\alpha,\beta} = N_{\alpha,\beta}/N$ are the fractions of types α and type β beads, $\boldsymbol{r}_{i,j}$ are the coordinates of beads α and β, and $\langle\ldots\rangle$ denotes an ensemble average. The time evolution of the partial RDFs of all the sacrificial beads, $g_{ss}(r, t)$, is then used in Ref. [73] to quantify the characteristic length scale of the domains formed by the sacrificial beads. It is worth noting that the total RDF in the system reads $g(r) = c_{ss}^2 g_{ss}(r, t) + 2c_{ss}c_{mm}g_{sm}(r, t) + c_{mm}^2 g_{mm}(r, t)$, where $g_{mm}(r, t)$ is the partial RDF for all the beads comprising PHMS chains. The partial RDF of the sacrificial component, $g_{ss}(r, t)$, exhibits primary peaks at the distance corresponding to the bond length in Eq. (2.2). The magnitude of $g_{ss}(r, t)$ increases with time until an equilibrium is reached, reflecting an effect of clustering the sacrificial beads into distinct domains.

The characteristic length scale within this system can be defined via zero crossing [82–84] of the shifted RDF, $g_{ss}(r, t) - 1$. It is worth noting that alternatively the characteristic length scale in the system can also be defined from the first minimum [84] of $g_{ss}(r, t)$ at the distances exceeding the short-distance correlations. It had been demonstrated that the characteristic length scales calculated from zero crossings of the shifted RDFs follow the expected scaling during spinodal decomposition [82, 84]. Notably, the ratio between the length scales defined by these two approaches (from the minima of RDF and via the zero crossing of the shifted RDF)

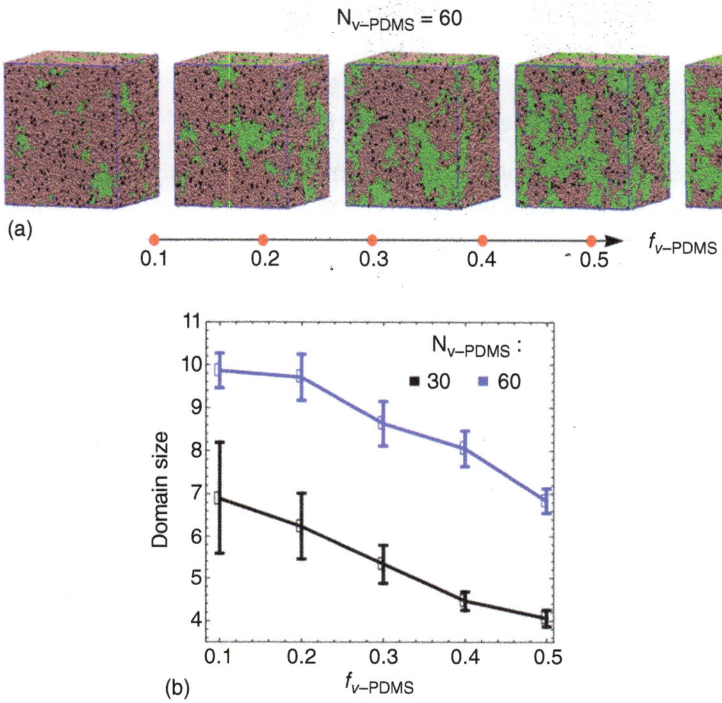

Figure 2.2 Effect of the fraction of the sacrificial component *v*-PDMS (denoted as f_{v-PDMS}) on the steady-state morphology and characteristic size of the sacrificial domains in PHMS/*v*-PDMS blends cross-linked via the hydrosilylation reaction. (a) Snapshots of blends with the fraction of *v*-PDMS from left to right as schematically marked on the arrow (f_{v-PDMS} = 0.1, 0.2, 0.3, 0.4, and 0.5). (b) The average characteristic length scale representing the average size of the sacrificial domain as a function of f_{v-PDMS} for two degrees of polymerization of *v*-PDMS: $N_{v-PDMS} = 30$ (in black) and $N_{v-PDMS} = 60$ (in blue); error bars represent standard deviations. Source: Reproduced from Ref. [73] MDPI/CC BY 4.0.

remains approximately constant [82, 84]. The characteristic length scale identified via zero crossing of $g_{ss}(r, t) - 1$ was used to identify characteristic length scales of sacrificial domains as a fraction of a sacrificial component, $f_{\nu-PDMS}$, for two degrees of polymerization of *ν*-PDMS: $N_{\nu-PDMS} = 30$ (black curve) and $N_{\nu-PDMS} = 60$ (blue curve) (Figure 2.2b). This plot shows that the characteristic average length scale in equilibrium decreases with an increase in $f_{\nu-PDMS}$ for both values of $N_{\nu-PDMS}$. This decrease is most pronounced for the highest values of $f_{\nu-PDMS}$, which in turn correspond to the highest number of reactive chain ends and correspondingly the highest number of cross-links formed in this blend upon reaching an equilibrium. Further, smaller sacrificial domains are formed for lower values of $N_{\nu-PDMS}$; similar trend showing smaller pore sizes for lower molecular weights of *ν*-PDMS had been demonstrated in experiments [85]. To summarize, results of the DPD simulations show that the phase separation between the PHMS and *ν*-PDMS is arrested due to the hydrosilylation reaction. These results clearly demonstrate [73] that the morphology of the sacrificial domains in the phase-separated network encompassing

PHMS can be tailored by tuning the composition, the degree of polymerization, and chemical nature of the sacrificial components.

2.2.4 Mesoscale Modeling of Degradation and Erosion

We reported mesoscale modeling of degradation and erosion in polymer networks in our recent work [68]. Here, degradation is defined as the chemical reaction that cleaves chemical bonds, while erosion is the process of mass loss during the degradation [86, 87]. We model degradation in tetra-arm polyethylene glycol (tetra-PEG) gels [88] which consist of two distinct tetra-PEG precursors. The two precursors have distinct and mutually reactive end functionalities; the end-linking reaction of these precursors leads to formation of nearly ideal homogenous networks as had been demonstrated by Sakai et al. [88]. A schematic of the starting structure of the tetra-PEG network is shown in Figure 2.3a.

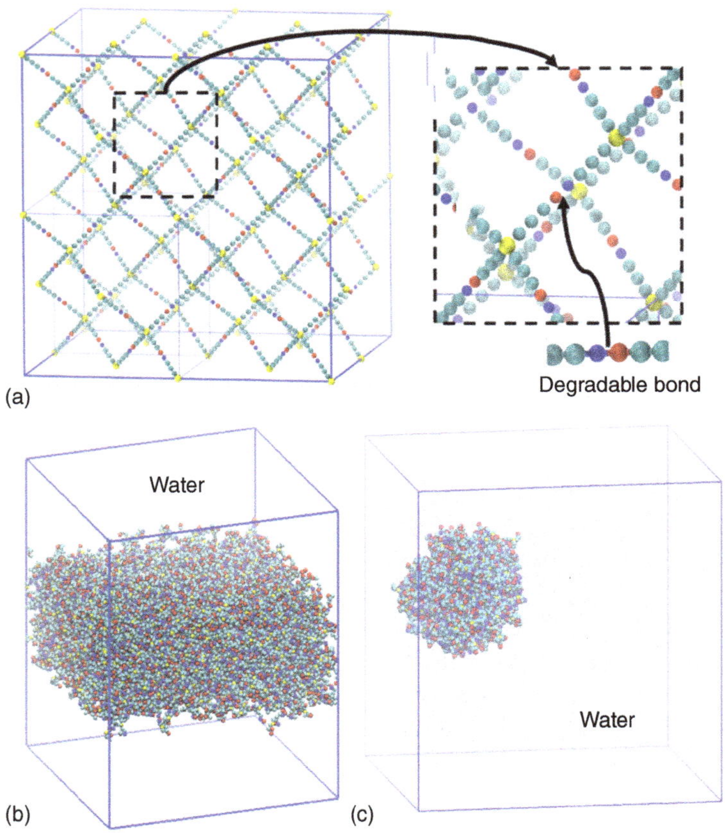

Figure 2.3 (a) Schematic of the initial diamond-like network topology for tetra-PEG gels. The inset shows a zoomed in view of the network highlighting the degradable bond. Beads are colored as follows: cyan for PEG beads, red and blue for end functionalities, and yellow for the centers of each tetra-arm precursor. Snapshots of (b) an equilibrated hydrogel film and (c) a nanogel particle in water. Beads representing water are hidden in (b) and shown as points in (c).

Beads representing the PEG monomers are shown as cyan, while the two end functionalities are shown as red and blue and the centers of tetra-arm precursors shown as yellow. Only bonds between end functionalities are considered for the degradation reaction corresponding to typical choice of cleavable end groups in experiments [89, 90]. The initial structure of the polymer network is modeled as a diamond-like lattice [36, 40, 91] with the centers of the tetra-arm precursors occupying lattice sites. Starting from an initial unit cell of the diamond-like lattice, hydrogel films [68] and nanogel particles [92, 93] can be constructed. The complete details of constructing the starting structure for each of these systems are provided in the respective original publications. Figure 2.3b shows an equilibrated hydrogel film that is swollen in a good solvent for the polymer (water). Correspondingly, Figure 2.3c shows a swollen nanogel particle suspended in water. Periodic boundary conditions are imposed in the simulations shown in the snapshots in Figure 2.3.

We use the stochastic protocol described in Section 2.2 to simulate degradation and erosion of these materials. The first-order degradation reaction can be simulated by setting appropriate values for the reaction probability, P, and the reaction time step τ_r. We simulated such first-order degradation of nanogel particles in our recent work [93]. The first-order degradation rate constant in these simulations is a function of P and τ_r as $k = P/\tau_r$ [68, 93]. The progress of the degradation reaction can be measured via the fraction of degradable bonds intact, p, which follows the first-order exponential relation $p = \exp(-kt)$. The evolution of fraction of degradable bonds intact for degradation of nanogels at various degradation rates is plotted in Figure 2.4a. Symbols in Figure 2.4a represent average measurements from five independent DPD simulations with error bars representing standard deviation. The degradation rate is controlled by adjusting P and τ_r and thus tuning k. Several combinations of P and τ_r values were used [93] as shown in Figure 2.4b. Solid lines in Figure 2.4a represent the analytical plot of $p = \exp(-kt)$. It can be seen that the first-order reaction model is followed in simulations with all tested combinations of parameters.

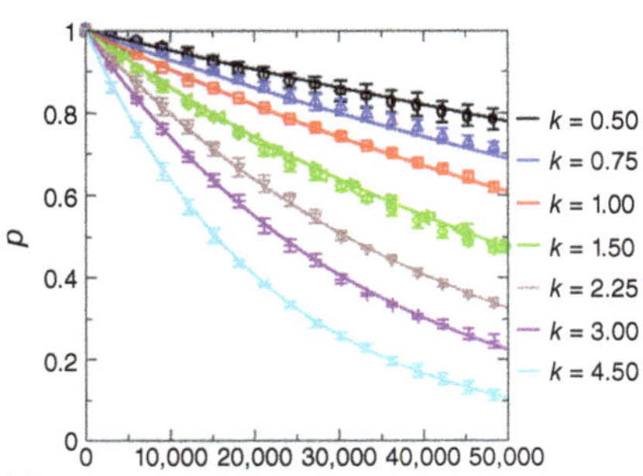

(a)

(b)

Symbol	k (units $10^{-5}\tau^{-1}$)	P	τ_R (units τ)
○	0.50	3×10^{-6}	0.6
△	0.75	3×10^{-6}	0.4
□	1.00	6×10^{-6}	0.6
◇	1.50	9×10^{-6}	0.6
◁	1.50	6×10^{-6}	0.4
▷	1.50	3×10^{-6}	0.2
▽	2.25	9×10^{-6}	0.4
+	3.00	6×10^{-6}	0.2
×	4.50	9×10^{-6}	0.2

Figure 2.4 (a) Evolution of the fraction of bonds intact as a function of time. Symbols represent average measurements over five DPD simulations with error bars representing standard deviations. Lines represent the function $p = \exp(-kt)$, with $k = P/\tau_r$ provided in the legend in the units of $10^{-5}\tau^{-1}$. (b) Table containing a list of P and τ_r values for each simulation in (a). Figure reproduced from ref. [93] with permission from Springer Nature.

In our recent work, we simulated the degradation and erosion of hydrogel films [68]. To track the progress of this process, we defined topological clusters or fragments as any set of bonded tetra-PEG precursors. Consequently, the cluster size is defined as the number of tetra-PEG precursors constituting that cluster. Before degradation commences, a single topological cluster encompasses all the precursors. Progress of degradation is tracked by following the evolution of both the size of the largest cluster, $N_T(t)$ (black curve, left axis) and the total number of clusters during degradation (red curve, right axis) as shown in Figure 2.5a.

Initially, relatively small fragments leave the largest cluster, and hence the size of the largest cluster only decreases slightly. The release of only small fragments is evident in the simulation snapshot in Figure 2.5b along with the corresponding distribution of cluster sizes in Figure 2.5e. It is evident that along with a large number of smaller fragments, only one large cluster exists in the system (inset in Figure 2.5b). The large cluster corresponds to the hydrogel film. During the entire degradation process, smaller clusters dominate the distribution (see Figure 2.5e–g). As shown in Figure 2.5b–c, several of these clusters leave from the hydrogel and diffuse away from the film. Apart from the clusters that diffused away, there are also some clusters that remain stuck either at the surface or within the bulk of the hydrogel film. The topological characterization discussed earlier does not allow one to differentiate between these clusters. Additional distance-based characterization to quantify the aspects such as the mass loss from the system is discussed in the following section. Multiple larger clusters appear as the degradation reaction continues (Figure 2.5c,d), which is accompanied by a sharp decrease in the size of the largest cluster (Figure 2.5a). At some point during the sharp decrease, the percolating hydrogel network vanishes, and many relatively small clusters are formed in the system. This is evident from the distributions in Figure 2.5f,g. Hence, at some point during this process, the reverse gelation transition occurred.

To quantify the reverse gel point, we track the evolution of molecular weights during the degradation process [68]. The weight average degree of polymerization is defined as $DP_{\mathrm{w}}(t) = \frac{\Sigma n_i(t)i^2}{\Sigma n_i(t)i}$, where i denotes the number of beads in a cluster and $n_i(t)$ the number of clusters with i beads. DP_{w} becomes infinitely large at the gel point in analytical gelation theories [94–96]; however, it does not diverge during finite-size simulations of gelation or reverse gelation. Hence, the reduced weight average degree of polymerization,

$$DP^{\mathrm{r}}_{\mathrm{w}}(t) = \frac{\Sigma' n_i(t)i^2}{\Sigma' n_i(t)i}, \tag{2.5}$$

where the summation is taken over all the clusters excluding the largest cluster, is typically used in finite-size simulations of gelation [97–101] and in our work on reverse gelation [68].

$DP^{\mathrm{r}}_{\mathrm{w}}$ is plotted in Figure 2.6a and exhibits a peak at the reverse gel point during the degradation process. This is in analogy to the peak in $DP^{\mathrm{r}}_{\mathrm{w}}$ observed for simulations of gelation in finite-size systems [98–101]. Figure 2.6a contains $DP^{\mathrm{r}}_{\mathrm{w}}$ from five independent simulations, each normalized by the corresponding maximum values, for three different degradation rate constants. $DP^{\mathrm{r}}_{\mathrm{w}}$ is initially low since the hydrogel

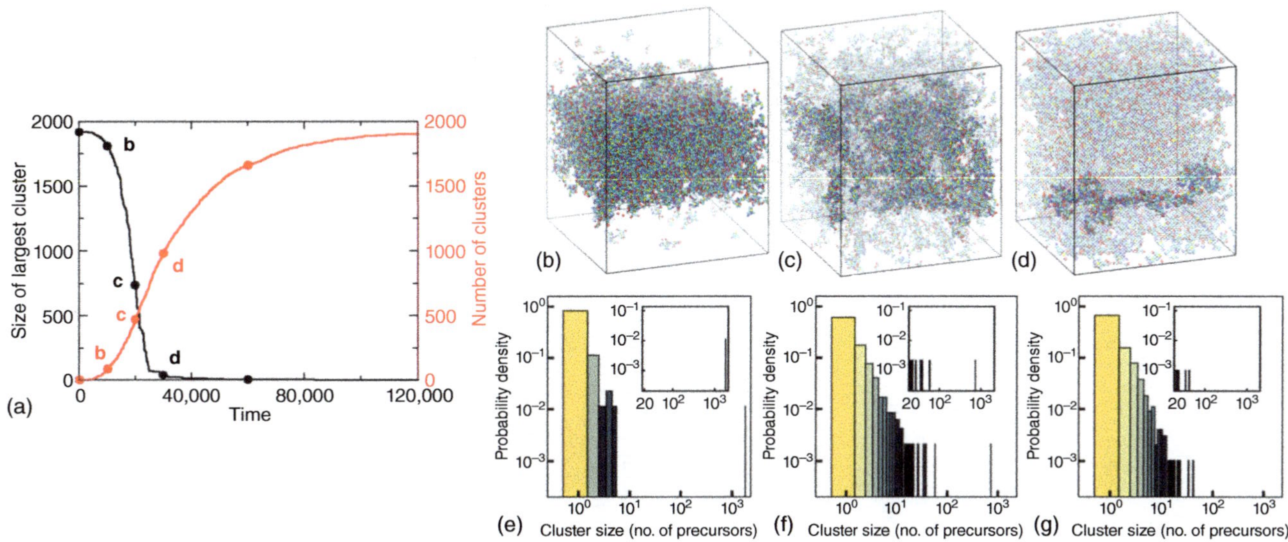

Figure 2.5 (a) Evolution of the size of the largest cluster measured in number of precursors (black curve) and total number of clusters (red curve) during degradation of a hydrogel film. Snapshots of the degradation process at (b) $t = 10{,}000$, (c) $t = 20{,}000$, and (d) $t = 30{,}000$. The largest cluster during degradation is highlighted with all fragments that broke off from the film shown as translucent. Points corresponding to snapshots are marked in (a). Probability density distributions of the cluster sizes during degradation at (e) $t = 10{,}000$, (f) $t = 20{,}000$, and (g) $t = 30{,}000$. Insets in (e–g) show zoomed in view for larger cluster sizes. Source: Reproduced with permission from ref [68]. Copyright 2022 American Chemical Society.

film constitutes the only large cluster in the system. As bonds break, DP_w^r increases up to the reverse gel point where it exhibits a peak value corresponding to the disintegration of the percolating hydrogel into rather large fragments. As these fragments further degrade, DP_w^r decreases with time. t_c^w, the critical time corresponding to the peak in DP_w^r, enables the definition of a reverse gel point via the corresponding critical value of p as $p_c^w = \exp\left(-kt_c^w\right)$. In addition to this definition of reverse gel point, in a similar manner the z-average degree of polymerization, $DP_z(t) = \frac{\Sigma n_i(t)i^3}{\Sigma n_i(t)i^2}$, can also be used to define a reverse gel point. In analogy to DP_w^r, we calculate the reduced z-average degree of polymerization, DP_z^r, 2.as:

$$DP_z^r(t) = \frac{\Sigma' n_i(t)i^3}{\Sigma' n_i(t)i^2}, \tag{2.6}$$

where the summation is again taken over all clusters excluding the largest cluster. Figure 2.6b shows both a representative DP_z^r and the corresponding DP_w^r curves for each of the degradation rates. In similar fashion to t_c^w, the time corresponding to the peak in DP_z^r, t_c^z, enables an alternate definition of the reverse gel point as $p_c^z = \exp\left(-kt_c^z\right)$. Detailed analysis of the measured reverse gel point and comparison with expected analytical values is provided in the following section. To further characterize the degradation process, we track the dispersity (Đ) (formerly rereferred to as Polydispersity Index (PDI)) in Figure 6c as:

$$Đ(t) = \frac{DP_w(t)}{DP_n(t)}, \tag{2.7}$$

where $DP_n(t) = \frac{\Sigma n_i(t)i}{\Sigma i}$ is the number average degree of polymerization. Although Đ also exhibits a peak similar to DP_w^r and DP_z^r, the peak in Đ is observed prior to the reverse gel point (the circles of corresponding color in Figure 2.6c represent averaged values of the reverse gel point). Previous gelation simulations have reported an analogous trend where the peak in Đ was observed after the gel point [101].

Averaged over five independent simulations each, we reported [68] the following values for the reverse gel point: $p_c^w = 0.425 \pm 0.014, 0.401 \pm 0.015$ and 0.408 ± 0.029

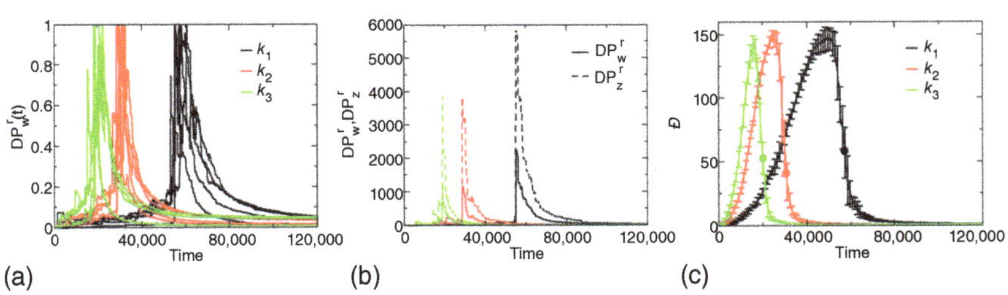

Figure 2.6 Evolution of (a) reduced weight average degree of polymerization DP_w^r, (b) example DP_w^r and corresponding reduced z-average degree of polymerization DP_z^r, and (c) dispersity for hydrogel films degrading with a first-order rate constants $k = 1.5 \cdot 10^{-5}\tau^{-1}$ (black curves), $k = 3.0 \cdot 10^{-5}\tau^{-1}$ (red curves), $k = 4.5 \cdot 10^{-5}\tau^{-1}$ (green curves). Plots in (a) and (b) contain data from five simulations at each of the three degradation rates. Each curve in (a) and (b) represents one simulation, while each curve in (c) represents an average over five simulations with error bars representing standard deviation. Source: Reproduced with permission from ref [68]. Copyright 2022 American Chemical Society.

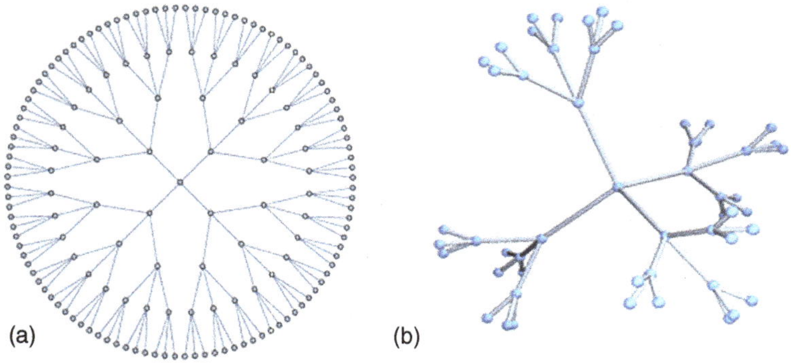

Figure 2.7 Bethe lattice with junction functionality 4 in 2D (in a) and in 3D (in b); images are generated using Wolfram Mathematica@ software.

and $p_c^z = 0.425 \pm 0.014, 0.402 \pm 0.019, 0.400 \pm 0.04$, correspondingly, for the three degradation rates discussed in Figure 2.6. These reported values are significantly higher than the value predicted by mean-field percolation theory for tetrafunctional networks [94, 95] ($p_c = 0.33$). This difference can be attributed to the difference between the initial diamond lattice-like network topology representing tetrafunctional network (Figure 2.3a) and the Bethe lattice structure (see Figure 2.7a,b) postulated in the mean-field theories. Notably, during the gelation processes, an increase in values of gelation points observed in a number of previous studies [101–107] with respect to those predicted by the mean-field theories was primarily contributed to the intramolecular reactions in polymer networks.

Since we use the diamond lattice topology as our starting structure (Figure 2.3a), the percolation problem closest to our simulations is that of bond percolation on a diamond lattice [108] ($p_c = 0.39$). A gel point close to $p_c = 0.39$ was also reported in the experiments on gelation of tetra-arm PEG precursors at close to overlap concentration [109, 110].

It is worth noting that the values of the reverse gel points measured in our simulations as listed earlier are close to but somewhat exceed the value $p_c = 0.39$ predicted theoretically for bond percolation on diamond lattice. This could be attributed to the finite number of precursors in the simulations. Notably, prior studies of gelation of finite-size network reported an increase in the gel point with the number of precursors within the system compared to the gel point for the infinite network [97, 108].

Having characterized the reverse gel point, we now focus on characterizing erosion, the process of loss of material during degradation [86, 87]. The evolution of fractional mass loss, $f(t) = 1 - m(t)/m_0$, where m_0 is the initial mass, and $m(t)$ is the mass at time t, is tracked in hydrogel degradation experiments [86]. Unlike the simple first-order degradation reaction, kinetics of the fractional mass loss in experiments demonstrate different regimes. At early times, slow mass loss is observed, which is followed by an accelerated mass loss attributed to reverse gelation [86, 111].

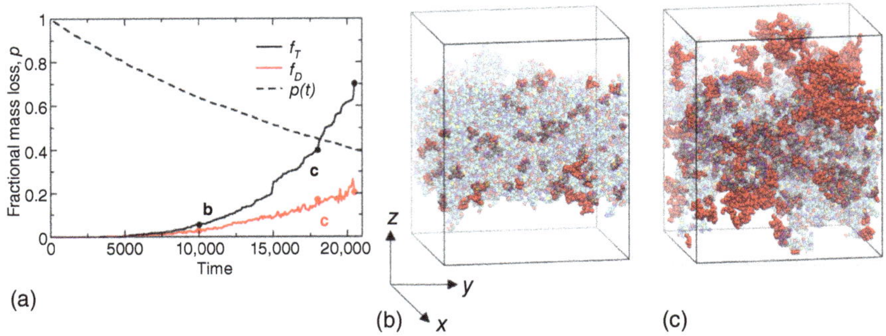

Figure 2.8 (a) Evolution of the fraction of bonds intact, p (dashed curve), topological mass loss (solid black curve), and distance-based mass loss (red curve) during degradation of a hydrogel film. Snapshots of the degrading hydrogel film, at (b) $t = 10{,}000$ and (c) $t = 18{,}000$. The hydrogel film is shown as translucent, and broken-off fragments that are stuck either within the bulk or at the surface of the film are shown as red. Fragments that broke off and diffused more than r_c from the film are not shown. Source: Reproduced with permission from ref [68]. Copyright 2022 American Chemical Society.

Based on the definition of topological cluster aforementioned, we measure mass loss up to the reverse gel point by measuring the mass of detached topological clusters that leave the hydrogel due to bond cleavage. As all DPD beads have the same mass, the fractional mass loss based on topological clusters is defined as $f_T(t) = 1 - N_T(t)/N_0$ (black curve in Figure 2.8a, $t_c = 20{,}500$), where $N_T(t)$ corresponds to the size of largest cluster at time t, measured via number of precursors, and $N_0 = N_T(0)$ is the total number of precursors in the system. The slow and fast mass loss regimes as discussed earlier are clearly identifiable in the mass loss data. Up to $t = 10{,}000$, for example, only ≈5% of the mass is lost ($f_T(t) \approx 0.05$) while ≈35% of the degradable bonds have broken (dashed curve in Figure 2.8a); this point is marked as (b) in Figure 2.8a. Some of the detached precursors do not diffuse away and remain stuck either at the surface or within the bulk of the hydrogel film. The mass loss trend accelerates notably before reverse gelation occurs, since larger fragments detach from the network as degradation proceeds.

$f_T(t)$ represents a purely topological definition of mass loss and does not consider spatial position of clusters. Such a definition overestimates the actual mass loss from the film as it does not differentiate between clusters that detached and diffused away and clusters that remain stuck within the film. To only consider fragments that no longer interact with the film as "being lost," we define a distance-based cluster or an agglomerate as a set of precursors each within r_c from another precursor. Correspondingly, mass loss from the largest agglomerate is defined as: $f_D(t) = 1 - N_D(t)/N_0$, where N_D is the number of precursors in the largest agglomerate. The evolution of f_D is provided in Figure 2.8a (red curve). The detached fragments that remain stuck with the largest topological cluster are now incorporated in the largest agglomerate. These stuck fragments are highlighted in dark red and shown through the hydrogel film (the film is shown as translucent) in the snapshots at $t = 10{,}000$ ($p = 0.64$) and at $t = 18{,}000$ ($p = 0.44$) in Figure 2.8b and c, respectively. Understanding the main features of degradation and erosion on the mesoscale could provide guidelines for designing degrading materials with controlled properties.

2.3 Continuum Modeling of Reactive Hydrogels

2.3.1 Modeling Chemo- and Photo-Responsive Reactive Hydrogels

While it is critically important to understand dynamics of reactive hydrogels on the mesoscale, the characteristic sizes of the hydrogel matrix in many relevant applications significantly exceed the length scales accessible via DPD modeling introduced in the section earlier. Herein, we first briefly review selected approaches to continuum modeling of the reactive hydrogels with the primary focus on modeling of degradation of various hydrogels on the continuum length scales.

A dynamic response of a broad range of chemo-responsive gels to variations in external stimuli including light and external confinements can be characterized using gel Lattice Spring Model (gLSM). The gLSM was originally developed by Yashin and Balazs [112, 113] to capture dynamics of the chemo-responsive gels undergoing Belousov–Zhabotinsky chemical reactions (referred to as BZ gels) in two dimensions; shortly after this approach was extended to model gel elastodynamics in three dimensions [114]. Within the gLSM, the total energy of the deformed polymer network encompasses elastic energy and Flory–Huggins mixing energy contributions, which in turn incorporates an additional term proportional to the volume fraction of the reactive component grafted onto the polymer matrix (such as a catalyst in its oxidized state in the original formulation of the model [112, 113]). This additional term captures the effect of chemical species on the degree of swelling of the hydrogel, i.e. mechanical response of the hydrogel to the chemical reactions taking place within the system. The results obtained using gLSM are in a good agreement with a number of experimental findings [115–121]; the details of this approach and its applications can be found in reviews by Yashin et al. [122] and Kuksenok et al. [117]. Most recent extensions of gLSM include development of a framework for modeling networks encompassing loops and dangling chain ends [123], modeling formation of 3D helices in gel–fiber composites [124], modeling directed and programmed locomotion of active chemo-responsive BZ gels [125–127], and modeling elastodynamics of thin azobenzene-functionalized hydrogels immersed into the α-cyclodextrin (α-CD) solution under various illumination conditions [128].

Current advances in modeling large deformation and fracture of hydrogels are surveyed in a recent review by Lei et al. [129], where a range of models from CGMD models of polymer network to constitutive models capturing viscoelasticity and various fracture criteria are considered. A range of simulation approaches suitable to model various aspects of dynamics of soft hydrogels, including finite element methods and meshless methods, are surveyed in the same review [129]. Further, contributions of ionization and polarization to the free energy required for constitutive model of polyelectrolyte gels, and free energy of photoexcitation in a gel filled with the light-absorbing nanoparticles are also discussed [129]. Finally, with respect to the photo-induced reactions within the hydrogels, recent finite element modeling and theoretical analysis of light-responsive spiropyran-functionalized hydrogels demonstrated different modalities of buckling and an importance of the effects of light attenuation across the sample thickness [130].

2.3.2 Continuum Modeling of Degradation of Polymer Network

Early one-dimensional approaches for degradation of polymer matrix employed the diffusion equation for drug release [131] via modeling polymer erosion by means of a constantly moving front [132]. Lattice-based two-dimensional models were later developed to account for microstructural changes during polymer erosion [133, 134]. With respect to modeling hydrogel degradation, analytical models have been developed capturing degradation and erosion [86, 89, 135–140] of bulk degrading hydrogels. In these models, the overall process of network degradation and erosion, i.e. breaking of fragments from the network and the erosion accompanied with it, is understood as the reverse of the gelation process, where smaller fragments combine to form a percolating network. Correspondingly, a reverse gel point is typically reported [86, 89], in analogy to the gel point during gelation, in terms of the fraction of bonds that need to be broken (or inversely the fraction of bonds that remain intact) and is reported to accompany a sudden and complete mass loss [86]. A lattice-based stochastic model, which employed analytical theories to relate degradation reaction kinetics with gel swelling, was also introduced [135]. Analytical and two-dimensional lattice-based Monte Carlo simulations have also been utilized in the analysis of drug release from hydrogel-based drug delivery platforms [141].

Degradation of polymer network can also occur under mechanical forcing. The classical constitutive models were recently extended by Lu et al. [142] to capture the deformation of reversible and irreversible tough double-network hydrogels; the calculated stress–strain curves agreed well with the concurrent experimental results. Lamont et al. [143] formulated a model based on the transient network theory to track the damage in dynamic polymer networks and to identify effects of both reversible bond exchange and irreversible chain rupture on materials properties, such as strain stiffening and self-healing; the modeling results were shown to be in a good agreement with the experimental data on cyclic loading and self-healing in dynamic polymer networks. A model capturing polymer chains rearrangement in the crack tip vicinity in hybrid double-network hydrogels under cyclic stretching was recently developed by Liu et al. [144]. During the loading, a fraction of physical bonds break resulting in water diffusion into the crack tip region, while during the unloading, some of these bonds reconnect, leading to localized compression and water escape from this region [144].

A modeling approach that captures homogenization or phase separation depending on the respective Flory–Huggins interaction parameter describing affinity between the polymer and solvent, isotropic swelling mechanics, and uniaxial tensile stress for a tetra-PEG gel was recently introduced by Wagner et al. [145]. The numerical framework for this model was implemented in two dimensions based on the appropriate choice of representative volume elements (RVEs). It had been shown that the model can be used to predict the mechanical response and failure of PEG-based gels with various functionalities of network junctions and molecular weights of macromers forming polymer network [145].

Most recently, Pan and Brassart [146] formulated a continuum framework and constitutive models for large deformation of hydrogels coupled with the hydrolytic degradation; the constitutive models captured the evolution of elastic modulus and mass loss of polymer based on network topology evolution during hydrolysis. The proposed approach was validated with respect to the experimental data available for the two types of gels including biodegradable tetra-PEG hydrogels. Importantly, fast diffusion of water molecules and reaction products was assumed in the this model [146] based on the fact that the characteristic time scale for chemical degradation significantly exceeds characteristic diffusion time scales.

As mentioned in the introduction, one of the important applications of degradable hydrogels is their use in tissue engineering as controllably degradable scaffolds. An important challenge in designing specific hydrogels promoting tissue growth is tailoring hydrogel degradation along with the transport of ECM so that sustained mechanical integrity of the construct incorporating hydrogel scaffold and growing tissue can be maintained [147]. Akalp et al. [147] developed a mathematical model and a respective numerical nonlinear finite element framework incorporating the following: (i) degradation of hydrogel around a single cell; (ii) ECM deposition and growth of the new tissue (neo-tissue); and (iii) characterization of the mechanical integrity of the construct incorporating degrading hydrogel and growing neo-tissue. This work focused on the scenario where cartilage cells (chondrocytes) are embedded into 8-arm PEG hydrogel with enzyme-sensitive peptide cross-links [147], in which degradation kinetics of the cross-links can be tailored by the selection of amino acids [148]. The cell encapsulated within the hydrogel scaffold releases enzymes and ECM molecules. While the diffusion of relatively large ECM molecules is essentially arrested within the intact hydrogel, sufficiently small enzymes can diffuse into the hydrogel and cleave a fraction of cross-links via Michaelis–Menten kinetics [147]. The relative size of the enzyme with respect to the gel mesh size, r^*, and normalized rate of degradation, k^*, which captures competition between degradation and diffusion, identify diffusion-limited and diffusion-dominated degradation regimes; r^* and k^* increase from top to bottom row in Figure 2.9. These results show that for the large network mesh size or small enzymes, the diffusive processes are close to those in pure solvent and scaffold degradation is a rate-limiting process (Figure 2.9a), while for the large enzymes or a small mesh size, the rate-limiting step is enzymes diffusivity so that sharp degradation front is observed (Figure 2.9c). A new tissue is formed when ECM percolate and can sustain mechanical load to maintain mechanical integrity of the construct; hence, the reverse gelation of the scaffold must be matched with the percolation threshold of the newly formed ECM [147], which in turn introduces restrictions on the realization of the construct which would indeed effectively promote neo-tissue growth. Three-dimensional continuum modeling by Sridhar et al. [149] demonstrated a key role of heterogeneity in defining mechanical

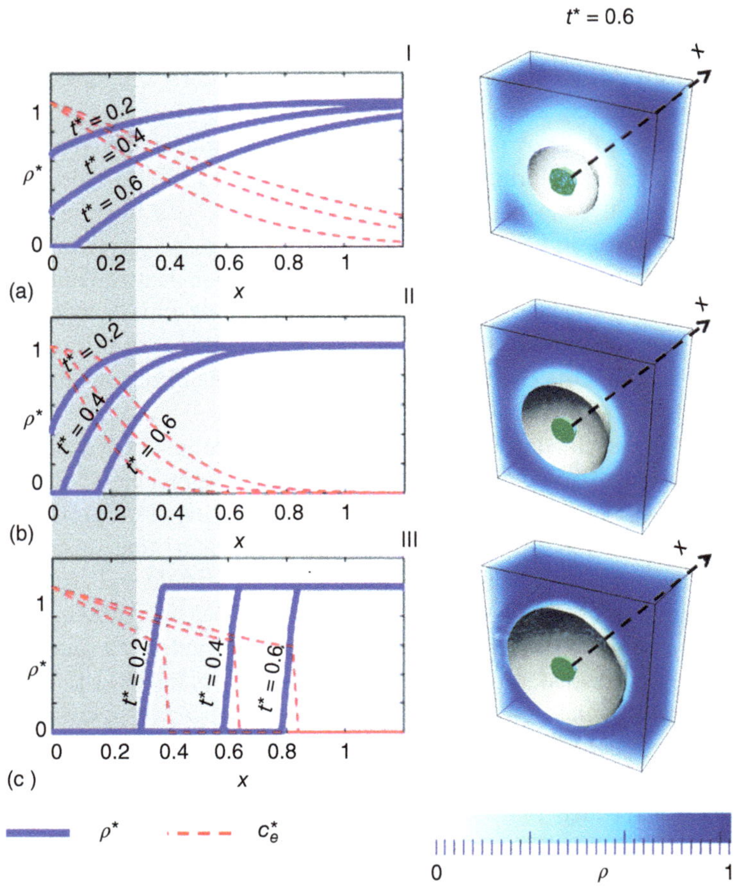

Figure 2.9 Modeling degradation of hydrogel scaffold around a single cell. Normalized cross-link density of the degrading hydrogel and normalized enzyme concentration are shown by solid blue and dashed red lines in (a–c). The distance x in (a–c) denotes the distance from the cell surface. The relative size of the enzyme with respect to the gel mesh size (r^*) and the normalized rate of degradation (k^*) increase from (a) to (c). The images on the right correspond to the latest normalized time instant shown in (a–c).
Source: Reproduced from ref. [147] with permission from the Royal Society of Chemistry.

properties of hydrolytically degradable cell-laden hydrogel scaffolds ensuring an effective neo-tissue growth. Current advances in computational modeling and experimental studies of cell-laden hydrogels, including mechanisms of hydrogel degradation coupled with ECM assembly and neo-tissue growths, are surveyed in recent comprehensive review by Vernerey et al. [150].

2.4 Conclusions

Depending on a particular application of reactive hydrogels, characteristic relevant length scales of the problem can range from tens of nanometers to microns and to significantly larger length scales often exceeding millimeter sizes. The respective choice of the relevant approach to model dynamic response of reactive hydrogels thereby needs to be informed by the relevant characteristic length scale of the problem. In this review, we surveyed most recent select approaches to model reactive processes in polymer networks with the emphasis on characterizing controlled degradation in swollen networks. Examples of catalyst-initiated reactions such as hydrosilylation reaction relevant in many practical applications and enzymatic degradation of hydrogels were also introduced. The mesoscale DPD-based approach described in this review combines an efficient mSRP approach allowing one to minimize topology violations in DPD simulation of polymers [67] with a stochastic approach for modeling chemical reactions. The mSRP DPD approach was used in simulations of the hydrosilylation reaction during synthesis of PDCs and for modeling degradation and erosion in hydrogel films. Measurements of domain size clearly demonstrated that the blends morphology can be controlled during the hydrosilylation reaction via polymer properties such as types of polymers used and the degree of polymerization. In mesoscale simulations of the degrading hydrogel film, the reverse gel point and the mass loss from the largest cluster were characterized. In addition to the mesoscale simulations, continuum approaches to model reactive polymer networks with the focus on modeling degradation, and in particular enzymatic degradation of hydrogels, were also introduced. Understanding reactivity and in particular degradation of hydrogels plays an important role in a plethora of applications and could provide guidelines for designing reactive hydrogel-based materials with dynamically controlled properties.

Acknowledgments

This work was supported in part by the National Science Foundation under NSF Award No. 2110309.

References

1 Zhang, Y.S. and Khademhosseini, A. (2017). Advances in engineering hydrogels. *Science* 356 (6337): eaaf3627.
2 Rosales, A.M. and Anseth, K.S. (2016). The design of reversible hydrogels to capture extracellular matrix dynamics. *Nature Reviews Materials* 1 (2): 1–15.
3 Lin, C.-C. and Metters, A.T. (2006). Hydrogels in controlled release formulations: network design and mathematical modeling. *Advanced Drug Delivery Reviews* 58 (12-13): 1379–1408.
4 Wang, Z., Zheng, X., Ouchi, T. et al. (2021). Toughening hydrogels through force-triggered chemical reactions that lengthen polymer strands. *Science* 374 (6564): 193–196.
5 Kloxin, A.M., Kasko, A.M., Salinas, C.N., and Anseth, K.S. (2009). Photodegradable hydrogels for dynamic tuning of physical and chemical properties. *Science* 324 (5923): 59–63.

6 Frey, M.T. and Wang, Y.L. (2009). A photo-modulatable material for probing cellular responses to substrate rigidity. *Soft Matter* 5 (9): 1918–1924.

7 Wong, D.Y., Griffin, D.R., Reed, J., and Kasko, A.M. (2010). Photodegradable hydrogels to generate positive and negative features over multiple length scales. *Macromolecules* 43 (6): 2824–2831.

8 Luo, Y. and Shoichet, M.S. (2004). A photolabile hydrogel for guided three-dimensional cell growth and migration. *Nature Materials* 3 (4): 249–253.

9 Griffin, D.R. and Kasko, A.M. (2012). Photoselective delivery of model therapeutics from hydrogels. *ACS Macro Letters* 1 (11): 1330–1334.

10 Ramanan, V.V., Katz, J.S., Guvendiren, M. et al. (2010). Photocleavable side groups to spatially alter hydrogel properties and cellular interactions. *Journal of Materials Chemistry* 20 (40): 8920–8926.

11 Kloxin, A.M., Tibbitt, M.W., Kasko, A.M. et al. (2010). Tunable hydrogels for external manipulation of cellular microenvironments through controlled photodegradation. *Advanced Materials* 22 (1): 61.

12 Azagarsamy, M.A., Alge, D.L., Radhakrishnan, S.J. et al. (2012). Photocontrolled nanoparticles for on-demand release of proteins. *Biomacromolecules* 13 (8): 2219–2224.

13 McKinnon, D.D., Brown, T.E., Kyburz, K.A. et al. (2014). Design and characterization of a synthetically accessible, photodegradable hydrogel for user-directed formation of neural networks. *Biomacromolecules* 15 (7): 2808–2816.

14 Hoogerbrugge, P.J. and Koelman, J.M.V.A. (1992). Simulating microscopic hydrodynamic phenomena with dissipative particle dynamics. *Europhysics Letters (EPL)* 19 (3): 155–160.

15 Espanol, P. and Warren, P. (1995). Statistical-mechanics of dissipative particle dynamics. *Europhysics Letters* 30 (4): 191–196.

16 Groot, R.D. and Warren, P.B. (1997). Dissipative particle dynamics: bridging the gap between atomistic and mesoscopic simulation. *The Journal of Chemical Physics* 107 (11): 4423–4435.

17 Espanol, P. and Warren, P.B. (2017). Perspective: dissipative particle dynamics. *The Journal of Chemical Physics* 146 (15): 150901.

18 Vishnyakov, A., Lee, M.T., and Neimark, A.V. (2013). Prediction of the critical micelle concentration of nonionic surfactants by dissipative particle dynamics simulations. *Journal of Physical Chemistry Letters* 4 (5): 797–802.

19 Mills, Z.G., Mao, W.B., and Alexeev, A. (2013). Mesoscale modeling: solving complex flows in biology and biotechnology. *Trends in Biotechnology* 31 (7): 426–434.

20 Shillcock, J.C. and Lipowsky, R. (2005). Tension-induced fusion of bilayer membranes and vesicles. *Nature Materials* 4 (3): 225–228.

21 Dutt, M., Kuksenok, O., Nayhouse, M.J. et al. (2011). Modeling the self-assembly of lipids and nanotubes in solution: forming vesicles and bicelles with transmembrane nanotube channels. *ACS Nano* 5 (6): 4769–4782.

22 Zhang, M., Larison, T., Tu, S. et al. (2022). Effect of fluorophobic character upon switching nanoparticles in polymer films from aggregated to dispersed

states using immersion annealing. *ACS Applied Polymer Materials* 4 (10): 7042–7053.

23 Tu, S., Choudhury, C.K., Giltner, M. et al. (2022). Mesoscale modeling of agglomeration of molecular bottlebrushes: focus on conformations and clustering criteria. *Polymers* 14 (12): 2339.

24 Wei, L., Caliskan, T.D., Tu, S. et al. (2020). Highly oil-repellent thermoplastic boundaries via surface delivery of CF3 groups by molecular bottlebrush additives. *ACS Applied Materials & Interfaces* 12 (34): 38626–38637.

25 Xu, Z., Yang, Y., Zhu, G. et al. (2019). Simulating transport of soft matter in micro/nano channel flows with dissipative particle dynamics. *Advanced Theory and Simulations* 2 (2): 1800160.

26 Xu, Z., Dai, X., Bu, X. et al. (2021). Enhanced heterogeneous diffusion of nanoparticles in semiflexible networks. *ACS Nano* 15 (3): 4608–4616.

27 Chen, P., Yue, H., Zhai, X. et al. (2019). Transport of a graphene nanosheet sandwiched inside cell membranes. *Science Advances* 5 (6): eaaw3192.

28 Xu, Z., Zhu, G., Chen, P. et al. (2019). Optimal ligand–receptor binding for highly efficient capture of vesicles in nanofluidic transportation. *Nanoscale* 11 (46): 22305–22315.

29 Chen, S.S. and Yong, X. (2018). Dissipative particle dynamics modeling of hydrogel swelling by osmotic ensemble method. *The Journal of Chemical Physics* 149 (9).

30 Nikolov, S., Fernandez-Nieves, A., and Alexeev, A. (2018). Mesoscale modeling of microgel mechanics and kinetics through the swelling transition. *Applied Mathematics and Mechanics-English Edition* 39 (1): 47–62.

31 Gumerov, R.A., Filippov, S.A., Richtering, W. et al. (2019). Amphiphilic microgels adsorbed at oil-water interfaces as mixers of two immiscible liquids. *Soft Matter* 15 (19): 3978–3986.

32 Gumerov, R.A., Rudov, A.A., Richtering, W. et al. (2017). Amphiphilic arborescent copolymers and microgels: from unimolecular micelles in a selective solvent to the stable monolayers of variable density and nanostructure at a liquid interface. *ACS Applied Materials & Interfaces* 9 (37): 31302–31316.

33 Karg, M., Pich, A., Hellweg, T. et al. (2019). Nanogels and microgels: from model colloids to applications, recent developments, and future trends. *Langmuir* 35 (19): 6231–6255.

34 Gumerov, R.A., Rumyantsev, A.M., Rudov, A.A. et al. (2016). Mixing of two immiscible liquids within the polymer microgel adsorbed at their interface. *ACS Macro Letters* 5 (5): 612–616.

35 Mourran, A., Wu, Y., Gumerov, R.A. et al. (2016). When colloidal particles become polymer coils. *Langmuir* 32 (3): 723–730.

36 Yong, X., Kuksenok, O., Matyjaszewski, K., and Balazs, A.C. (2013). Harnessing interfacially-active nanorods to regenerate severed polymer gels. *Nano Letters* 13 (12): 6269–6274.

37 Yong, X., Kuksenok, O., and Balazs, A.C. (2015). Modeling free radical polymerization using dissipative particle dynamics. *Polymer* 72: 217–225.

38 Singh, A., Kuksenok, O., Johnson, J.A., and Balazs, A.C. (2017). Photo-regeneration of severed gel with iniferter-mediated photo-growth. *Soft Matter* 13 (10): 1978–1987.

39 Singh, A., Kuksenok, O., Johnson, J.A., and Balazs, A.C. (2016). Tailoring the structure of polymer networks with iniferter-mediated photo-growth. *Polymer Chemistry* 7 (17): 2955–2964.

40 Choudhury, C.K., Palkar, V., and Kuksenok, O. (2020). Computational design of nanostructured soft interfaces: focus on shape changes and spreading of cubic nanogels. *Langmuir* 36 (25): 7109–7123.

41 Liu, Y., Aizenberg, J., and Balazs, A.C. (2021). Using dissipative particle dynamics to model effects of chemical reactions occurring within hydrogels. *Nanomaterials* 11 (10): 2764.

42 Zhang, T., Biswas, S., Cuthbert, J. et al. (2020). Understanding the origin of softness in structurally tailored and engineered macromolecular (STEM) gels: a DPD study. *Polymer* 208: 122909.

43 Zholudev, S.I., Gumerov, R.A., Larina, A.A., and Potemkin, I.I. (2023). Swelling, collapse and ordering of rod-like microgels in solution: computer simulation studies. *Journal of Colloid and Interface Science* 629: 270–278.

44 Lin, F., Yu, J., Tang, W. et al. (2013). Peptide-functionalized oxime hydrogels with tunable mechanical properties and gelation behavior. *Biomacromolecules* 14 (10): 3749–3758.

45 Zhu, J.M. (2010). Bioactive modification of poly(ethylene glycol) hydrogels for tissue engineering. *Biomaterials* 31 (17): 4639–4656.

46 Lin, C.C. and Anseth, K.S. (2009). PEG hydrogels for the controlled release of biomolecules in regenerative medicine. *Pharmaceutical Research* 26 (3): 631–643.

47 Santo, K.P. and Neimark, A.V. (2021). Dissipative particle dynamics simulations in colloid and Interface science: a review. *Advances in Colloid and Interface Science* 298: 102545.

48 Groot, R.D. and Rabone, K.L. (2001). Mesoscopic simulation of cell membrane damage, morphology change and rupture by nonionic surfactants. *Biophysical Journal* 81 (2): 725–736.

49 Sliozberg, Y.R., Gair, J.L., and Hsieh, A.J. (2020). Dissipative particle dynamics simulation of microphase separation in polyurethane urea nanocomposites. *Polymer* 193: 122339.

50 Gavrilov, A.A., Chertovich, A.V., Khalatur, P.G., and Khokhlov, A.R. (2013). Effect of nanotube size on the mechanical properties of elastomeric composites. *Soft Matter* 9 (15): 4067–4072.

51 Anderson, R.L., Bray, D.J., Ferrante, A.S. et al. (2017). Dissipative particle dynamics: systematic parametrization using water-octanol partition coefficients. *The Journal of Chemical Physics* 147 (9).

52 Lee, M.-T., Mao, R., Vishnyakov, A., and Neimark, A.V. (2016). Parametrization of chain molecules in dissipative particle dynamics. *The Journal of Physical Chemistry B* 120 (22): 4980–4991.

53 Sami, S., Yildirim, E., Yurtsever, M. et al. (2014). Understanding the influence of hydrogen bonding and diisocyanate symmetry on the morphology and properties of segmented polyurethanes and polyureas: computational and experimental study. *Polymer* 55 (18): 4563–4576.
54 Yildirim, E., Yurtsever, M., Yilgor, E. et al. (2018). Temperature-dependent changes in the hydrogen bonded hard segment network and microphase morphology in a model polyurethane: experimental and simulation studies. *Journal of Polymer Science: Polymer Physics* 56 (2): 182–192.
55 Kmiecik, S., Gront, D., Kolinski, M. et al. (2016). Coarse-grained protein models and their applications. *Chemical Reviews* 116 (14): 7898–7936.
56 Choudhury, C.K. and Kuksenok, O. (2020). Native-based dissipative particle dynamics approach for α-helical folding. *The Journal of Physical Chemistry B* 124 (50): 11379–11386.
57 Vishnyakov, A., Mao, R., Lee, M.T., and Neimark, A.V. (2018). Coarse-grained model of nanoscale segregation, water diffusion, and proton transport in Nafion membranes. *The Journal of Chemical Physics* 148 (2).
58 Vishnyakov, A., Talaga, D.S., and Neimark, A.V. (2012). DPD simulation of protein conformations: from alpha-helices to beta-structures. *Journal of Physical Chemistry Letters* 3 (21): 3081–3087.
59 Lee, M.T., Vishnyakov, A., and Neimark, A.V. (2015). Modeling proton dissociation and transfer using dissipative particle dynamics simulation. *Journal of Chemical Theory and Computation* 11 (9): 4395–4403.
60 Liu, H., Li, M., Lu, Z.Y. et al. (2009). Influence of surface-initiated polymerization rate and initiator density on the properties of polymer brushes. *Macromolecules* 42 (7): 2863–2872.
61 Lisal, M., Brennan, J.K., and Smith, W.R. (2006). Mesoscale simulation of polymer reaction equilibrium: combining dissipative particle dynamics with reaction ensemble Monte Carlo. I. Polydispersed polymer systems. *The Journal of Chemical Physics* 125 (16).
62 Gissinger, J.R., Jensen, B.D., and Wise, K.E. (2017). Modeling chemical reactions in classical molecular dynamics simulations. *Polymer* 128: 211–217.
63 Akkermans, R.L.C., Toxvaerd, S., and Briels, W.J. (1998). Molecular dynamics of polymer growth. *The Journal of Chemical Physics* 109 (7): 2929–2940.
64 Kawagoe, Y., Kikugawa, G., Shirasu, K., and Okabe, T. (2021). Thermoset resin curing simulation using quantum-chemical reaction path calculation and dissipative particle dynamics. *Soft Matter* 17 (28): 6707–6717.
65 Feng, Y.H., Zhang, X.P., Hu, L.F. et al. (2021). Mesoscopic simulation for the effect of cross-linking reactions on the drug diffusion properties in microneedles. *Journal of Chemical Information and Modeling* 61 (8): 4000–4010.
66 Kumar, S. and Larson, R.G. (2001). Brownian dynamics simulations of flexible polymers with spring-spring repulsions. *The Journal of Chemical Physics* 114 (15): 6937–6941.
67 Sirk, T.W., Slizoberg, Y.R., Brennan, J.K. et al. (2012). An enhanced entangled polymer model for dissipative particle dynamics. *The Journal of Chemical Physics* 136 (13): 11.

68 Palkar, V. and Kuksenok, O. (2022). Controlling degradation and erosion of polymer networks: insights from mesoscale modeling. *The Journal of Physical Chemistry B* 126 (1): 336–346.

69 Thompson, A.P., Aktulga, H.M., Berger, R. et al. (2022). LAMMPS-a flexible simulation tool for particle-based materials modeling at the atomic, meso, and continuum scales. *Computer Physics Communications* 271.

70 Plimpton, S. (1995). Fast parallel algorithms for short-range molecular dynamics. *Journal of Computational Physics* 117 (1): 1–19.

71 LAMMPS Molecular Dynamics Simulator. http://lammps.org (10 September 2023).

72 Vaibhav Palkar. LAMMPS pair style srp/react command .https://docs.lammps.org/pair_srp.html#pair-style-srp-react-command (10 September 2023).

73 Xiong, Y., Choudhury, C.K., Palkar, V. et al. (2022). Mesoscale modeling of phase separation controlled by hydrosilylation in polyhydromethylsiloxane (PHMS)-containing blends. *Nanomaterials* 12 (18).

74 Dibandjo, P., Dire, S., Babonneau, F., and Soraru, G.D. (2010). Influence of the polymer architecture on the high temperature behavior of SiCO glasses: a comparison between linear- and cyclic-derived precursors. *Journal of Non-Crystalline Solids* 356 (3): 132–140.

75 Zhang, C., Zhang, C., Cui, X.M. et al. (2015). Transparent and dense ladder-like alkylene-bridged polymethylsiloxane coating with enhanced water vapor barrier property. *ACS Applied Materials & Interfaces* 7 (40): 22157–22165.

76 Colombo, P. (2008). Engineering porosity in polymer-derived ceramics. *Journal of the European Ceramic Society* 28 (7): 1389–1395.

77 Vakifahmetoglu, C., Zeydanli, D., and Colombo, P. (2016). Porous polymer derived ceramics. *Materials Science and Engineering Reports* 106: 1–30.

78 Yu, Z.J., Feng, Y., Li, S., and Pei, Y.X. (2016). Influence of the polymer-polymer miscibility on the formation of mesoporous SiC(O) ceramics for highly efficient adsorption of organic dyes. *Journal of the European Ceramic Society* 36 (15): 3627–3635.

79 Nakajima, Y. and Shimada, S. (2015). Hydrosilylation reaction of olefins: recent advances and perspectives. *RSC Advances* 5 (26): 20603–20616.

80 Meister, T.K., Riener, K., Gigler, P. et al. (2016). Platinum catalysis revisited-unraveling principles of catalytic olefin hydrosilylation. *ACS Catalysis* 6 (2): 1274–1284.

81 Hansen, J.P. and McDonald, I.R. (2013). *Theory of Simple Liquids: with Applications to Soft Matter*. San Diego: Academic Press.

82 Yelash, L., Virnau, P., Paul, W. et al. (2008). Spinodal decomposition of polymer solutions: a parallelized molecular dynamics simulation. *Physical Review E* 78 (3): 031801.

83 Liu, H.J., Cavaliere, S., Jones, D.J. et al. (2018). Morphology of hydrated nafion through a quantitative cluster analysis: a case study based on dissipative particle dynamics simulations. *Journal of Physical Chemistry C* 122 (24): 13130–13139.

84 Ma, W.J., Maritan, A., Banavar, J.R., and Koplik, J. (1992). Dynamics of phase-separation of binary fluids. *Physical Review A* 45 (8): R5347–R5350.

85 Blum, Y., Soraru, G.D., Ramaswamy, A.P. et al. (2013). Controlled mesoporosity in SiOC via chemically bonded polymeric "spacers". *Journal of the American Ceramic Society* 96 (9): 2785–2792.

86 Metters, A.T., Bowman, C.N., and Anseth, K.S. (2000). A statistical kinetic model for the bulk degradation of PLA-b-PEG-b-PLA hydrogel networks. *The Journal of Physical Chemistry B* 104 (30): 7043–7049.

87 Göpferich, A. (1997). Polymer bulk erosion. *Macromolecules* 30 (9): 2598–2604.

88 Sakai, T., Matsunaga, T., Yamamoto, Y. et al. (2008). Design and fabrication of a high-strength hydrogel with ideally homogeneous network structure from tetrahedron-like macromonomers. *Macromolecules* 41 (14): 5379–5384.

89 Li, X., Tsutsui, Y., Matsunaga, T. et al. (2011). Precise control and prediction of hydrogel degradation behavior. *Macromolecules* 44 (9): 3567–3571.

90 Azagarsamy, M.A., McKinnon, D.D., Age, D.L., and Anseth, K.S. (2014). Coumarin-based photodegradable hydrogel: design, synthesis, gelation, and degradation kinetics. *ACS Macro Letters* 3 (6): 515–519.

91 Jha, P.K., Zwanikken, J.W., Detcheverry, F.A. et al. (2011). Study of volume phase transitions in polymeric nanogels by theoretically informed coarse-grained simulations. *Soft Matter* 7 (13): 5965–5975.

92 Choudhury, C.K. and Kuksenok, O. (2018). Modeling dynamics of polyacrylamide gel in oil-water mixtures: dissipative particle dynamics approach. *MRS Advances* 3 (26): 1469–1474.

93 Palkar, V., Choudhury, C.K., and Kuksenok, O. (2020). Development of dissipative particle dynamics framework for modeling hydrogels with degradable bonds. *MRS Advances* 5 (17): 927–934.

94 Flory, P.J. (1941). Molecular size distribution in three dimensional polymers. I. Gelation. *Journal of the American Chemical Society* 63: 3083–3090.

95 Macosko, C.W. and Miller, D.R. (1976). New derivation of average molecular-weights of nonlinear polymers. *Macromolecules* 9 (2): 199–206.

96 Rubinstein, M. and Colby, R.H. (2003). *Polymer Physics*. OUP Oxford.

97 Lin, T.S., Wang, R., Johnson, J.A., and Olsen, B.D. (2018). Topological structure of networks formed from symmetric four-arm precursors. *Macromolecules* 51 (3): 1224–1231.

98 Gupta, A.M., Hendrickson, R.C., and Macosko, C.W. (1991). Monte-Carlo description of af homopolymerization – diffusional effects. *The Journal of Chemical Physics* 95 (3): 2097–2108.

99 Shy, L.Y., Leung, Y.K., and Eichinger, B.E. (1985). Critical exponents for off-lattice gelation of polymer-chains. *Macromolecules* 18 (5): 983–986.

100 Somvarsky, J. and Dusek, K. (1994). Kinetic Monte-Carlo simulation of network formation. 2. Effect of system size. *Polymer Bulletin* 33 (3): 377–384.

101 Polanowski, P., Jeszka, J.K., Li, W.W., and Matyjaszewski, K. (2011). Effect of dilution on branching and gelation in living copolymerization of monomer and divinyl cross-linker: modeling using dynamic lattice liquid model (DLL) and Flory-Stockmayer (FS) model. *Polymer* 52 (22): 5092–5101.

102 Flory, P.J. (1953). *Principles of Polymer Chemistry*. Cornell University Press.

103 Rankin, S.E., Kasehagen, L.J., McCormick, A.V., and Macosko, C.W. (2000). Dynamic Monte Carlo simulation of gelation with extensive cyclization. *Macromolecules* 33 (20): 7639–7648.

104 Spouge, J.L. (1986). Equilibrium ring formation in polymer-solutions. *Journal of Statistical Physics* 43 (1-2): 143–196.

105 Wang, R., Lin, T.S., Johnson, J.A., and Olsen, B.D. (2017). Kinetic Monte Carlo simulation for quantification of the gel point of polymer networks. *ACS Macro Letters* 6 (12): 1414–1419.

106 Tanaka, Y., Stanford, J.L., and Stepto, R. (2012). Interpretation of gel points of an epoxy-amine system including ring formation and unequal reactivity: measurements of gel points and analyses on ring structures. *Macromolecules* 45 (17): 7197–7205.

107 Lang, M. and Muller, T. (2020). Analysis of the gel point of polymer model networks by computer simulations. *Macromolecules* 53 (2): 498–512.

108 Stauffer, D. and Aharony, A. (1994). *Introduction To Percolation Theory*. Taylor & Francis.

109 Sakai, T., Katashima, T., Matsushita, T., and Chung, U.I. (2016). Sol-gel transition behavior near critical concentration and connectivity. *Polymer Journal* 48 (5): 629–634.

110 Nishi, K., Fujii, K., Chung, U. et al. (2017). Experimental observation of two features unexpected from the classical theories of rubber elasticity. *Physical Review Letters* 119 (26).

111 Ashley, G.W., Henise, J., Reid, R., and Santi, D.V. (2013). Hydrogel drug delivery system with predictable and tunable drug release and degradation rates. *Proceedings of the National Academy of Sciences of the United States of America* 110 (6): 2318–2323.

112 Yashin, V.V. and Balazs, A.C. (2006). Pattern formation and shape changes in self-oscillating polymer gels. *Science* 314 (5800): 798–801.

113 Yashin, V.V. and Balazs, A.C. (2007). Theoretical and computational modeling of self-oscillating polymer gels. *The Journal of Chemical Physics* 126 (12): 124707.

114 Kuksenok, O., Yashin, V.V., and Balazs, A.C. (2008). Three-dimensional model for chemoresponsive polymer gels undergoing the Belousov-Zhabotinsky reaction. *Physical Review E* 78 (4): 041406.

115 Yoshida, R., Tanaka, M., Onodera, S. et al. (2000). In-phase synchronization of chemical and mechanical oscillations in self-oscillating gels. *The Journal of Physical Chemistry. A* 104 (32): 7549–7555.

116 Chen, I.C., Kuksenok, O., Yashin, V.V. et al. (2011). Shape- and size-dependent patterns in self-oscillating polymer gels. *Soft Matter* 7: 3141–3146.

117 Kuksenok, O., Yashin, V.V., Kinoshita, M. et al. (2011). Exploiting gradients in cross-link density to control the bending and self-propelled motion of active gels. *Journal of Materials Chemistry* 21 (23): 8360–8371.

118 Yuan, P., Kuksenok, O., Gross, D.E. et al. (2013). UV patternable thin film chemistry for shape and functionally versatile self-oscillating gels. *Soft Matter* 9: 1231–1243.

119 Yashin, V.V., Suzuki, S., Yoshida, R., and Balazs, A.C. (2012). Controlling the dynamic behavior of heterogeneous self-oscillating gels. *Journal of Materials Chemistry* 22 (27): 13625–13636.

120 Xiong, Y. and Kuksenok, O. (2021). Mechanical adaptability of patterns in constrained hydrogel membranes. *Langmuir* 37 (16): 4900–4912.

121 Xiong, Y., Dayal, P., Balazs, A.C., and Kuksenok, O. (2018). Phase transitions and pattern formation in chemo-responsive gels and composites. *Israel Journal of Chemistry* 58 (6-7): 693–705.

122 Yashin, V.V., Kuksenok, O., and Balazs, A.C. (2010). Modeling autonomously oscillating chemo-responsive gels. *Progress in Polymer Science* 35 (1-2): 155–173.

123 Biswas, S., Yashin, V.V., and Balazs, A.C. (2021). Dynamic behavior of chemically tunable mechano-responsive hydrogels. *Soft Matter* 17 (47): 10664–10674.

124 Zhang, T., Yashin, V.V., Waters, J.T., and Balazs, A.C. (2021). Formation of helices with controllable chirality in gel-fiber composites. *Polymer* 212: 123191.

125 Ren, L., Wang, L., Gao, Q. et al. (2020). Programmed locomotion of an active gel driven by spiral waves. *Angewandte Chemie* 132 (18): 7172–7178.

126 Ren, L., Yuan, L., Gao, Q. et al. (2020). Chemomechanical origin of directed locomotion driven by internal chemical signals. *Science Advances* 6 (18): eaaz9125.

127 Wang, J., Ren, L., Teng, R. et al. (2021). Rotational locomotion of an active gel driven by internal chemical signals. *The Journal of Physical Chemistry Letters* 12 (50): 11987–11991.

128 Xiong, Y. and Kuksenok, O. (2022). Photocontrol of pattern formation and hysteresis loops in polymer gels with host-guest interactions. *Iscience* 105606.

129 Lei, J., Li, Z., Xu, S., and Liu, Z. (2021). Recent advances of hydrogel network models for studies on mechanical behaviors. *Acta Mech. Sin.* 37 (3): 367–386.

130 Aggarwal, A., Li, C., Stupp, S.I., and de la Cruz, M.O. (2022). Controlling the shape morphology of origami-inspired photoresponsive hydrogels. *Soft Matter* 18 (11): 2193–2202.

131 Paul, D.R. and McSpadden, S.K. (1976). Diffusional release of a solute from a polymer matrix. *Journal of Membrane Science* 1 (1): 33–48.

132 Lee, P.I. (1980). Diffusional release of a solute from a polymeric matrix – approximate analytical solutions. *Journal of Membrane Science* 7 (3): 255–275.

133 Gopferich, A. and Langer, R. (1993). Modeling of polymer erosion. *Macromolecules* 26 (16): 4105–4112.

134 Zygourakis, K. (1990). Development and temporal evolution of erosion fronts in bioerodible controlled release devices. *Chemical Engineering Science* 45 (8): 2359–2366.

135 Jahanmir, G., Abdekhodaie, M.J., and Chau, Y. (2018). Stochastic modeling of degradation behavior of hydrogels. *Macromolecules* 51 (11): 3941–3952.

136 Metters, A. and Hubbell, J. (2005). Network formation and degradation behavior of hydrogels formed by Michael-type addition reactions. *Biomacromolecules* 6 (1): 290–301.

137 Metters, A.T., Anseth, K.S., and Bowman, C.N. (2001). A statistical kinetic model for the bulk degradation of PLA-b-PEG-b-PLA hydrogel networks: Incorporating network non-idealities. *The Journal of Physical Chemistry B* 105 (34): 8069–8076.

138 Reddy, S.K., Anseth, K.S., and Bowman, C.N. (2005). Modeling of network degradation in mixed step-chain growth polymerizations. *Polymer* 46 (12): 4212–4222.

139 Martens, P., Metters, A.T., Anseth, K.S., and Bowman, C.N. (2001). A generalized bulk-degradation model for hydrogel networks formed from multivinyl cross-linking molecules. *Journal of Physical Chemistry B* 105 (22): 5131–5138.

140 Tibbitt, M.W., Kloxin, A.M., and Anseth, K.S. (2013). Modeling controlled photodegradation in optically thick hydrogels. *Journal of Polymer Science* 51 (9): 1899–1911.

141 Reid, R., Sgobba, M., Raveh, B. et al. (2015). Analytical and simulation-based models for drug release and gel-degradation in a tetra-PEG hydrogel drug-delivery system. *Macromolecules* 48 (19): 7359–7369.

142 Lu, T., Wang, Z., Tang, J. et al. (2020). A pseudo-elasticity theory to model the strain-softening behavior of tough hydrogels. *Journal of the Mechanics and Physics of Solids* 137: 103832.

143 Lamont, S.C., Mulderrig, J., Bouklas, N., and Vernerey, F.J. (2021). Rate-dependent damage mechanics of polymer networks with reversible bonds. *Macromolecules* 54 (23): 10801–10813.

144 Liu, D., Ma, S., Yuan, H., and Markert, B. (2022). Computational modelling of poro-visco-hyperelastic effects on time-dependent fatigue crack growth of hydrogels. *International Journal of Plasticity* 155: 103307.

145 Wagner, R.J., Dai, J., Su, X., and Vernerey, F.J. (2022). A mesoscale model for the micromechanical study of gels. *Journal of the Mechanics and Physics of Solids* 104982.

146 Pan, Z. and Brassart, L. (2022). Constitutive modelling of hydrolytic degradation in hydrogels. *Journal of the Mechanics and Physics of Solids* 167: 105016.

147 Akalp, U., Bryant, S.J., and Vernerey, F.J. (2016). Tuning tissue growth with scaffold degradation in enzyme-sensitive hydrogels: a mathematical model. *Soft Matter* 12 (36): 7505–7520.

148 Lutolf, M.P., Lauer-Fields, J.L., Schmoekel, H.G. et al. (2003). Synthetic matrix metalloproteinase-sensitive hydrogels for the conduction of tissue regeneration: engineering cell-invasion characteristics. *Proceedings of the National Academy of Sciences* 100 (9): 5413–5418.

149 Sridhar, S.L., Schneider, M.C., Chu, S. et al. (2017). Heterogeneity is key to hydrogel-based cartilage tissue regeneration. *Soft Matter* 13 (28): 4841–4855.

150 Vernerey, F.J., Lalitha Sridhar, S., Muralidharan, A., and Bryant, S.J. (2021). Mechanics of 3d cell–hydrogel interactions: experiments, models, and mechanisms. *Chemical Reviews* 121 (18): 11085–11148.

3

Dynamic Bonds in Associating Polymer Networks

Jiayao Chen[1], Xiao Zhao[2], and Peng-Fei Cao[1]

[1]State Key Laboratory of Organic-Inorganic Composites, College of Materials Science and Engineering, Beijing University of Chemical Technology, Beisanhuan Road East, Beijing 100029, China
[2]GCP Applied Technologies Inc., Wilmington, Massachusetts 01887, USA

3.1 Introduction of Dynamic Bonds

Polymeric materials with covalent cross-links have advantages of facile synthesis, high mechanical performance, and decent chemical resistance, despite the emerging problems regarding the reprocess ability and recyclability [1–3]. There are basically two ways to introduce dynamic bonds into polymers: (i) within the polymer backbone and (ii) as reversible cross-links between polymer chains. Polymers with dynamic bonds, often termed associating polymers, have been attracting great research interest in recent years because of their unique viscoelastic properties, self-healing ability, and recyclability. A variety of dynamic covalent and noncovalent chemistries have been explored with respect to their capabilities to form transient bonding in polymer networks. Recently, the micro phase-separated dynamic aggregates have also been developed due to their unique architectures and performances in comparison to associating networks that are cross-linked merely by binary associating groups. Moreover, various dynamic bonds that differ by the bond-dissociation energy were broadly investigated to reveal the mechanisms that control the characteristic time of the network topology rearrangements. In this section, two typical types of dynamic bonds, i.e. dynamic covalent bonds and dynamic noncovalent bonds, are presented with regard to their categories, rearrangement mechanisms, and characteristics.

3.1.1 Dynamic Covalent Bonds

Dynamic covalent bonds typically cross-link polymer chains to form the dynamic covalent networks, which is also known as covalent adaptive networks (CANs) [3]. Relaxation dynamics of CANs essentially depend on dynamics of the polymer chains and nature of the covalent bonds that construct the polymer networks. The intermolecular forces are relatively negligible in comparison to the strong covalent bonds. The dynamic covalent bonding mechanisms employed in polymer networks

Dynamics and Transport in Macromolecular Networks: Theory, Modeling, and Experiments,
First Edition. Edited by Li-Tang Yan.

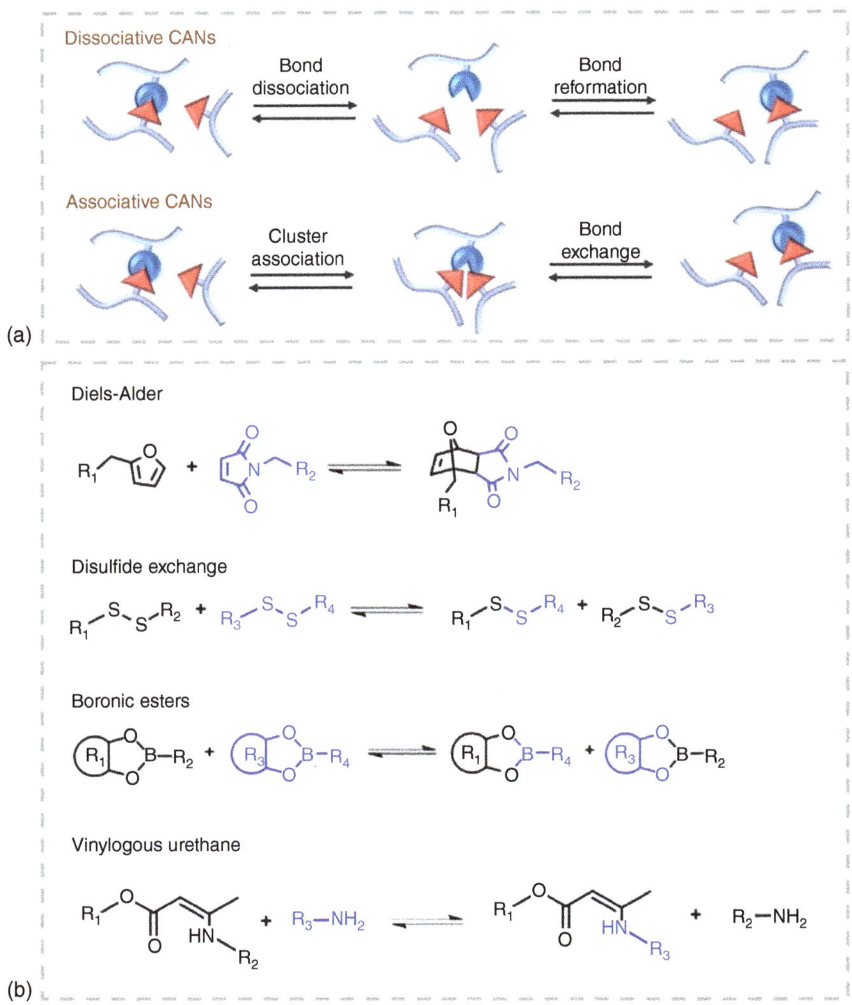

Figure 3.1 (a) Schematics of dynamic covalent mechanisms of dissociative and associative CANs. (b) A few types of dynamic covalent bonds [3].

are categorized as dissociative and associative, of which both bond reformation or exchange are governed by the bond dissociation energy barrier. In dissociative CANs, an existing bond is dissociated prior to a new bond being reformed, whereas in associative CANs, a new bond is associated to form a cluster prior to a bond rupture, i.e. bond exchange (Figure 3.1a). In other words, the bond exchange behavior allows the stress to relax and the topology to rearrange without compromising the network connectivity and degree of cross-linking. Such polymers are also known as vitrimers that demonstrate an Arrhenius temperature dependence of molecular dynamics, like the vitrification of silica glass [4]. Vitrimers possess the advantages of cross-linked networks as thermosets at the ambient using temperature and the ability to be reprocessed as thermoplastics at elevated temperatures.

Accordingly, the dynamic covalent bonds can be classified into two types: inherently reversible and extrinsically reversible. The former allows materials to spontaneously adapt the network even under mild and room temperature conditions (e.g. Diels–Alder reactions). On the other hand, the extrinsically reversible bonds govern the exchange rate that depends on external conditions, such as pH, temperature, or the presence of a soluble competitor (e.g. diol–boronic acid bonds). Specifically, the major types of dynamic covalent bonds are shown in Figure 3.1b: [2] (i) Diels–Alder reaction is attached to a dissociative mechanism and involves a conjugated diene reacting with a dienophile, a thermally triggered reversible covalent bond. For example, upon application of physical damage, heat could be introduced to trigger the reverse Diels–Alder reaction to regenerate the dienes and dienophiles, and these reactants could then react to reform chemical bonds and heal the damage. (ii) Reversible disulfide bonds are widely used to design the self-healing materials via the exchange reaction of S–S linkages that often occur at an elevated temperature [5–7]. (iii) The formation of boroxines from the dehydration of boronic acids is reversible upon hydrolysis [8]. (iv) Vinylogous urethane is another type of associative dynamic bonds, which undergoes transamination reactions at an elevated temperature even without a catalyst [9].

3.1.2 Dynamic Noncovalent Bonds

Noncovalent bonds, another type of interactions to design the dynamic associative networks, are physical interactions that often involve among self-associating pendant groups [10, 11]. These physical interactions include hydrogen bonding, ionic bonding, metal–ligand coordination, guest–host interaction, and π–π stacking (Figure 3.2) [2]. (i) Hydrogen bonding (H-bond) is one of the most widely utilized dynamic noncovalent bonds. Dominated by the electrostatic force, a hydrogen (H) atom could be attracted by an adjacent atom or group with higher electronegativity, such as atoms of nitrogen (N), oxygen (O), or fluorine (F). H-bond forms between the two atoms or groups once the pairing between the donor and acceptor occurs. Such bonds with relatively low strength are able to achieve faster bond breakage and formation, of which the feature can be used to create a polymer network with relatively fast molecular rearrangements [12]. (ii) Ionic bonds are based on electrostatic interactions, where two oppositely charged ionic species attract each other. (iii) Metal–ligand coordination bonds include a metallic core and at least one π-donor ligand.

The polymers cross-linked by dynamic noncovalent bonds can be referred to as supramolecular polymeric networks (SPNs) [13, 14]. Supramolecular motifs normally contribute additional order and directionality to gain physical cross-linking and chain extension, while remaining dynamic interactions [13, 15, 16]. Among various types of noncovalent interactions, macrocyclic host–guest complexes have earned great attention because of the macrocycle-based association and tunability of the underlying binding imparts (i.e. various stimuli-responsive functions) to the resulting complexes [13]. As the widely accessible dynamic noncovalent interaction, H-bond is still the most promising candidate in the supramolecular design.

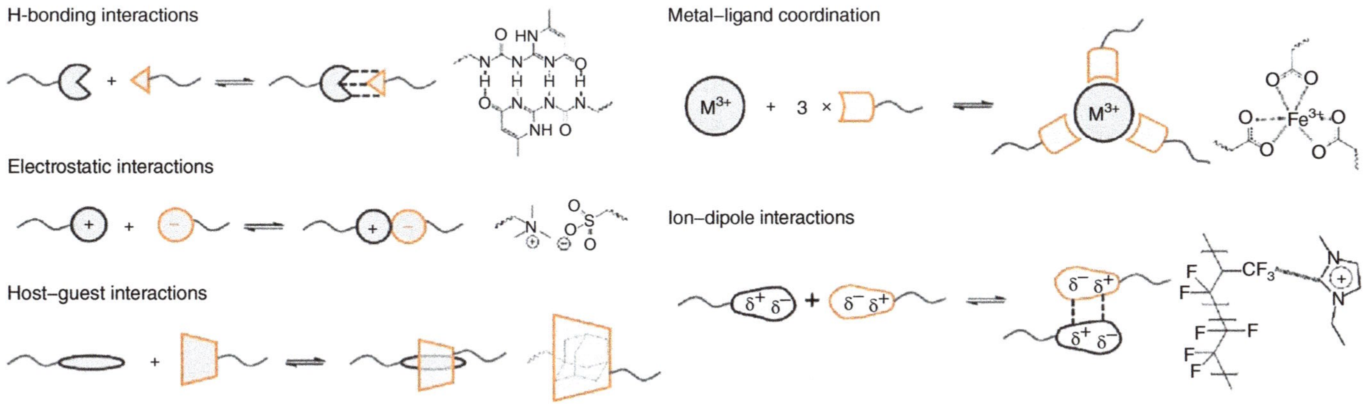

Figure 3.2 Schematics of various dynamic noncovalent bonds [2].

3.2 Physical Insight of Dynamic Bonds

The relaxation dynamics and structural configurations of dynamic associative networks determine their advanced performance and have been studied by means mainly including rheology, scattering techniques, dielectric spectroscopy, as well as modeling. These endeavors are of great importance to understand the time scales of dynamic processes in the dynamic networks and to shed a light on tailoring the dynamic bonds for prolonged material lifetime and advanced applications.

The rate of polymer dynamics is usually quantified using a characteristic relaxation time, which describes the overall time required for the intra- or intermolecular rearrangement and stress relaxation. In this section, the structural configuration of dynamic networks and the pertinent theoretical or phenomenological models have been reviewed, with a focus on the topology rearrangement mechanism and the state-of-the-art tools/methods/techniques for the characterization of the network dynamics.

3.2.1 Segmental and Chain Dynamics

Dynamic associative networks are often applied to take advantages of their mechanical and viscoelastic properties, and these properties can be tuned by modifying the interaction strength, morphology of the associating groups, and the polymer architectures. Telechelic polymers are mostly studied to understand the physics of associating dynamics since they have the simplest molecular structure, i.e. two reactive end groups on a linear backbone [17].

Xing et al. carried out the study on molecular dynamics in a telechelic poly(dimethylsiloxane) (PDMS) with different chain lengths that form supramolecular networks, and two types of end groups (amine-terminated PDMS-NH_2 and carboxylic-terminated PDMS-COOH) were investigated [18]. Despite the similarities in backbone chemistry, the designed PDMS with variant molecular weight and end groups exhibited some qualitative differences in their supramolecular network formation. Figure 3.3a shows Arrhenius plots of the mean dielectric relaxation times of various relaxation modes for PDMS-COOH. The fastest dielectric relaxation mode in all samples was assigned to the segmental or α-relaxation, because its extrapolation to ~100 seconds coincides with the glass transition temperature (T_g) from the differential scanning calorimetry (DSC) measurements. This process slowed down as the degree of polymerization (DP) of a single PDMS chain decreased, consistent with the increase in T_g, because of the increased number of associating chain ends per unit. The α-relaxation spectrum displayed an increased broadening with decreasing DP, indicating a dynamical constraint and frustrated packing (Figure 3.3b). The PDMS-COOH samples also exhibited a trace of second relaxation (α^*-relaxation) that varied in several characteristics from its α-relaxation counterpart but were partially superimposed for different DP. For example, the α^*-relaxation shows typical Arrhenius temperature dependence, whereas the α-relaxation exhibits a super Arrhenius behavior. The extrapolation of the well-separated process in PDMS-COOH matched with the second glass transition observed in these systems

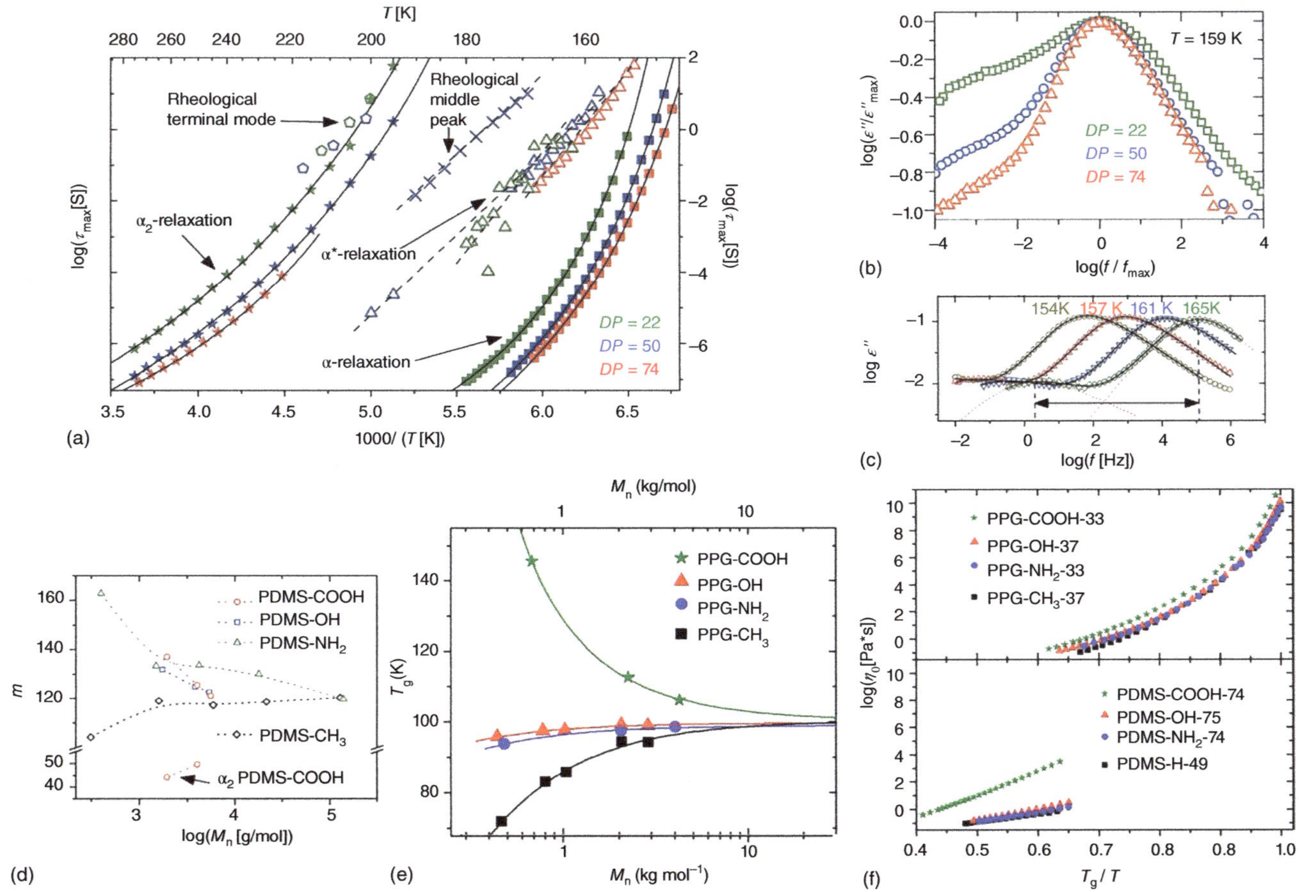

Figure 3.3 (a) Arrhenius plot of the dielectric mean relaxation times, the rheological relaxation time of the middle peak in the sample with DP = 50, and the rheological terminal mode for PDMS-COOH systems of different DP as indicated. (b) Normalized dielectric loss vs. normalized frequency of the α-relaxation of PDMS-NH_2. (c) Dielectric loss spectra at several temperatures as indicated of PDMS-COOH samples with DP = 74. (d) Fragility index m vs. total number averaged molecular weight M_n (including end groups) of the α-relaxation [18]. (e) Calorimetric T_g vs. M_n (including end groups) in telechelic PPG. (f) Zero shear viscosity η_0 vs. T_g/T for telechelic PPG and PDMS with different end groups as indicated with DP [19].

and thus referred as the α_2-process (Figure 3.3c). A major difference between the α- and the α_2-process was unraveled by the fragility index m which quantifies the steepness of the temperature dependence of the relaxation time at T_g. It can be estimated using the Vogel–Fulcher–Tamman (VFT) fit parameters according to

$$m \equiv \left[\frac{\partial \log_{10}\tau}{\partial(T_g/T)}\right]_{T=T_g} = \frac{BT_{g,diel}}{\ln 10(T_{g,diel} - T_0)^2}. \tag{3.1}$$

The m of the α-relaxation in both PDMS-NH_2 and PDMS-COOH took values of 120–140 and increased with decreasing molecular weight (MW) or DP, respectively (Figure 3.3d). On the other hand, m of the α_2-process was less than 50, which is consistent with the structural relaxation of a local phase that mainly consists of (H-bond rich) chain ends rather than polymer chains. In conclusion, the PDMS-COOH revealed an outstanding feature: though consisting of only one species of terminal group, a dual network with two types of topologies was resembled, namely the effectively permanent and transient (temporary) bonds. The former refers to the formation of clustered end groups that work as the multi-arm cross-linking point and even create phases separated from the PDMS domain. On the other hand, the latter are the rather loose, nonsegregated associates (possibly dimers) from the telechelic head-to-end associated chains, where the transient stress relaxation can be achieved.

The research group further studied the impact of types of H-bond end groups on segmental and chain dynamics of telechelic polymers mainly via broadband dielectric spectroscopy (BDS), DSC, and rheology [19]. Polypropylene glycol (PPG) with three types of H-bond end groups (hydroxyl-terminated PPG-OH, amine-terminated PPG-NH_2, and carboxylic-terminated PPG-COOH) that have different interaction strength were compared with a reference, i.e. non-H-bonding end group (methyl-terminated PPG-CH_3). The DSC result indicated that the T_g of PPG-CH_3 increased with increasing MW (Figure 3.3e), following the normal trend for polymers [20]. In the case of PPG-OH and PPG-NH_2, the T_g was almost independent of MW with only a slight reduction for small MW. The T_g of PPG-COOH increased strongly with decrease in MW, consistent with the previously observed increase in the T_g of linear associating polymers with strong intermolecular H-bonds [21, 22]. These results suggested the significant influence of the chain-end interaction strength on T_g at relatively low MWs. In the meantime, viscosity appeared to be almost independent of the chain-end interactions once the shift of T_g was accounted for (by scaling the temperature with T_g) (Figure 3.3f). By comparing with the previous work [18], it revealed that the difference in the backbone flexibility might play a significant role in the chain-end association. A qualitative explanation considering the lifetime of chain-end dissociation in comparison to the characteristic chain and segmental relaxation time was proposed. If the dissociation time is much longer than the segmental time, T_g turns out to reach a limiting value with no significant dependence on the strength of chain-end association. If the dissociation time is shorter than the chain relaxation time, no significant influence on viscosity is expected. Hence, the ratio of these two characteristic time scales should be considered in the design of associating polymers for desired properties.

Since the activation energy (E_a) determines the temperature dependence of the bond relaxation time, its calculation needs to be accurately accounted. Chen et al. presented a new method to accurately estimate the bond exchange dynamics that excluded the often-neglected backbone chain dynamics [23]. The synthesized PDMS-based recyclable elastic network (REN) combined the permanent chemical cross-links with the high-density dynamic physical cross-links via side chain reactions. The rheology test of the REN elastomer was conducted, and the representative relaxation time, τ_{rheo}, was estimated using the time–temperature superposition (TTS) shift factor as well as the reference time scale, as shown in Figure 3.4a. To calculate the E_a, the stress relaxation experiment was conducted at 60, 70, and 80 °C (Figure 3.4b), and the master curve using 60 °C as the reference temperature was constructed to eliminate the influence of initial stress upon varying temperature. The stress relaxation time, τ_{SR}, at these three temperatures were obtained and plotted against temperature. The apparent activation energy, E_{app}, being denoted from τ_{SR}, was calculated to be 112.825 kJ mol^{-1}. This energy accounted the H-bond rearrangement dynamics that depended on both bond dissociation energy E_d and segmental dynamics. Therefore, the E_d was estimated without the chain dynamics using the following equation:

$$\frac{\tau_{\mathrm{SR}}}{\tau_\alpha} = \exp\left(\frac{E_d}{RT}\right). \tag{3.2}$$

The corresponding results were plotted as the inset in Figure 3.4a, and the E_d was calculated to be 65 kJ mol^{-1} using τ_{rheo} for τ_α. The calculated E_d was within the range of the typical associating dynamic bonds used in vitrimers, indicating that the developed REN exhibited a vitrimer-like rheological behavior.

3.2.2 Phase-Separated Aggregate Dynamics

There is an interesting scenario that dynamic noncovalent bonds such as ionic or hydrogen bonds form local microphase segregated aggregations in the polymer matrix. The limited miscibility with the polymer matrix drives the formation of

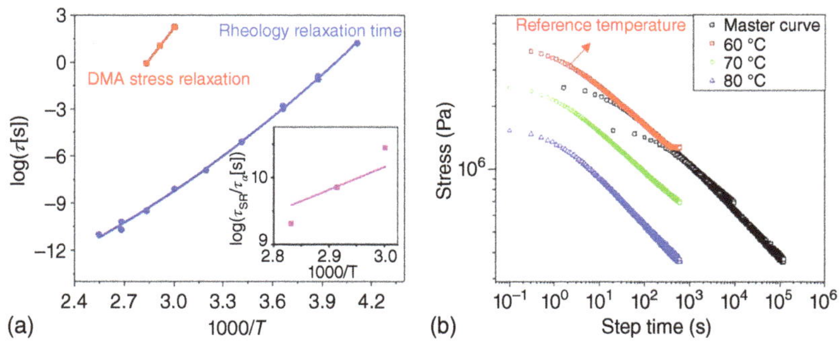

Figure 3.4 (a) Plots of log (τ) vs. 1000/T from stress relaxation time τ_{SR} and rheology relaxation time τ_{Rheo} (TTS and reference time from rheology). Inset: the τ_{SR} excluding the effect of τ_{Rheo}. (b) Stress relaxation curves at different temperatures and constructed master curve [23].

the microphase separation of stickers [18]. The structural relaxation of these phase-separated clusters dominates the viscosity of polymers rather than the polymer chain relaxation. Moreover, an interfacial polymer layer with restricted segmental mobility and higher modulus can be formed around the microphase-separated clusters. Therefore, the polymers with these structures exhibit unique viscoelastic properties with relatively high modulus and extensibility, leading to exceptional toughness [18, 24], analogous to spider silk, where partially crystalline structures and strong hydrogen bonding also lead to extremely high tensile stress and extensibility [2, 25].

The interfacial layer that exists in associating polymers with microphase separation is similar to the interfacial polymer layer formed around the nanoparticles of the polymer nanocomposites [26, 27]. This layer changes rheological behaviors such as the glass-to-rubber transition regime and the rubbery plateau level [28]. Ge et al. utilized small-angle X-ray scattering (SAXS), BDS, and rheology to demonstrate the structure and viscoelastic properties of telechelic PDMS polymers with the microphase-separated stickers (carboxy- and double-urea-terminated) [29]. Analysis of the SAXS data illustrated that aggregates of the associating end groups were formed. A direct estimate of the average distance between the aggregates was shown to be dominated by the chain end-to-end distance. They found that aggregates containing ~30–55 associating groups with the size of 1.4–1.8 nm was proportional to the reciprocal of backbone length.

The analysis of the BDS spectra confirmed the formation of an interfacial polymer layer surrounding these clusters. Considering the interfacial layer process and α^*-relaxation process that is additional to the PDMS segmental relaxation, three Havriliak–Negami (HN) functions were additively used for fitting the loss spectra:

$$\varepsilon''(\upsilon) = -\mathrm{Im}\left\{ \frac{\Delta\varepsilon_{\mathrm{bulk}}}{\left[1+(2\pi i\upsilon\tau_{\mathrm{bulk}})^{\alpha_1}\right]^{\gamma_1}} + \frac{\Delta\varepsilon_{\mathrm{int}}}{\left[1+(2\pi i\upsilon\tau_{\mathrm{int}})^{\alpha_2}\right]^{\gamma_2}} + \frac{\Delta\varepsilon_{\alpha^*}}{\left[1+\left(2\pi i\upsilon\tau_{\alpha^*}\right)^{\alpha_3}\right]^{\gamma_3}} \right\}, \tag{3.3}$$

where $\Delta\varepsilon_{\mathrm{bulk}}$, $\Delta\varepsilon_{\mathrm{int}}$, and $\Delta\varepsilon_{\alpha^*}$ are the dielectric relaxation strengths of bulk-like PDMS segments, segments in the interfacial layer, and of the binary association process (α^* process), respectively. τ_{bulk}, τ_{int}, and τ_{α^*} are the corresponding HN relaxation times; α and γ represent the shape parameters for their respective loss spectra. The fit results are shown in Figure 3.5a–c. The segmental relaxation in the interfacial layer was approximately one order of magnitude slower than in the bulk-like polymer. The amplitude of the bulk-like relaxation decreased and that of the interfacial layer increased gradually with decreasing DP, indicating an increasing volume fraction of the interfacial region with a decrease in the chain length. The interfacial layer volume fraction and thickness (~0.7–0.9 nm) in these materials were fully verified against the SAXS results.

The binding/dissociation energy and the lifetime of the dynamic bonds are important parameters that determine the viscoelastic properties of polymers with sticky bonds. At shorter time scale, material with these sticky bonds performs as a network, while at the time scale longer than that of the bond dissociation, the

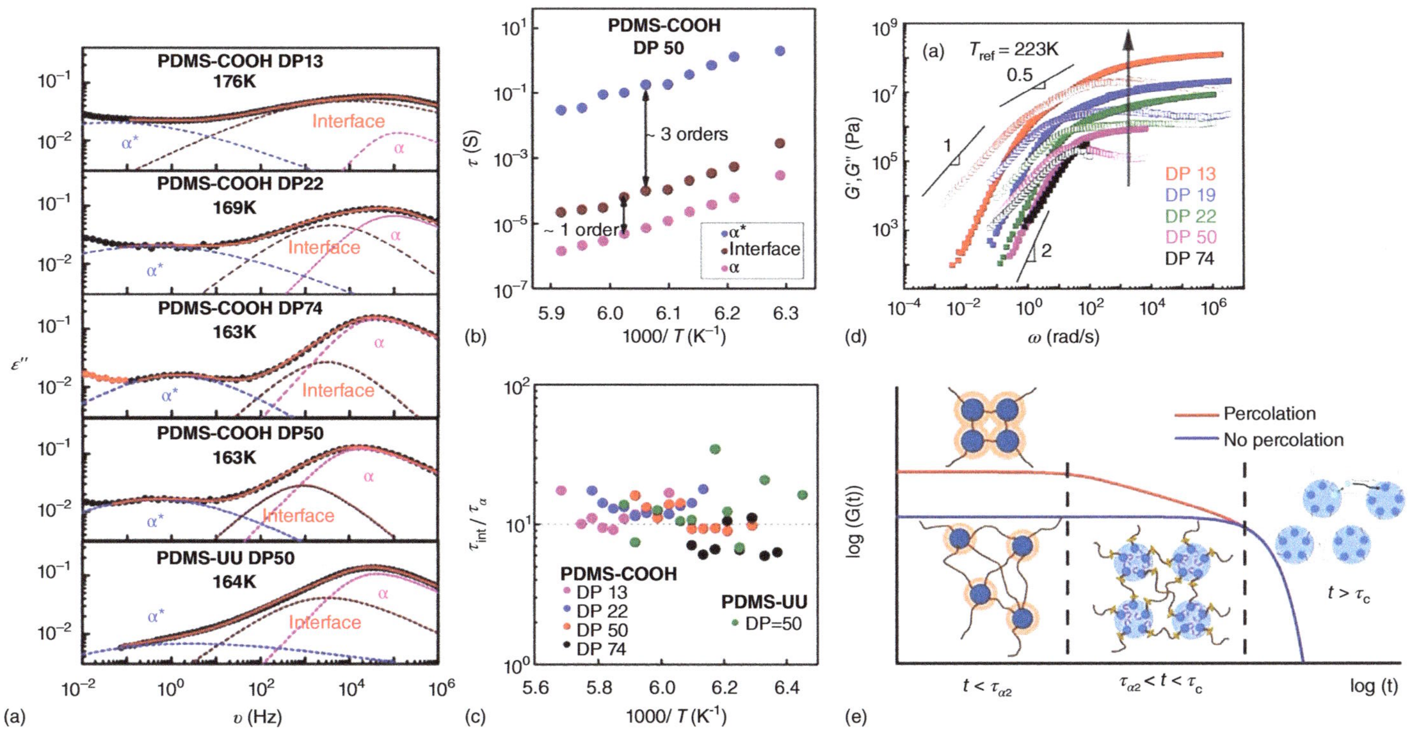

Figure 3.5 (a) Dielectric loss spectra $\varepsilon''(\nu)$ for PDMS with different associating end groups and DP. (b) Activation plots of the segmental relaxation time in the bulk-like PDMS (pink), in the interfacial layer (brown), and the α^* (blue) relaxation time in PDMS-COOH DP 50 sample. (c) Ratio between the relaxation times of the interfacial layer and bulk-like PDMS segments [29]. (d) Shear modulus master curves for the telechelic PDMS-COOH measured at temperatures higher than second T_g. (e) Qualitative picture of the shear modulus variation with time for a phase-separated associated polymer network (bottom panel) [30].

restriction to material flow is eliminated [31]. Clusters with higher glass transition or melting temperature exhibit a prolonged rubbery plateau compared to that of polymer matrix [18]. Based on the result of previous COOH-terminated PDMS system, the group presented a mechanical percolation model to comprehensively explain the unusually high rubbery plateau modulus (Figure 3.5d) [30]. More importantly, they proposed a mechanism of the network rearrangements via single-chain hopping between clusters, which were governed by a thermodynamic energy barrier related to the immiscibility of the end groups in the polymer matrix. Namely, the chain end in a phase-separated cluster needs to be pulled out and diffuse to another phase-separated cluster to achieve a rearrangement process. Borrowing the idea of chain exchange kinetics in block copolymer micelles [32, 33], relationship between τ_c (terminal relaxation time) and τ_{α_2-M} (relaxation time of cluster motion from additional dielectric process) was expressed as:

$$\tau_c(T) = \tau_{\alpha_2-M}(T)\exp\left(\frac{E_a}{RT}\right) = \tau_{\alpha_2-M}(T)\exp(\alpha\chi N_{core}), \tag{3.4}$$

where E_a is related to the thermodynamic penalty for the sticker to be placed in the polymer matrix $\sim\alpha\chi N$core. Once the sticker pulled out from the cluster, it will diffuse through the polymer matrix to another cluster with a time scale of $\tau_{diffusion}$, which will be mainly governed by Rouse subdiffusive motion:

$$\tau_{diffusion} \approx \tau_\alpha\left(\frac{d}{l_k}\right)^4, \tag{3.5}$$

where d is the distance between clusters (~5 nm) and l_k is the Kuhn segment length of PDMS (~1 nm), resulting in $\tau_{diffusion} \approx 10^3\ \tau_\alpha$. At the onset of second T_g, where stickers moved inside the clusters, τ_α was ~10^{-9} seconds and $\tau_{diffusion}$ was ~10^{-6} seconds. This time scale was many orders faster than terminal relaxation time at the same temperatures. Hence, the process that a sticker was pulled out from the cluster was the rate-determining step in the single-chain hopping process (Figure 3.5e). A general picture of viscoelasticity in polymers with phase-separated dynamic bonds was proposed. Functional end group clusters had a higher T_g value than the polymer matrix, and their dynamics were governed by a time scale of structural relaxation in these clusters, τ_{α_2}. This in-depth understanding of viscoelasticity in polymers with phase-separated dynamic clusters can provide design rules for developing functional materials with tunable viscoelastic properties, enhanced mechanical properties, or better recyclability.

By controlling the distribution or topology of the clusters, the mechanical properties of the polymer materials can be tailored on demand. Zheng et al. designed a series of high-performance cross-linked PDMS elastomers via controlling the density and morphology of the hydrogen-bonding clusters, surpassing the traditional stiffness-extensibility trade-off [24]. Two approaches were used for the network design: (i) incorporation of the elastic chemical network (ECN) with H-bonds to form a physical-chemical interpenetrated network (pc-IPN); (ii) chemically grafting of efficient hydrogen-bonding units, having quadruple H-bonds such as 2-ureido-4 [1H]-pyrimidinone (UPy), on the elastic chemical network (cg-ECN), and tuning

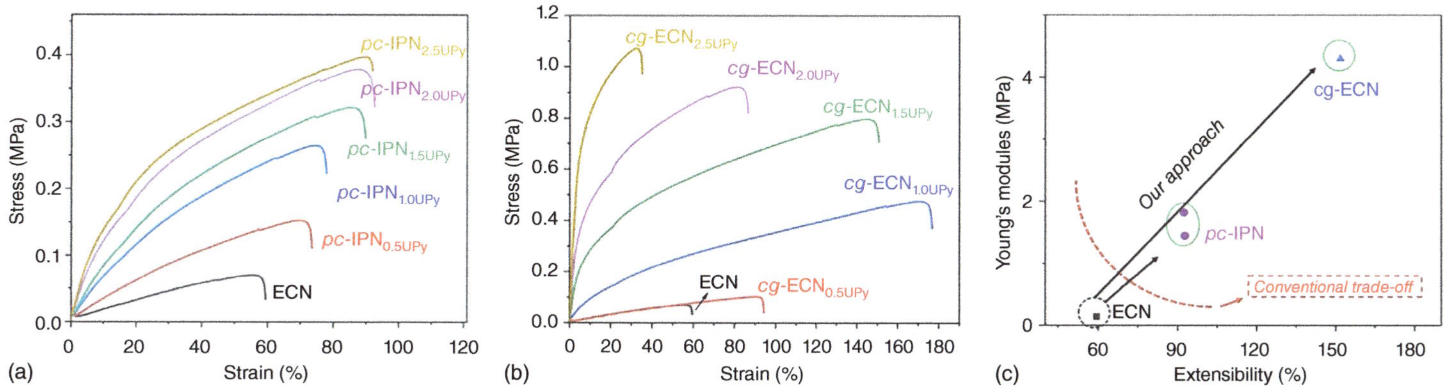

Figure 3.6 Stress–strain curves of (a) pc-IPNs and (b) cg-ECNs. (c) Simultaneously increased extensibility and Young's modulus of the pc-IPN and cg-ECN compared with the conventional trade-off [24].

the distribution of UPy units and topology of H-bond clusters without varying the overall UPy concentrations.

Compared with those of the ECN, the elongation at break and Young's modulus of the pc-IPN were improved, achieving a 10-fold increase of toughness from 0.027 to 0.260 MJ m^{-3} for the interpenetrated network with 2.5 equivalents of UPy, i.e. pc-$IPN_{2.5UPy}$ (Figure 3.6a). The improvement for cg-ECN in mechanical performances was much more significant (Figure 3.6b). For example, from ECN to cg-$ECN_{2.5UPy}$, Young's modulus increased by 150-fold. Figure 3.6c shows that the significant improvement in both performances successfully surpassed the traditional stiffness-extensibility trade-off in elastic networks, which is beyond most of state-of-the-art approaches that could not break through the reciprocal relationship of these two parameters. In addition to the effect of concentration, the effect of distribution of the H-bond units on mechanical properties was also studied by comparing the cg-$ECN_{1.5UPy}$ and cg-$ECN_{1.5UPy'}$. The cg-$ECN_{1.5UPy'}$ was prepared with minor modifications, where more UPy units were selectively grafted on part of AMS-PDMSs instead of homogenously distributing them on each PDMS chain. The result showed that the Young's modulus of cg-$ECN_{1.5UPy'}$ was about four times higher than that of cg-$ECN_{1.5UPy}$ without sacrificing the extensibility of the elastomers. Given the same H-bond concentration in both cases, the higher small-angle intensity and smaller Porod exponent in cg-$ECN_{1.5UPy'}$ suggested larger network of clusters that aggregated in this elastomer in comparison with cg-$ECN_{1.5UPy}$. It was concluded that the simultaneously increased Young's modulus and elongation at break were attributed to the synergistic effect of H-bond interactions and rearrangement, along with the formation of hydrogen-bonding clusters and their aggregation. The success of beyond the state-of-the-art approaches in the improvement of both Young's modulus and extensibility in a model elastic network light a shed for designing other mechanically robust elastic materials for various applications such as automotive industries and medical devices.

3.3 Properties and Applications

With the development in macromolecular design employing dynamic bonds, many attractive functionalities such as extreme stretchability, self-healability, recyclability, high adhesion property, and 3D printability have been presented over the recent decades. Such functionalities are promising in applications on gas separation [34–37], sealants and adhesives [38], 3D printing [39, 40], and polymer electrolytes of Li-metal batteries (LMBs) [41, 42], etc. In this section, several typical reported dynamic polymeric materials with attractive performance are highlighted among various application fields, as well as their state-of-the-art synthesis approaches and characterization. Some significant systems without dynamic bonds are also reviewed as the references for readers/researchers to design polymer networks with desired properties.

3.3.1 Gas Separation

To date, many efforts have been devoted to developing novel polymer membranes with increased gas selectivity [43, 44]. However, the overall gas separation performance and lifetime of membrane are still negatively influenced by the weak mechanical performance, low plasticization resistance, and poor physical aging tolerance. To utilize the mechanical improvement of chemical cross-linking and plasticization resistance of physical interactions, Cao et al. designed a series of urethane-rich PDMS-based polymer networks (U-PDMS-NW) with improved mechanical performance for gas separation application [34]. The PDMS membranes were fabricated by the in situ reaction between the amine-terminated PDMS and the multifunctional isocyanate cross-linker (Figure 3.7a). The cross-link density of U-PDMS-NWs was tailored by varying the molecular weight (M_n) of PDMS, and it was expressed as:

$$c_x = \frac{2G'}{3RT} = \frac{2\rho}{3M_x}, \tag{3.6}$$

where c_x is the moles of cross-links per volume, G' is the plateau value of real part of shear modulus, R is the gas constant, T is the temperature in Kelvin, ρ is the polymer density, and M_x is the number-average molecular weight of the polymer segment between the cross-links. The U-PDMS-NWs exhibits up to 400% elongation before break and tunable Young's modulus (1.3–122.2 MPa), ultimate tensile strength (1.1–14.3 MPa), and toughness (0.7–24.9 MJ m^{-3}) (Figure 3.7b). All of the U-PDMS-NWs showed salient gas separation performance with excellent thermal resistance and aging tolerance, high gas permeability (>100 Barrer), and tunable gas selectivity (up to $\alpha[P_{CO2}/P_{N2}] \approx 41$ and $\alpha[P_{CO2}/P_{CH4}] \approx 16$) (Figure 3.7c). With well-controlled mechanical properties and gas separation performance, these U-PDMS-NW can be used as a polymer membrane platform for gas separation and other functional potential applications such as microfluidic channels and stretchable electronic devices.

Utilization of reversible chemical/physical bonding to heal mechanical fractures and recover functionalities of the polymer materials, such as extensibility and separation capability, is an effective way to prolong the lifetime of polymer membranes [45, 46]. On the basis of the aforementioned work, the same group also explored a series of urea functionalized PDMS-based elastomers (U-PDMS-Es) to further improve the performance of the system including extremely high stretchability (up to 5500%), self-healing mechanical properties, and recoverable gas separation performance [35]. A cross-linker *tetra*-diethylenetri-amine-terminated PDMS (TDA-PDMS), amine-terminated PDMS (PDMS-NH_2), and urea were reacted in a certain ratio to synthesize a series of U-PDMS-Es with four types of associating units (Figure 3.7d). Amine-terminated PDMS (PDMS-NH_2), urea, and TDA-PDMS as cross-linker were reacted in a certain ratio to synthesize a series of U-PDMS-Es with four types of associating units (Figure 3.7d). Among different associating units, type I exhibited the strongest supramolecular interactions and shortest hydrogen bond length, while type II showed the weakest interactions. By tailoring the MW of PDMS or weight ratio of elastic cross-linker, the mechanical properties of the

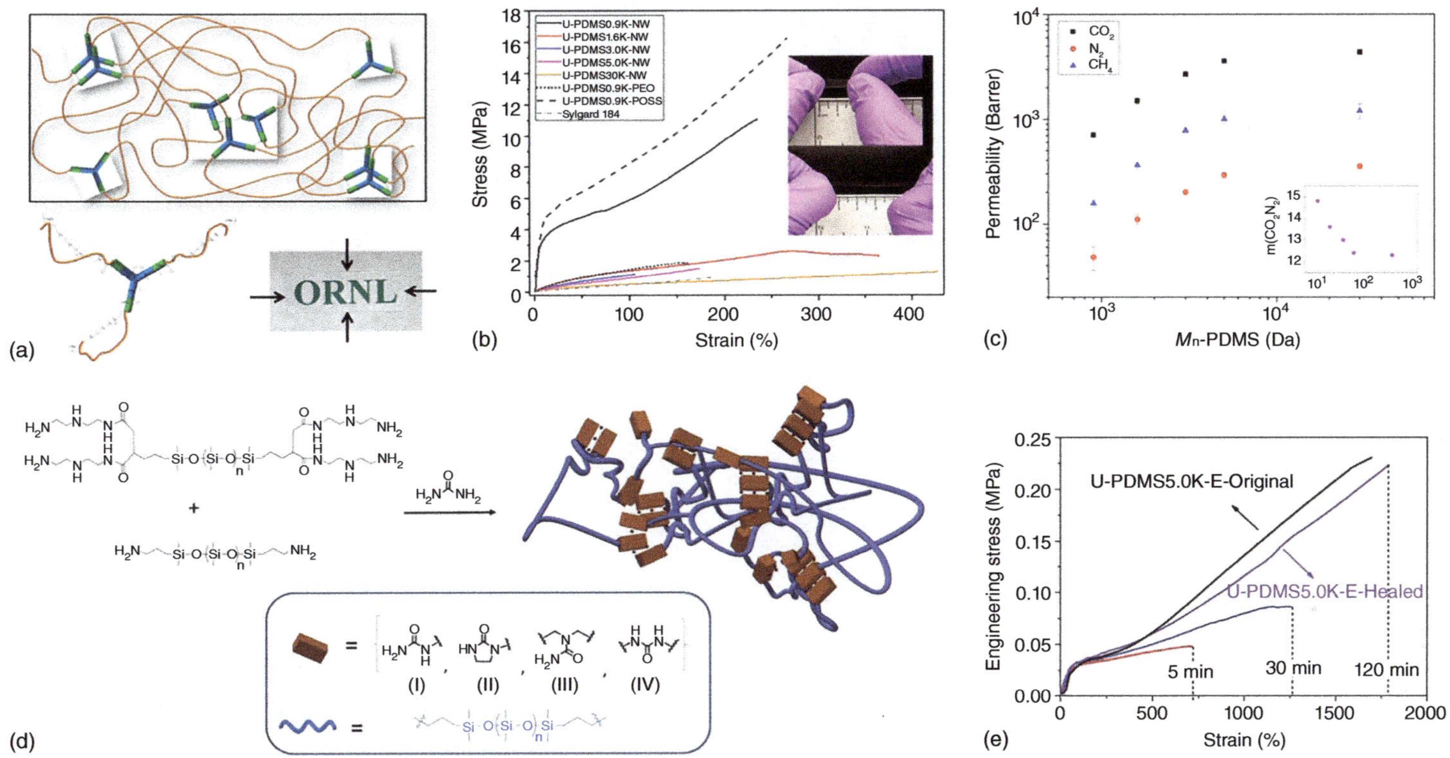

Figure 3.7 (a) Schematics of the U-PDMS-NW and the repeating units, and picture of the obtained U-PDMS-NW film. (b) Stress–strain curves of the U-PDMS-NWs. (c) Gas permeability of the U-PDMS-NW vs. the M_n of the feeding PDMS [34]. Peng-Fei Cao et al. 2017/Reproduced with permission from American Chemical Society. (d) Synthesis of the U-PDMS-Es. (e) Comparative tensile test of the original U-PDMS-E membrane and after cut and healing process [35].

obtained U-PDMS-Es were tuned in wide ranges, namely, ultimate elongation before break from 984% to 5600%, Young's modulus from 230 to 27.3 kPa, ultimate tensile strength from 61.1 to 1110 kPa, and toughness from 1.46 to 7.14 MJ m^{-3}. Due to the high density of hydrogen bonding and fast segmental dynamics, the U-PDMS-Es showed efficient self-healing capacity with completely restored mechanical properties under two hours at ambient temperature or 20 minutes at 40 °C (Figure 3.7e). As such, the U-PDMS-Es was applied to recoverable gas separation membranes with retained permeability/selectivity after being damaged and healed.

There are two main strategies for fabrication of stretchable functional polymeric materials with physical adhesion between functional components and the elastic polymer matrix: (i) utilization of the stretchable polymer as it interconnects between rigid island components with specific functionality and (ii) embedding functional particles/molecules into elastic polymer matrices [47]. Poly(ionic liquids) (PILs) are usually known as "brittle" functional polymers due to their "glassy" nature at ambient temperature. Li et al. developed a unique approach to endow glassy PILs with high flexibility and good elasticity via a rational molecular design of chemical composition and polymer architectures (Figure 3.8a) [36]. First, the reversible addition/fragmentation chain transfer agents (RAFT-CTAs) were chemically attached to the flexible PDMS backbones. The polymerization of functional ionic liquid monomers, e.g. 1-vinylimidazole, from RAFT-CTAs offered the grafted copolymers with the functional side chains, which were further cross-linked by difunctional PDMS. The ionic liquid with larger counter ions was reported to have a lower T_g because of the weaker ion associations [48, 49]. However, this study showed that only increasing the size of counterions in PIL-based membranes did not significantly improve their extendibility. Hence, the poly(ethylene glycol) methacrylate (PEGMA) was copolymerized with ionic liquid monomers to reduce the T_g, providing higher chain mobility and elasticity at ambient temperature. Such structure design allows soft PDMS as the polymer matrix for extensibility while it has PILs as the main chemical component for desired functionality.

PILs have been used as gas separation membranes due to their potential in improving gas permeation and selectivity for certain gas pairs (CO_2/N_2, CO_2/CH_4, etc.) by tuning solubility and utilizing optimized cation–anion pairs [50, 51]. With the elastic PDMS backbones providing fast segmental mobility at ambient temperature, as well as the ionic liquid and PEG segments serving as CO_2-philic functional groups, the elastic PILs resulted in high gas permeability and CO_2/N_2 selectivity (Figure 3.8b).

In addition to the cross-link density, associating units, and segmental mobility of the backbones, the features of the cross-linkers and side chains are also important to the gas separation performance of the polymer membranes through altering the crystalline regime. Hong et al. synthesized a series of multigrafted PDMS elastomers via a one-pot thiol–ene reaction for the application of gas separation membranes with tailorable CO_2-philicity [37]. Varying the cross-linkers, such as acryloxy-terminated block copolymer (EOPDMS), vinyl-terminated PDMS (VTPDMS), and poly(ethylene glycol) diacrylate (PEGDA), and grafting

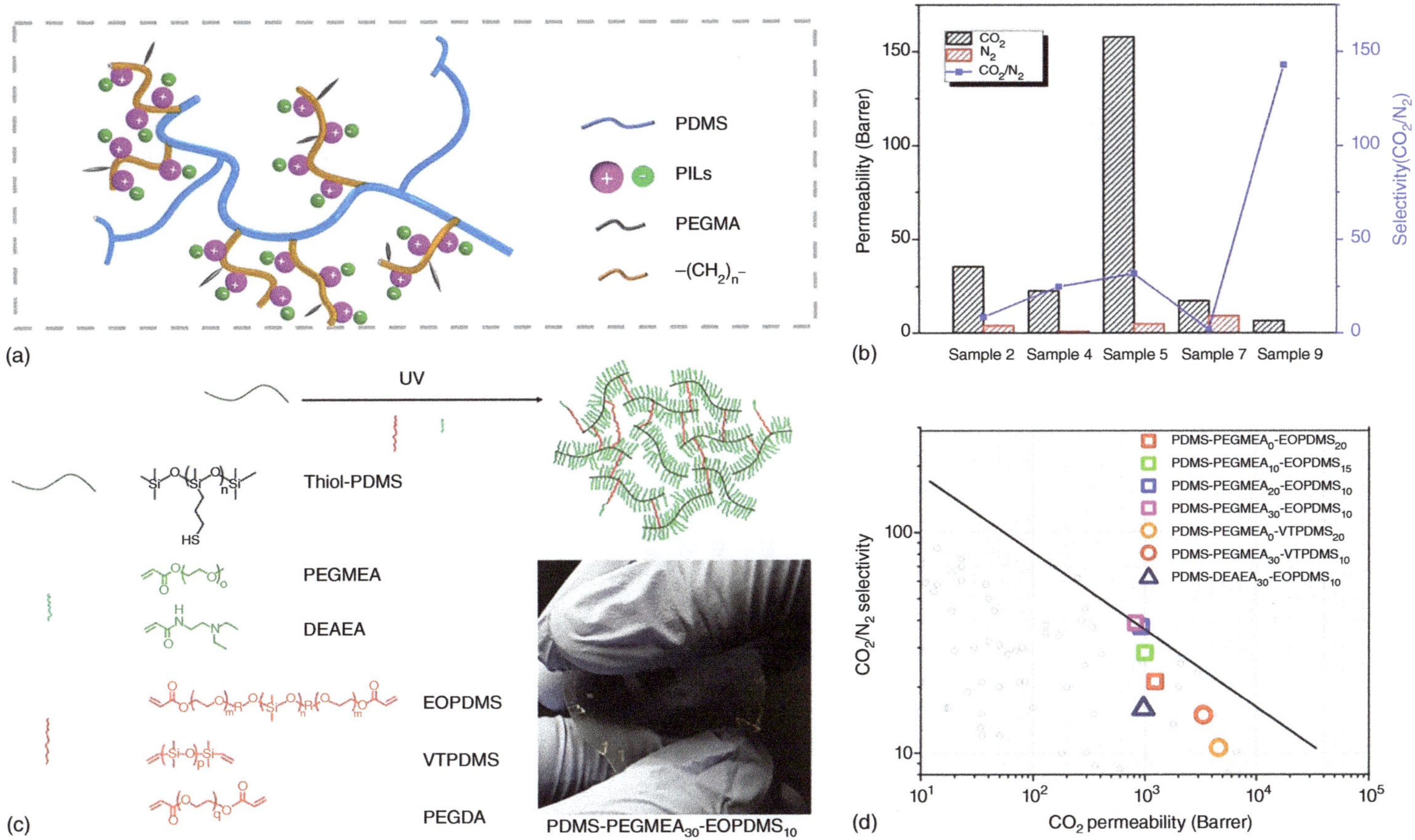

Figure 3.8 (a) Schematic of the designed PILs. (b) Gas permeation data of E-PIL membranes [36]. (c) Synthesis of PDMS elastomer membranes via thiol–ene click reaction. (d) Summary of the gas separation performance of synthesized membranes in 2008 Robeson plot [37]. Tao Hong et al. 2019/Reproduced with permission from American Chemical Society.

side chains including poly(ethylene glycol) methyl ether acrylate (PEGMEA), *N*-(2-(diethylamino)ethyl)acrylamide (DEAEA) allowed tuning of their thermal/mechanical properties and gas separation performance (Figure 3.8c). The relatively low MWs of PEGMEA and the presence of the cross-linked network prevented the formation of the crystalline regime at room temperature, achieving the minimum loss of gas permeabilities [34, 52]. The slightly cross-linked networks were designed by controlling the MW and reaction ratio of the starting materials to afford highly permeable rubbery membranes. The incorporation of CO_2-philic PEGMEA in both cross-linkers and side chains contributed to the significant improvement of gas selectivity. The membranes containing PEG moiety in the cross-linker exhibited more than three times improvement in terms of CO_2/N_2 selectivity. The copolymer membrane with weaker CO_2-philic DEAEA showed a much lower selectivity of CO_2 over N_2 and CH_4. The gas separation performance of such synthesized membranes was compared in 2008 Robeson plot, and the PDMS–PEGMEA$_{30}$–EOPDMS$_{10}$ reached the Robeson upper bound (Figure 3.8d). This work provides macromolecular design principles involving segmental mobility, cross-linking density, gas-philic functional groups, etc., for desired gas separation performance.

3.3.2 Adhesives and Additives

As discussed earlier, utilization of dynamic networks is capable of introducing functionalities such as self-healing and high stretchability. Self-healable elastomers are of great interest due to their ability to prolong product lifetime, which are promising in applications on sealants and adhesives. For adhesive applications, an important functionality is the high adhesion force to clean dusty surfaces. Zhang et al. reported a series of autonomous self-healable and highly adhesive elastomers (ASHA-Elastomers) that were composed of self-healable poly(2-[[(butylamino)carbonyl]oxy]ethyl acrylate) (Poly(BCOE)) and curable elastomers (C-Elastomers, i.e. silicon-based or polyurethane-based elastomers) (Figure 3.9a) [38]. The low glass transition temperature ($T_g \approx -3\,°C$) of the Poly(BCOE) and thus fast segmental dynamics at ambient temperature along with the numerous sacrificial hydrogen bonds of urethane groups offered the self-healing property to the obtained elastomer system. The C-Elastomers possessed high adhesive force due to the following two processes: (i) good contact of the elastomer precursors (liquid-like) with the substrate; and (ii) the subsequent moisture-triggered chemical cross-linking process that resulted in high-toughness elastomers. The shear modulus data showed that compared with the high G' value of C1-elastomer, the G' values of ASHA-C1-Elastomers decrease with the increased ratios of Poly(BCOE) because of the gradually decreased cross-linking densities of ASHA-C1-Elastomers. The ASHA-C1-Elastomer-30 showed outstanding mechanical properties with elongation at breakup to 2102% and toughness of $1.73\,MJ\,m^{-3}$. The damaged ASHA-Elastomer can autonomously self-heal with full recovery of functionalities, and the healing process was not affected by the presence of water.

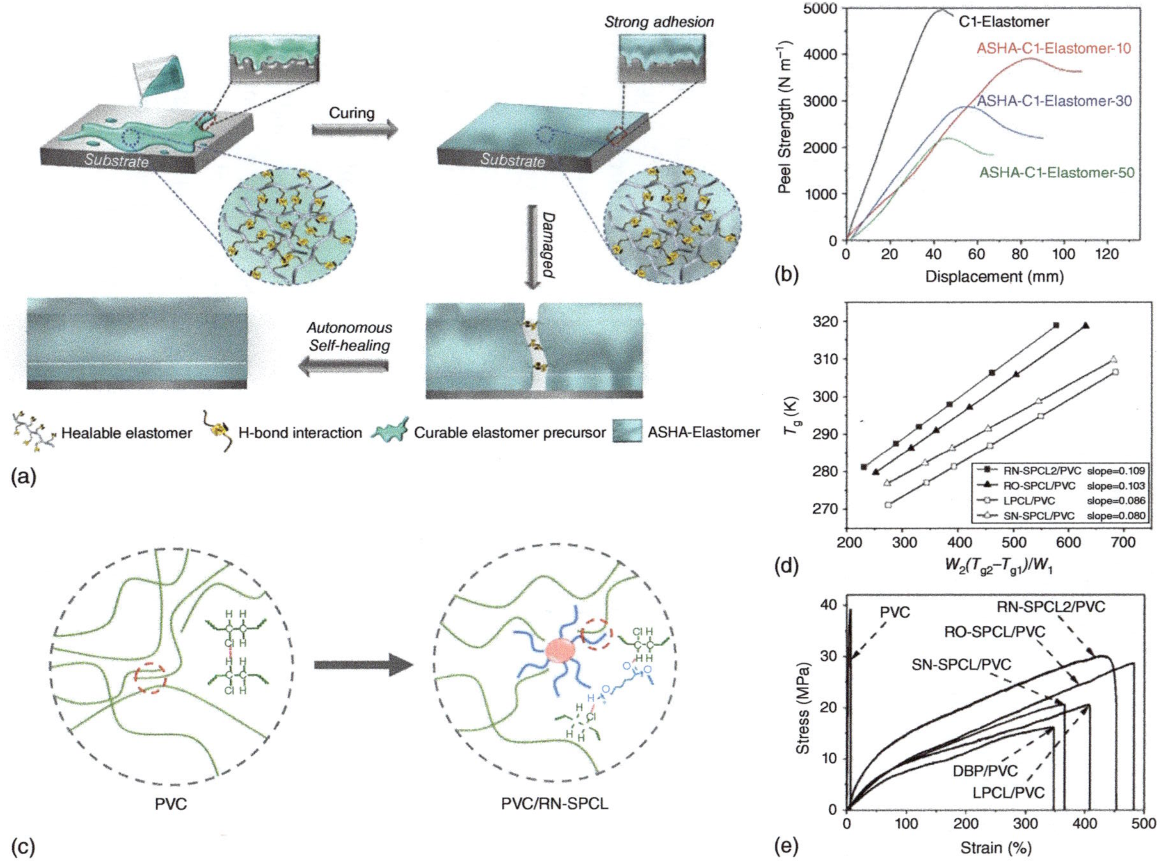

Figure 3.9 (a) Schematic of the fabrication process of the ASHA-Elastomer on an aluminum substrate and its self-healing process through H-bond interactions and molecular dynamics of the self-healable polymer Poly(BCOE). (b) Peel tests of ASHA-C1-Elastomers [38]. (c) Schematic of the plasticization mechanism by RN-SPCLs. (d) Gordon–Taylor plots of different PCL-based composites with PVC. (e) Stress–strain curves of the neat PVC and PVC-based elastomers [53].

The adhesive test results indicated that an adhesion force achieved up to 3488 N m^{-1}, outperforming previously reported self-healing adhesive elastomers (Figure 3.9b). Generally, the adhesion strengths of regular adhesive polymers are extremely sensitive to dust on the substrate because the presence of dust leads to reduced contact area. However, the adhesion force of the ASHA-Elastomer by this study was negligibly affected by dust on the surface. A possible explanation for the higher adhesion of ASHA-Elastomers on dusty surface was that the ASHA-C1-Elastomers were able to "encapsulate" the dust particles owe to the efficient physical interactions and fast polymer dynamic of Poly(BCOE) at ambient condition. The elastomer design guidance is promising in fabricating high-performance elastomers for applications with enhanced longevity and versatility, including their use in sealants, adhesives, and stretchable devices.

On the basis of the previous studies, the group further explored toughening approaches for different polymer systems. To address the mechanical brittleness and low toughness for poly(vinyl chloride) (PVC), Chen et al. designed a stretchable and ultratough PVC-based plastic by a star-shaped poly(ε-caprolactone) (PCL) copolymer (RN-SPCLs) constructed with a rigid, amino-containing, branched polylactide (N-BPLA) core and a soft PCL shell (Figure 3.9c) [53]. The N-BPLAs with a branching degree of 9% were employed as macroinitiators for the ring-opening polymerization of CL. With an optimal feed ratio of the CL monomer and N-BPLA core, the RN-SPCL2 efficiently lowered the T_g of PVC plastics, achieving the transition from the "glassy" to "rubbery" state at ambient temperature. The T_g of PVC systems could be reduced gradually through increasing the content of SPCL, and the data were fitted to the Gordon–Taylor equation [54]:

$$T_g = T_{g1} + KW_2(T_{g2} - T_{g1})/W_1, \tag{3.7}$$

where 1 and 2 in this work represent the PCL and PVC, respectively, W_i is the weight fraction of component i, and K is the Gordon–Taylor parameter that semiquantitatively determined the interaction strength between two polymers [55]. The fitted data in Figure 3.9d demonstrated a clear linear relationship for all PCL-based polymer-plasticized PVC systems that exhibited good miscibility.

The verified efficient physical interaction between the plasticizer and PVC matrix enabled the suppressed crystallinity of PCLs and the reduced T_g of PVCs, improving the stretchability, toughness, and migration resistance. The obtained RN-SPCL2/PVC exhibited high extensibility (453%) and maintained near 80% of tensile strength of the neat PVC (30.1 MPa) (Figure 3.9e). Its toughness reached 92.7 MJ m^{-3}, more than 50-fold higher than that of the neat PVC and two to three times higher than that of the linear PCL or dibutyl phthalate plasticized PVCs.

Additionally, the tertiary amine groups endowed the branched PLAs intrinsic photoluminescence nature and guest encapsulation capacity. Under UV light irradiation at a wavelength of 365 nm, the RN-SPCL2/PVC film emitted cyan light, and under the visible light irradiations at wavelength of 400–435 nm and 450–490 nm, it resulted in green light emission. The RN-SPCL could also serve as an efficient dispersant of TiO_2 nanoparticles, which significantly improved the anti-UV aging capability of PVC plastics.

3.3.3 3D Printing

With the combined advantages from both physical and chemical interactions, the polymer materials with dual-type dynamic bonds typically exhibit superior performances in terms of mechanical strength and extensibility, recyclability, self-healing, etc. The extrusion-based 3D printing such as direct ink writing (DIW) and fused deposition modeling (FDM) techniques employs liquid-phase inks with suitable viscosity (or resin filament) that are to be fed into the nozzle and then deposited layer-by-layer to fabricate the designated 3D shape of the objects [56]. The dynamic polymer materials with tuned printability have been applied in DIW or FDM to achieve self-healing parts.

Inspired by some organic solvents with high chain transfer constant that react with and terminate the radical species, Zhu et al. designed a poly(thioctic acid, TA) system with dual dynamic bonds (disulfide bonds and H-bonds) via an organic solvent-quenched polymer synthesis as adhesives and 3D printing inks [39]. The unique reaction mechanism was investigated by the solid-state proton nuclear magnetic resonance and first-principle simulations (Figure 3.10a). The result showed that the chlorinated solvent efficiently stabilized and mediated the depolymerization of poly(TA), which was more kinetically favorable upon lowering the temperature. Attributed to the numerous dual types of dynamic bonds, the obtained poly(TA) showed high extensibility (1000%, Figure 3.10b), self-healability (complete recovery of tensile strength within 24 hours, Figure 3.10c), and reprocessable properties (no significant morphological change after reprocessing). The poly(TA) also possessed promising adhesive property, and a good lap-shear strength of 1.15 and 1.63 MPa was achieved under the adhered aluminum and steel slices, respectively. Even on hydrophobic and low surface energy Teflon substrate, it still exhibited an excellent adhesive capability with a shear strength of 0.35 MPa. Different temperatures were conducted for the melting process during 3D printing. The printed parts showed slight degradation with a long heating time at 120 °C and visible sinkage upon decreasing the temperatures to 90 or 100 °C. Hence, 110 °C was selected as the optimal printing temperature with the printed parts shown in Figure 3.10d.

Vitrimers possessing robust chemical networks with associative dynamic covalent bonds are promising in 3D printing of parts with enhanced interlayer adhesion. Niu et al. designed a synthetic polyurea-based elastic vitrimer with vinylogous urethane cross-links for extrusion-based 3D printing to achieve elastomer parts with highly recyclable, mechanically isotropic, and healable properties (Figure 3.10e) [40]. Different types of polyurea vitrimers were synthesized by varying the types of isocyanate and chain length of the linear polyurea to optimize the printability. The vitrimer using 4,4′-methylenebis and 2 : 1 feed ratio of PDMS and di-isocyanates to 2 : 1 was selected for the printing because of the suitable printing temperature (70 °C) and mechanical strength. The tensile bars with raster angles of 0° and 90° were printed, and the mechanical isotropy was achieved in both with and without thermal annealing (70 °C for 20 hours, Figure 3.10f). In addition, the vitrimer could be recycled for five generations with retained mechanical properties, and the printed sample could be basically repaired after damage (Figure 3.10g).

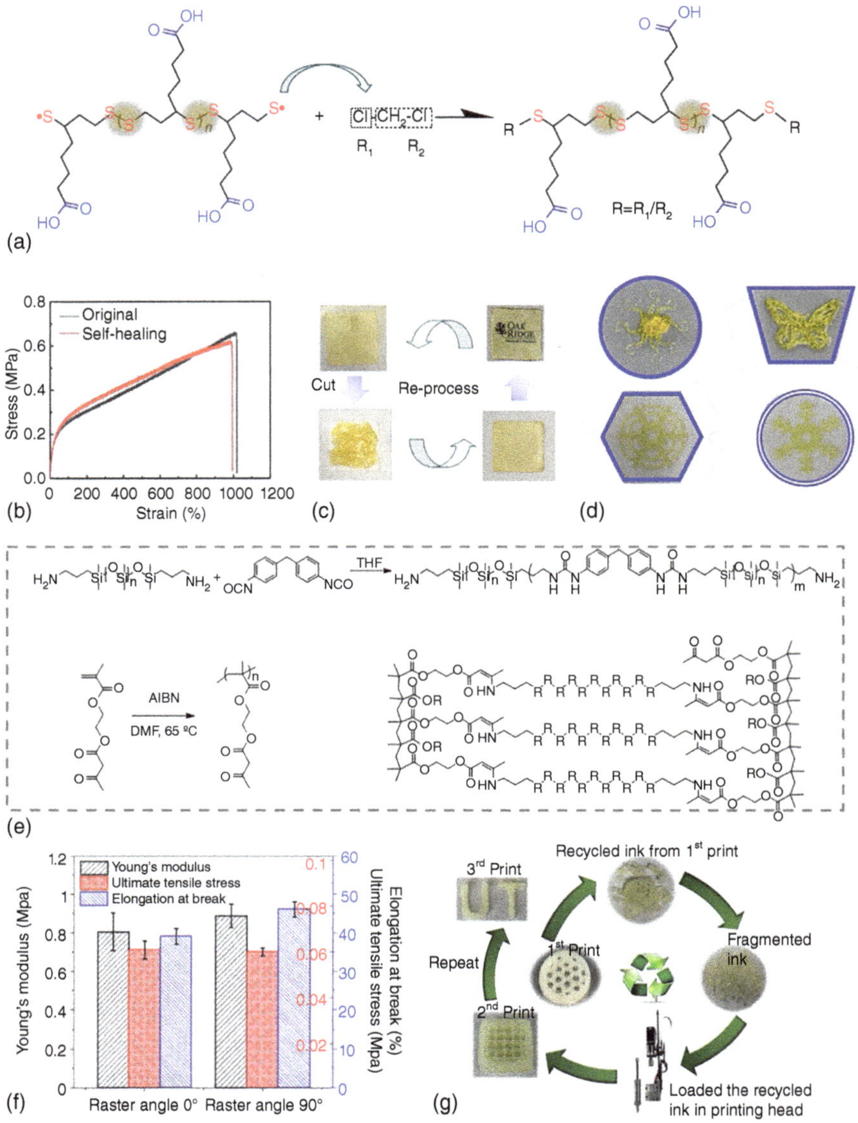

Figure 3.10 (a) Schematic of the reaction mechanism in the presence of dichloromethane (DCM). (b) Stress–strain curves of the poly(TA) before and after self-healing. (c) Reprocess of the poly(TA). (d) FDM-printed parts with solid infill [39]. (e) Synthesis scheme of the polyurea vitrimer. (f) The tensile properties of the polyurea vitrimer samples printed along different printing orientations with thermal annealing. (g) Images of 3D-printed parts through recycle of the polyurea vitrimer ink [40].

3.3.4 Polymer Electrolytes

The stretchable functional polymeric electrolytes may induce interfacial resistance issues, where the physical adhesion between functional components and elastic polymers is insufficient. Cao et al. reported a versatile approach on the molecular-level intrinsically stretchable polymer materials with defined functionality [47]. PDMS having a large number of reactive amine groups was selected as the polymer backbone and reacted with the coupling reaction of the carboxylic acid-terminated reversible addition–fragmentation chain transfer agent (RAFT–CTA–COOH) (Figure 3.11a). PEGMA and lithium (4-styrenesulfonyl) (trifluoromethane-sulfonyl) imide (STF–Li^+) with delocalized negative charge distribution were grafted from the PDMS backbone. The single-ion conducting polymer electrolytes (SICPEs) were obtained after being cross-linked (Figure 3.11b), and the rational design allowed tunable mechanical properties of the elastic membranes by changing the molar ratio of the elastic PDMS cross-linker and grafted block copolymers. Figure 3.11c shows that the galvanostatic test of the assembled cell was performed between 3.8 and 2.5 V with the capacity retention was calculated to be 81.5% after 100 cycles. In addition, the elastic polymer membranes with PEGMA and PEGMEMA as the side chains exhibited decent gas separation

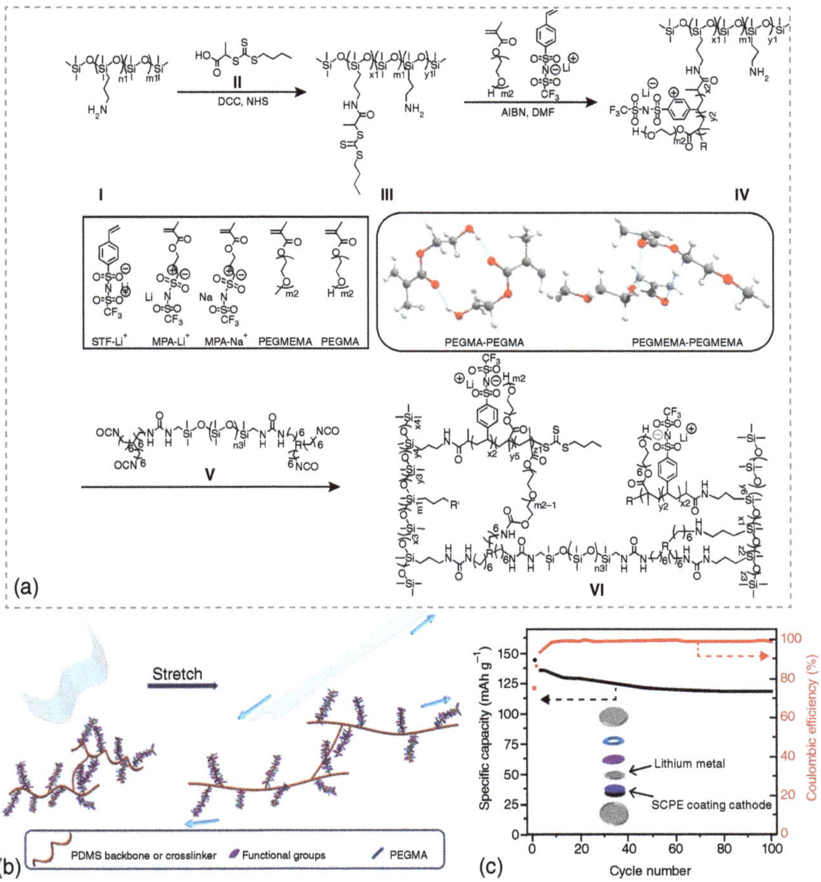

Figure 3.11 (a) Synthesis of the SICPEs. (b) Schematic of molecular-level intrinsically stretchable functional polymers. (c) Specific capacity and coulombic efficiency of the cell with the SICPE at 0.1 C under 65 °C [47].

performance ($\alpha[PCO_2/PCH_4] = 36.5$). The versatile molecular-level design of the stretchable functional materials is insightful for both fundamental understanding of rational polymer architecture design and various functional applications, where high stretchable performance is needed.

The intrinsic self-healable, stretchable polymer films with well-designed dynamic networks can be used as a protective coating on lithium (Li) anodes, which is a well-recognized practice to suppress dendrite growth and improve the safety of LMBs. Combined the chemical cross-linking and H-bonds, Sun et al. reported a highly stretchable, autonomous, self-healable, and ionic-conducting polymer network (SHIPN) via the polymerization of 2-[[(butylamino)carbonyl]oxy]ethyl acrylate (BCOE) and poly(ethylene glycol)-mono-methacrylate (PEGMMA) followed by chemical cross-linking with diisocyanate (HDI) (Figure 3.12a) [41]. Due to the presence of large amounts of H-bonds and their fast molecular dynamics at ambient temperature, the mechanical performance of the damaged SHIPN film was fully recovered after being completely cut and a one-hour healing process at an ambient condition. Such self-healing capability of SHIPN also worked even under a low temperature, which was of great significance for using as an efficient protecting layer of the Li-metal electrode that operated at low temperatures.

The full cells with the SHIPN-protected Li electrode exhibited a capacity retention of 85.6% after 500 cycles even at 5 °C (Figure 3.12b). The SHIPN@Li/$LiFePO_4$ (LFP) full cell using a commercial standard LFP cathode (96.8 wt% active material) retained a high discharge capacity up to 90 mAh g^{-1} over 100 cycles at 1 C. The impedance analysis and surface characterization indicated that the SHIPN film facilitated as a stable solid electrolyte interphase (SEI) layer effectively suppressed the growth of dendritic Li and thus reduced accumulation of "dead" lithium.

In a related way, He et al. proposed a chemically grafted hybrid dynamic network (CHDN) which was synthesized by the 4,4′-thiobisbenzenamine (4-AFD) cross-linked poly(poly(ethylene glycol) methyl ether methacrylate-*r*-glycidyl methacrylate) (PEGMEMA-*r*-GMA) and (3-glycidyloxypropyl) trimethoxysilane-functionalized SiO_2 nanoparticles (GLYMO@SiO_2). It has been utilized as the protective layer and hybrid solid-state electrolyte (HSE) of LMBs (Figure 3.12c) [42]. The dynamic covalent network was capable of self-healing and chemical stability in liquid electrolyte, and the inorganic fillers served as an agent to suppress the crystallinity of poly(ethylene oxide) for improving the ionic conductivity and the mechanical robustness of the developed dynamic network. The covalent attachment of the GLYMO@SiO_2 with the polymer network prevented the aggregation of such fillers even after the cycling process, ensuring the homogeneous ionic conducting channels.

With assembling the CHDN-based protective layers, excellent electrochemical performance was presented in Li/Cu half cells (coulombic efficiency of ~97.9% over 300 cycles at a current density of 0.5 mA cm^{-2}), Li/Li symmetric cells (over 1000 hours at the current density of 1 mA cm^{-2}), and Li/LFP full cells (capacity retention of 83.7% over 400 cycles at 1 C). The Li/HSE/LFP solid-state battery manifested high cycling stability with a discharge capacity of 145.2 mAh g^{-1} and capacity retention of 89.8% over 500 cycles at 0.5 C (Figure 3.12d). This work

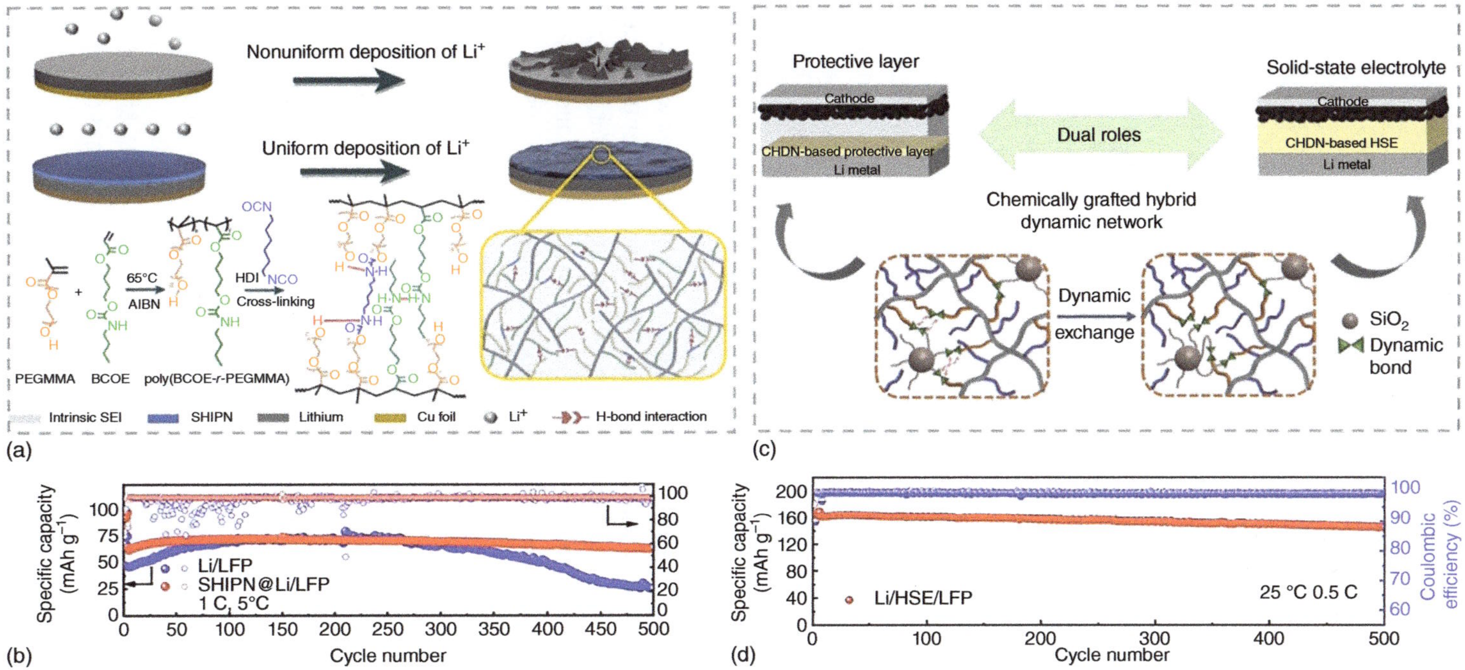

Figure 3.12 (a) Schematics of the SHIPN protective layer during the Li plating/stripping process and its chemical synthesis. (b) Cycling performance of the Li/LFP cells using the SHIPN@Li and bare Li as the anode at 1 C under 5 °C, respectively [41]. (c) Schematics of CHDN-based HSE and dynamic exchange of CHDN. (d) Cycling performance of the Li/HSE/LFP full cell at 0.5 C under 25 °C [42].

provided a design guideline for the fabrication of efficient protective layers and solid-state electrolytes of LMBs with recyclable and repairable abilities.

3.4 Conclusion

This chapter presents the recent progress achieved in studying the dynamic bonds in associating polymer networks, including fundamental understanding of the relaxation and morphology of these bonds, as well as the emergent design of dynamic polymer materials with the superior performance and energy-related applications. Models that explain the mechanisms of network rearrangement have been demonstrated for the telechelic polymer systems, along with the discussion of the effects from backbone and end group chemistry, chain lengths, molecular weight, and cluster topology. In addition to the conventional dynamic covalent and noncovalent bonding, a few new forms of dynamic networks have been developed including dual-/multi-dynamic bonding (e.g. combination of disulfide and hydrogen bonds), hybrid chemical/physical interactions (e.g. hybrid cross-linking of chemical and physical bonds), and phase-separated cluster interactions (e.g. hydrogen-bonding clusters), empowering the family of dynamic polymer materials. With precisely controlling the dynamic interactions in molecular level via synthesis, the systems with extreme stretchability, self-healing, recyclability, high adhesion property, or/and 3D printability can be achieved for many advanced applications.

This chapter is expected to interest and inspire readers/researchers to develop new dynamic polymer systems with advanced dynamic features, as well as to synergize with the presented dynamic networks. Furthermore, opportunities arise to improve understanding of the structure–properties relationships for these materials with respect to their dynamics and mechanics, especially how molecular-scale structure designs give rise to advanced functionalities and applications.

References

1. Webber, M.J. and Tibbitt, M.W. (2022). Dynamic and reconfigurable materials from reversible network interactions. *Nature Reviews Materials* 7 (7): 541–556.
2. Li, B., Cao, P.F., Saito, T., and Sokolov, A.P. (2023). Intrinsically self-healing polymers: from mechanistic insight to current challenges. *Chemical Reviews* 123 (2): 701–735.
3. Luo, J., Demchuk, Z., Zhao, X. et al. (2022). Elastic vitrimers: beyond thermoplastic and thermoset elastomers. *Matter* 5 (5): 1391–1422.
4. Montarnal, D., Capelot, M., Tournilhac, F., and Leibler, L. (2011). Silica-like malleable materials from permanent organic networks. *Science* 334 (6058): 965–968.
5. Guo, H., Han, Y., Zhao, W. et al. (2020). Universally autonomous self-healing elastomer with high stretchability. *Nature Communications* 11 (1): 2037.

6 Chen, J., Gao, Y., Shi, L. et al. (2022). Phase-locked constructing dynamic supramolecular ionic conductive elastomers with superior toughness, autonomous self-healing and recyclability. *Nature Communications* 13 (1): 4868.

7 Wang, J., Wu, B., Wei, P. et al. (2022). Fatigue-free artificial ionic skin toughened by self-healable elastic nanomesh. *Nature Communications* 13 (1): 4411.

8 Korich, A.L. and Iovine, P.M. (2010). Boroxine chemistry and applications: a perspective. *Dalton Transactions* 39 (6): 1423–1431.

9 Denissen, W., Rivero, G., Nicolaÿ, R. et al. (2015). Vinylogous urethane vitrimers. *Advanced Functional Materials* 25 (16): 2451–2457.

10 Jiang, G., Liu, C., Liu, X. et al. (2010). Network structure and compositional effects on tensile mechanical properties of hydrophobic association hydrogels with high mechanical strength. *Polymer* 51 (6): 1507–1515.

11 Jiang, H., Duan, L., Ren, X., and Gao, G. (2019). Hydrophobic association hydrogels with excellent mechanical and self-healing properties. *European Polymer Journal* 112: 660–669.

12 Cordier, P., Tournilhac, F., Soulie-Ziakovic, C., and Leibler, L. (2008). Self-healing and thermoreversible rubber from supramolecular assembly. *Nature* 451 (7181): 977–980.

13 Xia, D., Wang, P., Ji, X. et al. (2020). Functional supramolecular polymeric networks: the marriage of covalent polymers and macrocycle-based host-guest interactions. *Chemical Reviews* 120 (13): 6070–6123.

14 Amabilino, D.B., Smith, D.K., and Steed, J.W. (2017). Supramolecular materials. *Chemical Society Reviews* 46 (9): 2404–2420.

15 Appel, E.A., del Barrio, J., Loh, X.J., and Scherman, O.A. (2012). Supramolecular polymeric hydrogels. *Chemical Society Reviews* 41 (18): 6195–6214.

16 Mantooth, S.M., Munoz-Robles, B.G., and Webber, M.J. (2019). Dynamic hydrogels from host-guest supramolecular interactions. *Macromolecular Bioscience* 19 (1): e1800281.

17 Yan, T. and Guironnet, D. (2021). Synthesis of telechelic polyolefins. *Polymer Chemistry* 12 (36): 5126–5138.

18 Xing, K., Tress, M., Cao, P. et al. (2018). Hydrogen-bond strength changes network dynamics in associating telechelic PDMS. *Soft Matter* 14 (7): 1235–1246.

19 Xing, K., Tress, M., Cao, P.-F. et al. (2018). The role of chain-end association lifetime in segmental and chain dynamics of telechelic polymers. *Macromolecules* 51 (21): 8561–8573.

20 Kunal, K., Robertson, C.G., Pawlus, S. et al. (2008). Role of chemical structure in fragility of polymers: a qualitative picture. *Macromolecules* 41: 7232–7238.

21 Lewis, C.L., Stewart, K., and Anthamatten, M. (2014). The influence of hydrogen bonding side-groups on viscoelastic behavior of linear and network polymers. *Macromolecules* 47 (2): 729–740.

22 Osterwinter, C., Schubert, C., Tonhauser, C. et al. (2014). Rheological consequences of hydrogen bonding: linear viscoelastic response of linear polyglycerol and its permethylated analogues as a general model for hydroxyl-functional polymers. *Macromolecules* 48 (1): 119–130.

23 Chen, Q., Zhao, X., Li, B. et al. (2023). Exceptionally recyclable, extremely rtough, vitrimer-like pdms elastomers via combined physical and chemical networks. *Matter* .

24 Zhang, Z., Luo, J., Zhao, S. et al. (2022). Surpassing the stiffness-extensibility trade-off of elastomers via mastering the hydrogen-bonding clusters. *Matter* 5 (1): 237–252.

25 Nova, A., Keten, S., Pugno, N.M. et al. (2010). Molecular and nanostructural mechanisms of deformation, strength and toughness of spider silk fibrils. *Nano Letters* 10 (7): 2626–2634.

26 Eisenberg, A., Hird, B., and Moore, R.B. (1990). A new multiplet-cluster model for the morphology of random ionomers. *Macromolecules* 23: 4098–4107.

27 Yan, T., Schröter, K., Herbst, F. et al. (2017). What controls the structure and the linear and nonlinear rheological properties of dense, dynamic supramolecular polymer networks? *Macromolecules* 50 (7): 2973–2985.

28 Wu, S., Liu, S., Zhang, Z., and Chen, Q. (2019). Dynamics of telechelic ionomers with distribution of number of ionic stickers at chain ends. *Macromolecules* 52 (6): 2265–2276.

29 Ge, S., Samanta, S., Tress, M. et al. (2021). Critical role of the interfacial layer in associating polymers with microphase separation. *Macromolecules* 54 (9): 4246–4256.

30 Ge, S., Samanta, S., Li, B. et al. (2022). Unravelling the mechanism of viscoelasticity in polymers with phase-separated dynamic bonds. *ACS Nano* 16 (3): 4746–4755.

31 Ge, S., Tress, M., Xing, K. et al. (2020). Viscoelasticity in associating oligomers and polymers: experimental test of the bond lifetime renormalization model. *Soft Matter* 16 (2): 390–401.

32 Ma, Y. and Lodge, T.P. (2016). Chain exchange kinetics in diblock copolymer micelles in ionic liquids: the role of χ. *Macromolecules* 49 (24): 9542–9552.

33 Lu, J., Bates, F.S., and Lodge, T.P. (2013). Chain exchange in binary copolymer micelles at equilibrium: confirmation of the independent chain hypothesis. *ACS Macro Letters* 2 (5): 451–455.

34 Cao, P.F., Li, B., Hong, T. et al. (2017). Robust and elastic polymer membranes with tunable properties for gas separation. *ACS Applied Materials & Interfaces* 9 (31): 26483–26491.

35 Cao, P.F., Li, B., Hong, T. et al. (2018). Superstretchable, self-healing polymeric elastomers with tunable properties. *Advanced Functional Materials* 28 (22): 1800741.

36 Li, B., Zhao, S., Zhu, J. et al. (2020). Rational polymer design of stretchable poly(ionic liquid) membranes for dual applications. *Macromolecules* 54 (2): 896–905.

37 Hong, T., Cao, P.-F., Zhao, S. et al. (2019). Tailored CO_2-philic gas separation membranes via one-pot thiol–ene chemistry. *Macromolecules* 52 (15): 5819–5828.

38 Zhang, Z., Ghezawi, N., Li, B. et al. (2020). Autonomous self-healing elastomers with unprecedented adhesion force. *Advanced Functional Materials* 31 (4): 2006298.

39 Zhu, J., Zhao, S., Luo, J. et al. (2022). A novel dynamic polymer synthesis via chlorinated solvent quenched depolymerization. *CCS Chemistry* 1–13.

40 Niu, W., Zhang, Z., Chen, Q. et al. (2021). Highly recyclable, mechanically isotropic and healable 3D-printed elastomers via polyurea vitrimers. *ACS Materials Letters* 3 (8): 1095–1103.

41 Sun, F., Li, Z., Gao, S. et al. (2022). Self-healable, highly stretchable, ionic conducting polymers as efficient protecting layers for stable lithium-metal electrodes. *ACS Applied Materials & Interfaces* 14 (22): 26014–26023.

42 He, Y., Ma, M., Li, L. et al. (2023). Hybrid dynamic covalent network as a protective layer and solid-state electrolyte for stable lithium-metal batteries. *ACS Applied Materials & Interfaces* 15 (19): 23765–23776.

43 Wang, S., Li, X., Wu, H. et al. (2016). Advances in high permeability polymer-based membrane materials for CO_2 separations. *Energy & Environmental Science* 9 (6): 1863–1890.

44 Luo, S., Stevens, K.A., Park, J.S. et al. (2016). Highly CO_2-selective gas separation membranes based on segmented copolymers of poly(ethylene oxide) reinforced with pentiptycene-containing polyimide hard segments. *ACS Applied Materials & Interfaces* 8 (3): 2306–2317.

45 van der Kooij, H.M., Susa, A., Garcia, S.J. et al. (2017). Imaging the molecular motions of autonomous repair in a self-healing polymer. *Advanced Materials* 29 (26): 1701017.

46 Zhang, P. and Li, G. (2016). Advances in healing-on-demand polymers and polymer composites. *Progress in Polymer Science* 57: 32–63.

47 Cao, P.-F., Li, B., Yang, G. et al. (2020). Elastic single-ion conducting polymer electrolytes: toward a versatile approach for intrinsically stretchable functional polymers. *Macromolecules* 53 (9): 3591–3601.

48 Scovazzo, P., Kieft, J., Finan, D. et al. (2004). Gas separations using non-hexafluorophosphate [PF6]− anion supported ionic liquid membranes. *Journal of Membrane Science* 238 (1-2): 57–63.

49 Bocharova, V., Wojnarowska, Z., Cao, P.F. et al. (2017). Influence of chain rigidity and dielectric constant on the glass transition temperature in polymerized ionic liquids. *The Journal of Physical Chemistry. B* 121 (51): 11511–11519.

50 Zarca, G., Horne, W.J., Ortiz, I. et al. (2016). Synthesis and gas separation properties of poly(ionic liquid)-ionic liquid composite membranes containing a copper salt. *Journal of Membrane Science* 515: 109–114.

51 Vollas, A., Chouliaras, T., Deimede, V. et al. (2018). New pyridinium type poly(ionic liquids) as membranes for CO(2) separation. *Polymers* 10 (8): 912.

52 Hong, T., Lai, S., Mahurin, S.M. et al. (2017). Highly permeable oligo(ethylene oxide)-co-poly(dimethylsiloxane) membranes for carbon dioxide separation. *Advanced Sustainable Systems* 2 (4): 1700113.

53 Chen, W.-G., Wei, H.-J., Luo, J. et al. (2021). Highly stretchable, ultratough, and multifunctional poly(vinyl chloride)-based plastics via a green, star-shaped macromolecular additive. *Macromolecules* 54 (7): 3169–3180.

54 Gordon, M. and Taylor, J.S. (2007). Ideal copolymers and the second-order transitions of synthetic rubbers. i. non-crystalline copolymers. *Journal of Applied Chemistry* 2 (9): 493–500.

55 Belorgey, G., Aubin, M., and Prud'homme, R.E. (1982). Studies of polyester/chlorinated poly(vinyl chloride) blends. *Polymer* 23: 1051–1056.

56 Chen, J., Liu, X., Tian, Y. et al. (2022). 3D-printed anisotropic polymer materials for functional applications. *Advanced Materials* 34 (5): e2102877.

4

Direct Observation of Polymer Reptation in Entangled Solutions and Junction Fluctuations in Cross-linked Networks

Fengxiang Zhou[1,2] *and Lingxiang Jiang*[1,2]

[1] *South China Advanced Institute for Soft Matter Science and Technology, School of Emergent Soft Matter, South China University of Technology, Guangzhou 510640, China*
[2] *Guangdong Provincial Key Laboratory of Functional and Intelligent Hybrid Materials and Devices, South China University of Technology, Guangzhou 510640, China*

4.1 Introduction

A polymer network is a collection of macromolecular chains interlocked beyond the percolation point that can sustain stress by elastic deformation [1]. The interlocking can take places via physical entanglements with a finite relaxation time of disentanglement or chemical cross-linking with an infinite relaxation time. Without the entanglements or cross-linkers, polymer chains can move or reconfigure to relax any imposed stress, making a viscous fluid; with them, the chains are locked into position to form a percolated network that drastically turns the fluid into an elastic solid to withstand stress. Featuring unlimited versatility, polymer networks can manifest themself as a plastic or a rubber in the unswollen state or an organogel or a hydrogel in the solvent-swollen state, and they can be anything from weak to strong, soft to rigid, brittle to tough, and shear-thinning to self-recovery [2]. It is thus not surprising to notice the strong presence of polymer networks in almost every aspect of everyday life, industrial processes, and scientific endeavors [3, 4]. In particular, there has been a recent surge of interests in tough hydrogels and their applications in energy storage or harvesting, tissue engineering, and flexible and wearable materials, to name a few [2]. Nevertheless, there are still many challenges that must be addressed to attain a complete understanding of these materials [3].

Reptation model [5, 6], rubber theory [7, 8], and their variations or extensions [9, 10] have long been established, with recognizable success, to comprehend the polymer networks in threefold: static structures from micro- to microscales, molecular dynamics, and bulk properties such as elasticity. Experimental measurements of the static structures and bulk properties are relatively straightforward, and the results are generally in accordance with the theories [1]. In terms of molecular dynamics, however, experimental progress is seriously outpaced by the theoretical models on, for example, the reptation of polymer chains in entangled solutions (Figure 4.1a) and the fluctuation dynamics of junctions in cross-linked networks

Dynamics and Transport in Macromolecular Networks: Theory, Modeling, and Experiments, First Edition. Edited by Li-Tang Yan.

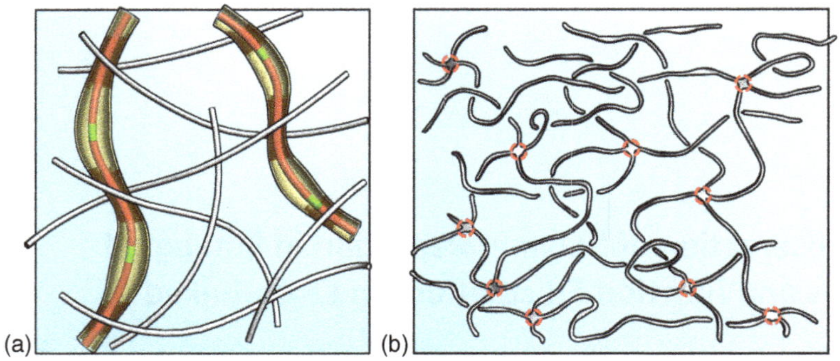

Figure 4.1 Schematic cartoons showing two labeled polymer chains in their virtual tubes (a) and a cross-linked network with highlighted junctions (b).

(Figure 4.1b). Traditionally, dynamic information was extracted by several experimental techniques such as nuclear magnetic resonance (NMR) [11, 12], neutron and light scattering [13–15], and fluorescence correlation spectroscopy (FCS) [16]. These ensemble-averaged means cannot provide spatially resolved data nor report on single-molecule dynamics. It has not been possible to directly observe reptating polymers or fluctuating cross-linkers in the real space on the molecular level until more recently [17–19], thanks to the implementation of state-of-art imaging technology [20, 21]. In this chapter, we shall recapitulate the recent progress on single-molecule visualization in polymer networks with emphasis to compare the theoretical predictions and experimental findings.

4.2 Reptation in Entangled Solutions

Proposed by de Gennes [5] and extended by Edwards and Doi [6], the famous reptation (or tube) model can account for the microscopic dynamics and macroscopic rheology of polymers in melts and concentrated solutions that arise from chain entanglement effects. In this model, a polymer chain is confined by the surrounding matrix and moves inside an imaginary tube defined by the transient network of entangled neighboring chains. The polymer molecule cannot make transverse movements and may relax only along the tube in a snake-like fashion (reptation). Despite the challenges in the detailed observation of polymer dynamics, researchers have managed to image and track individual polymers in real space to confirm and even to enrich the reptation model in quite a few notable publications.

In a typical experiment, a trace amount of fluorescently labeled polymers is mixed with a semidilute, entangled solution of unlabeled polymers such that only a few polymer chains are visible under the microscope. Consecutive images of the tracer polymers are then recorded by sCMOS or EMCCD, and the movie analyzed by customized codes to identify polymer positions and conformations and to produce statistics on polymer dynamics. A super-resolution optical setup based on 3D

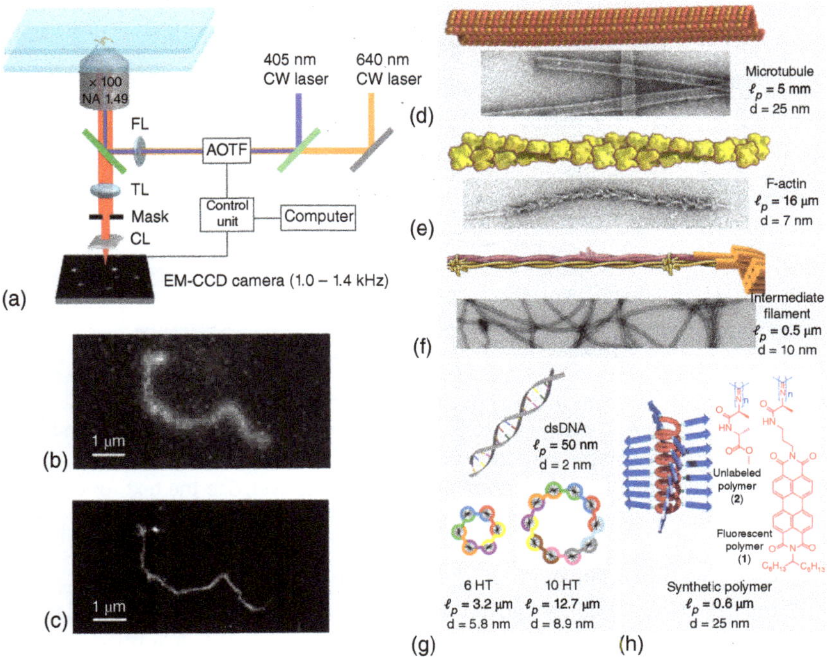

Figure 4.2 (a) Schematic illustration describing a particular optical setup for 3D storm imaging. AOTF, acousto-optic tunable filter; FL, focusing lens; FC, focus correction system; TL, tube lens; CL, cylindrical lens. (b) Conventional fluorescence image of Cy5-labeled lambda DNA in an aqueous solution containing 10 mg ml^{-1} concentration of non-labeled lambda DNA. (c) 2D-projected super-resolution fluorescence image of the DNA molecule displayed in (b). (d–h) Images, diameters, and persistence lengths of microtubules (d), F-actin (e), intermediate filaments (f), DNAs (g), and synthetic polymers (h). Source: (a–c) [22]. Maram Abadi et al. 2018/Reproduced with permission from Springer Nature/CC BY 4.0, (d–f) [23]. Shuying Yang & Lingxiang Jiang. 2020/Reproduced with permission from Royal Society of Chemistry, (g) from Ref. [19] with permissions from APS, (h) from Ref. [18] with permissions from ACS.

storm is depicted in Figure 4.2a [22]. Notably, semiflexible polymers were often chosen for direct visualization possibly due to their length (1–20 μm), extended conformations (easy to track the contour), and higher labeling density (against photobleaching, Figure 4.2b,c). These polymers include cytoskeletal filaments (microtubules, F-actin, and intermediate filaments) [23], DNAs (lambda DNA and DNA nanotubes) [19], and polyisocyanopeptides [18]. See Figure 4.2d–h for their diameters, d, and persistence lengths, ℓ_p.

In the framework of reptation model, there are several time scales and associated length scales (Figure 4.3) [6, 24] that deserve attentions before we delve into specific experimental systems. (i) $t < \tau_e$, at times less than the entanglement onset time τ_e, macromolecular filaments behave as thin, dilute, weakly bending threads of length L and diameter b that experience solvent-mediated hydrodynamic interactions. Filament segments move within lengths less than the static geometric mesh length ξ_m, and they cannot feel the constraints caused by nearby filaments or entanglements. This time scale is usually too short to be accessible by optical tracking.

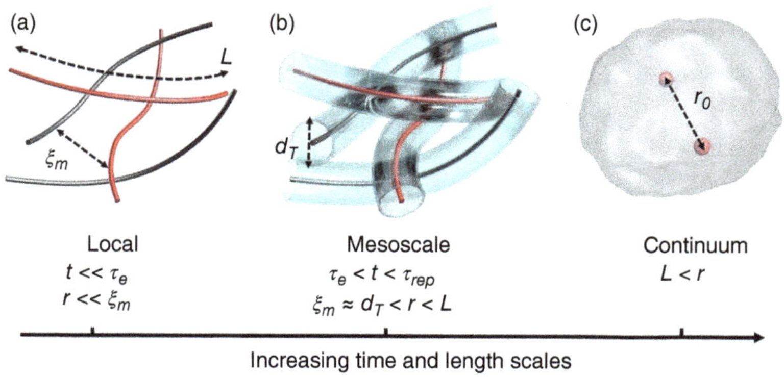

Figure 4.3 Schematic diagram showing the different time and length scales in an entangled system. (a) $t < \tau_e$ corresponds to local/microscopic scale. (b) $\tau_e < t < \tau_{rep}$ and $\tau_{rep} < t < \tau_d$ correspond to mesoscale when reptation is the dominant dynamics. (c) $\tau_d < t$ corresponds to continuum scale when normal diffusion takes over. See the text for details. Source: [24]. Boyce Tsang et al. 2017/Reproduced from National Academy of Science (NAS).

(ii) $\tau_e < t < \tau_{rep}$, at times longer than τ_e but less than the reptation time τ_{rep}, the filaments are effectively densely packed, filling at a nearly constant average density a tube of diameter d_T ($d_T \approx \xi_m$). They can transversely fluctuate only within the tube and reptate along their contours with diffusion constant $D_{\parallel}$. (iii) $\tau_{rep} < t < \tau_d$, at times longer than τ_{rep} but less than the disentanglement time τ_{rep}, the filament can longitudinally creep through the tube until fully move out of its original tube. (iv) $\tau_d < t$, at times longer than τ_d, normal diffusive motion sets in as the tubes are irrelevant.

4.2.1 Direct Confirmation of the Reptation Model

As early as 1994, Kas et al. [17] followed the motion of labeled actin filaments and imaged the confining tube in which the polymer chain could move by superimposing a large number of images and plotting the contours of the observed polymer chains. The restricted motion could be seen as undulations of the actin filament in a tube formed by its neighbors. The tube diameter was determined and shown to decrease with increasing concentration (Figure 4.4a). They also observed the characteristic thermally excited sliding of the filament out of the end of the tube, with the chain self-diffusion coefficient $D_{\parallel}$ decreasing approximately linearly as the filament length increases, in agreement with the reptation model.

Keshavarz et al. tracked the reptation of a synthetic polymer, perylenedimide-grafted polyisocyanopeptide, in the entangled solution (Figures 4.2h and 4.4b) [18]. They determined mean square displacements in the longitudinal direction ($\text{MSD}_{\parallel}$) and identified two regimes: roughly $t^{1/4}$ for $\tau_e < t < \tau_{rep}$ and $t^{1/2}$ for $\tau_{rep} < t < \tau_d$, both in line with the theoretical predictions (Figure 4.4c). They also observed the power laws τ_e and $\tau_{rep} \propto l^{2.07\pm0.05}$, which are in good agreement with the theory of reptation (the relationship should scale with l^2, Figure 4.4d). Plotting τ_d as a function of ℓ reveals a power law $\tau_d \propto l^{2.9\pm0.01}$, again in full agreement with reptation theory, which predicts $\tau_d \propto l^3$ (Figure 4.4e).

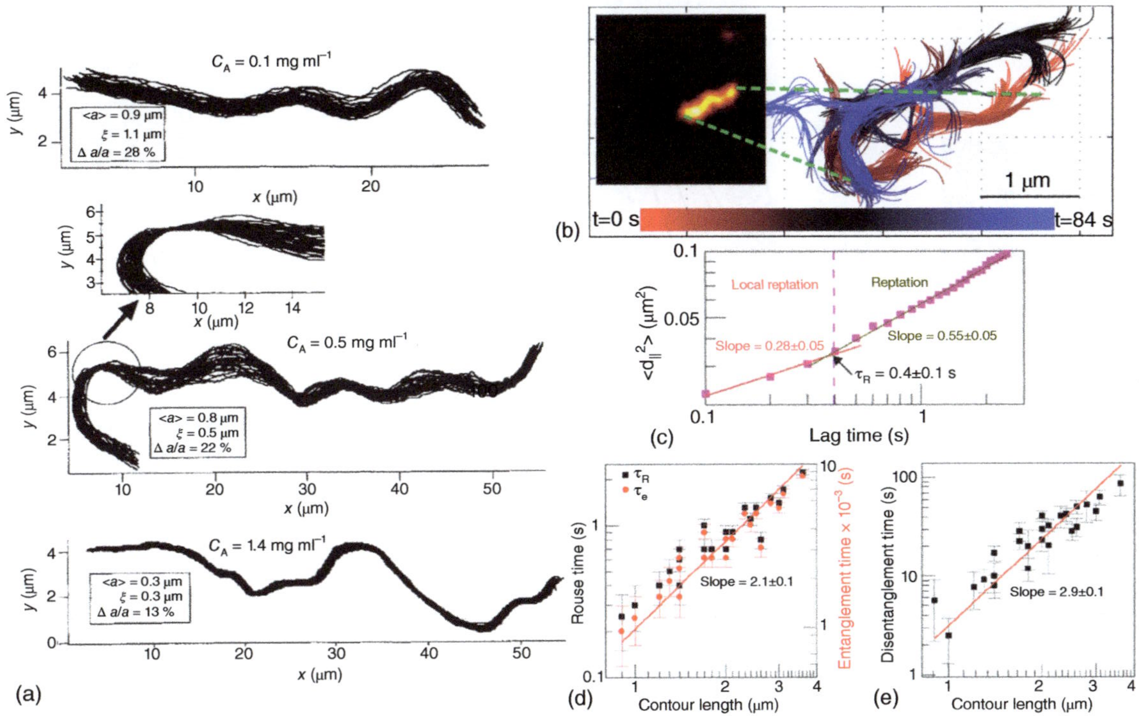

Figure 4.4 (a) Virtual tubes around labeled actin filaments were imaged by the superimposition of 60 traces of the contour taken at 0.1-second intervals. The mean tube width is determined by averaging the width of the dark bands along the contour. Above each tube the actin monomer concentration C_A of the surrounding F-actin solution is noted. Source: Reproduced from Ref. [17] with permissions from NPG. (b) Superimposed conformations of the fluorescent synthetic polymer chain contour for the entire movie from 0 to 84 seconds. The color code indicates the time. (c) $MSD_{\|}$ as a function of lag time, indicating the regimes of local reptation (solid line with a slope of 0.28 ± 0.05) and reptation (solid line with a slope of 0.55 ± 0.05). (d) τ_e and τ_{rep} as a function of contour length confirming the predictions that τ_e and $\tau_{rep} \propto l^2$. (e) Relationship between disentanglement time and the contour or primitive path length. The slope of the curve is in agreement with reptation theory $\tau_d \propto l^3$. Source: (b–e) [18]. Masoumeh Keshavarz et al. 2016/Reproduced with permission from American Chemical Society.

4.2.2 Tube Width Fluctuations

The tube width d_T is one of the most important parameters in the reptation model. While experimental results are in general agreement with the model, we present several important deviations from or complements to the model in Sections 2.2–2.4. The reptation model assumes a harmonic potential to confine the target polymer chain within the tube, so it predicts the distribution of transverse displacements to be Gaussian. Bo et al. [25] tested this assumption in entangled actin solutions (Figure 4.5a1). They, instead of tracking the whole polymer chain, marked one or

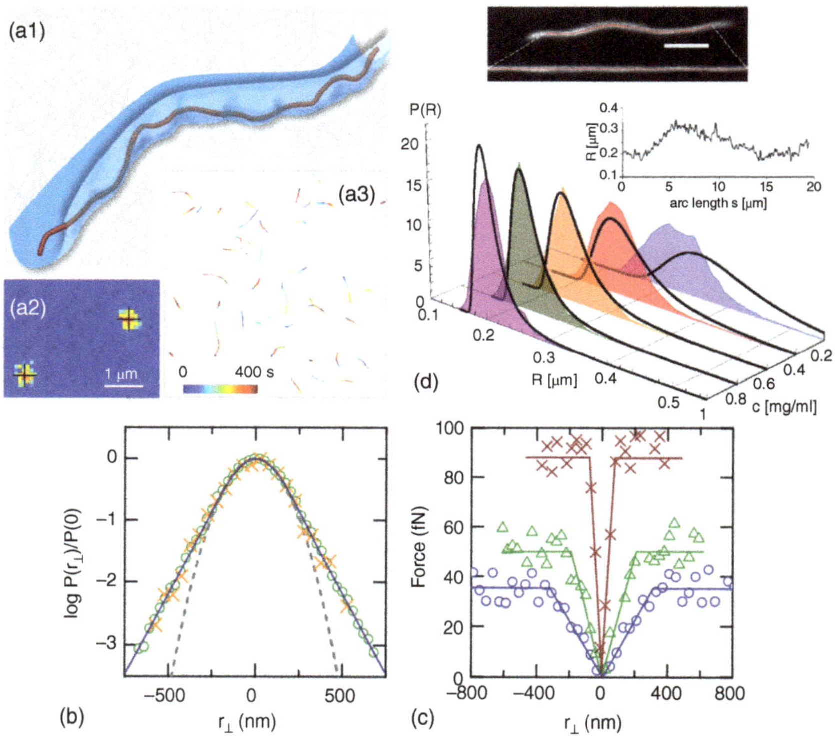

Figure 4.5 (a1) Schematic representation of a chain constrained into a tube-like region by an effective harmonic potential, the collective topological effect of surrounding chains. Minima of this potential define the primitive path. (a2) Representative image showing two labeled segments of a long actin filament. (a3) Overview of the variety of curvilinear-like diffusive motions of labeled actin segments in entangled F-actin networks. (b) Transverse distance probability distributions, plotted against distance normal to the primitive path, for an ensemble of hundreds of filaments (circles) and for the trajectory of a single filament (crosses). The dashed line is the corresponding Gaussian fit to the center part of the distribution. (c) Effective restoring forces, inferred from a Boltzmann distribution law, plotted against transverse distance. The actin concentrations are 1 (crosses), 0.5 (triangles), and 0.2 mg ml^{-1} (circles). (d) Superimposed confocal microscopy images of a fluorescent actin filament in a background solution and a spline representing the tube backbone; scale bar = 5 μm. Normalized tube radius distribution $P(R)$ obtained from cumulative contour lengths for different actin concentrations. Source: (a–c) [25]. Bo Wang et al. 2010/Reproduced with permission from American Physical Society and (d) [26]. J. Glaser et al. 2010/Reproduced with permission from American Physical Society.

two segments on an actin filament by sparse labeling to follow a specific point on a chain (Figure 4.5a2). In doing so, they counted the segment fluctuations transverse to the primitive path and further calculated the confining potential imposed by the virtual tube (Figure 4.5a3). The chain localization presents a small-displacement harmonic regime (Gaussian regime in Figure 4.5b) followed by a large-amplitude regime (exponential regime in Figure 4.5b) where the effective restoring force is independent of displacement (Figure 4.5c). The shortening on the tube periphery was attributed to the heterogeneity of physical and mechanical mesh sizes that may arise from the concentration fluctuations. A mean field model on the basis of binary collision approximation was proposed to quantify such heterogeneity. In the same year of 2010, Glaser et al. also reached similar conclusions on the tube width fluctuations (Figure 4.5d) [26].

4.2.3 Dependence of Tube Width on Chain Position

In the time scale of $\tau_{rep} < t < \tau_d$, large segmental freedom at the chain ends is expected within the reptation model because the two ends keep retracting from "old" tubes and entering "new" tubes. This was indeed observed in synthetic polymer systems [18]. Abadi et al. however, discovered unexpected, strong dependence of tube width on segment positions on a shorter time scale of $\tau_e < t < \tau_{rep}$ [22]. They employed super-resolution tracking and cumulative area analysis to identify the segmental (local) and whole-chain motions of DNA chains in entangled conditions (Figure 4.6a–c), and they observed that tube width is much larger at the chain ends than at the chain middle (Figure 4.6d). Larger and faster motion of entangle segments toward the chain ends caused the gradual change in the confinement level of the chain in the tube. Mutual relaxation of the entangled chains, which is not considered in the reptation theory, might be responsible for this behavior.

4.2.4 Tube Width under Shear

While most experimental measurements are conducted in equilibrium states, the polymer networks often serve under shear or strain. Early theory predicted that because stretched chains rapidly retract within a sheared or stretched tube, macroscopic deformations do not modify the tube diameter. Subsequently, tube expansion was predicted owing to deformation-induced chain orientation, while other theory predicted tube compression when chains are stretched affinely. Kaikai et al. confronted this common but poorly understood situation with single-molecule imaging over a wide range of shear rate [27]. They tracked alignment and stretching of entangled aqueous actin filaments in a homebuilt rheo-microscope (Figure 4.7a). Comparing filament alignment and shear thinning, they observed that the longest F-actin chains in a polydisperse mixture are most flexible, while the shortest F-actin chains are most rigid (Figure 4.7b). Increased shear rate causes dynamical tube nonlinearity, tube softening followed by hardening (Figure 4.7c). Physically, this means that dynamical localization first weakens owing to molecular alignment, then strengthens owing to filament stretching, even for semiflexible biopolymers shorter than their persistence length.

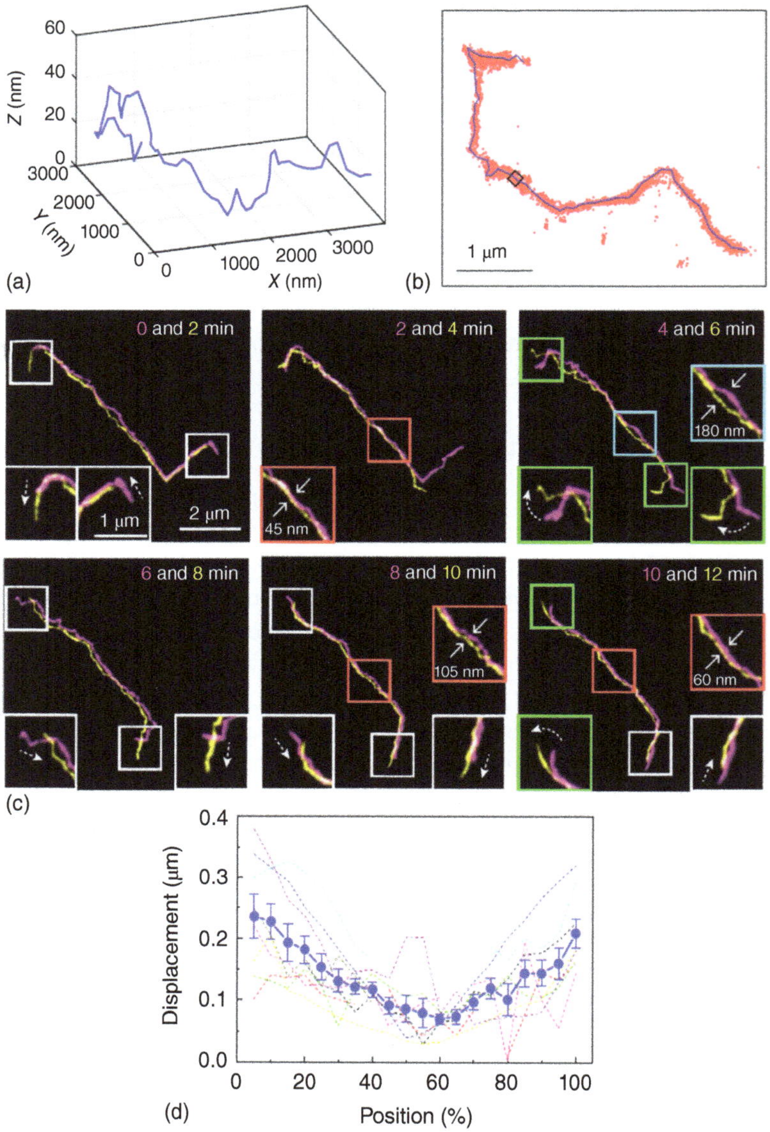

Figure 4.6 (a) 3D contour of a DNA molecule. (b) 2D-projected contour of the DNA molecule shown in (a). (c) Motion of the linear DNA molecules under entangled conditions. (d) Mean position-dependent displacement of the Cy5-labeled lambda DNA occurring along the direction perpendicular to the contours during the time gap between the adjacent time frames in an aqueous solution containing 7–10 mg ml^{-1} concentration of non-labeled lambda DNA. Source: [22]. Maram Abadi et al. 2018/Reproduced from Springer Nature/CC BY 4.0.

4.2.5 Interactions Between Reptating Polymer Chains

Recall that any polymer chain in an entangled solution may easily encounter tens or even hundreds of neighboring chains along its contour, making explicit modeling of polymer–polymer interactions impractical if not impossible. Success of the

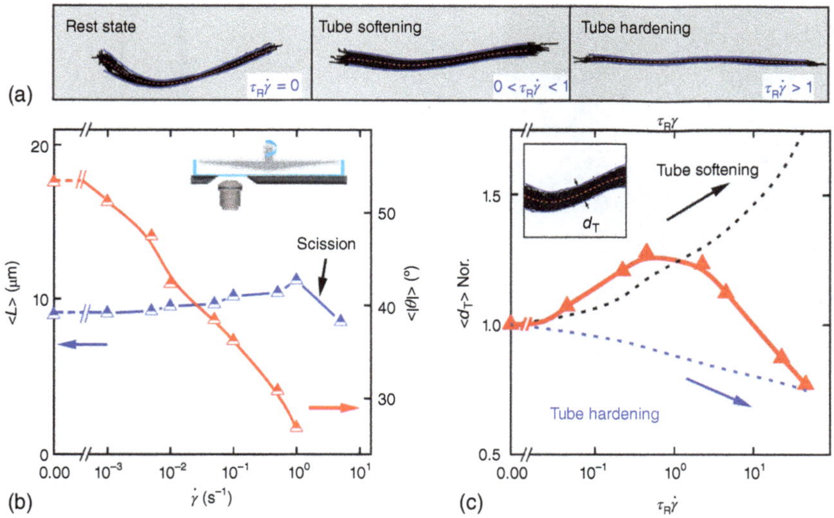

Figure 4.7 (a) "Tube" trajectories, each of a different single filament, after removing the mass center velocity, compared in the rest state, in the tube-softening regime (shear rate 0.1 s^{-1}) and in the tube-hardening regime (shear rate 1 s^{-1}). (b) As a function of logarithmic shear rate, this graph shows the mean filament length $\langle L \rangle$ and the orientation angle $\langle \theta \rangle$ with respect to shear. (c) Mean tube diameter of medium-length filaments plotted against the logarithmic product of shear rate and Rouse relaxation time, normalized to the mean tube diameter in the rest state. The two dotted lines compare to theoretical predictions for "tube softening" and "tube hardening." Source: Reproduced from Ref. [27] with permissions from APS.

tube model thus lies on the mean field approximation that treats all the nearby polymers and entanglements around the target polymer chain as a confining virtual tube, effectively turning an overcomplicated, multi-polymer problem to a tractable single-polymer problem [6]. Boyce et al., however, took a shift in the paradigm to study on the open question of how the motion of a given reptating macromolecule is coupled in space and time with others that reptate within its pervaded volume [24].

Their experiments were based on fluorescence tracking of dilutely labeled biopolymer, actin filaments (Figure 4.8a,b). The entangled pairs of aqueous actin filaments diffuse with separation-dependent dynamic cross-correlations that exceed those expected from continuum hydrodynamics up to strikingly large spatial distances of ≈15 μm (Figure 4.8c), which is more than 10^4 times the size of the solvent water molecules in which they are dissolved, and is more than 50 times the dynamic tube diameter, but is almost equal to the filament length. Modeling this entangled system as a collection of rigid rods, they presented a statistical mechanical theory that predicts these long-range dynamic correlations as an emergent consequence of an effective long-range interpolymer repulsion due to the de Gennes correlation hole, which is a combined consequence of chain connectivity and uncrossability. The key physical assumption needed to make theory and experiment agree is that solutions of entangled biofilaments localized in tubes that are effectively dynamically incompressible over the relevant intermediate time and length scales.

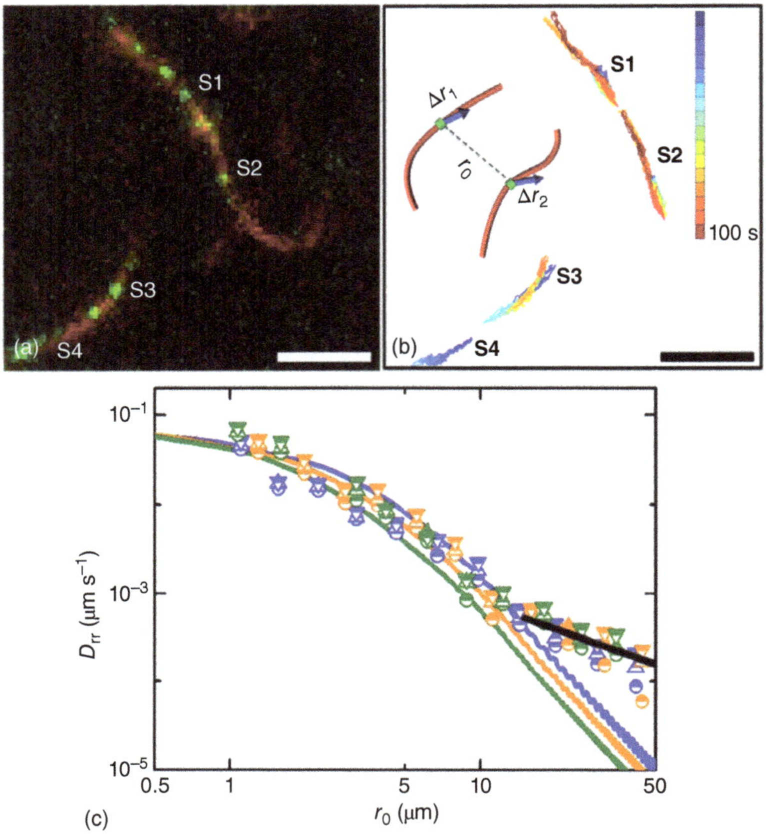

Figure 4.8 (a) Neighboring filaments both lie in the 2D focal plane, the time-dependent diffusion of red and green colors false-colored is tracked parallel to and transverse to the line between them at numerous separations r_0. (b) Image processing eliminates the background fluorescence and generates experimentally the time-dependent mutual trajectories of point sources S1, S2, S3, and S4 on the two molecules. (c) On log–log scales, the relative diffusivity data are plotted as points against separation for the actin concentrations 0.5 mg ml^{-1} (blue), 1 mg ml^{-1} (yellow), and 2 mg ml^{-1} (green) and times 0.1 seconds (circles), 0.2 seconds (upright triangles), and 0.4 seconds (inverted triangles). The solid curves show the corresponding theoretical predictions for the actin concentrations 0.5 (blue), 1 (yellow), and 2 (green) mg ml^{-1} and $L = 15$ μm. The straight line with slope −1 is the continuum hydrodynamic prediction. Source: Reproduced from Ref. [24] with permissions from NAS.

4.3 Dynamic Fluctuations of Cross-links

At the very heart of any cross-linked polymer networks are the junctions that interconnect neighboring chains and lock them in position to resist stress (Figure 4.1b). The dynamics of cross-links are expected to dictate network architecture, origin of elasticity, and transmission from macroscopic forces to local responses. There has been extensive theoretical work to correlate the network modulus to the entropic elasticity of polymer chains, where the key difference among varies models is how they treat the thermal fluctuations of cross-links around their equilibrium positions [7–9]. On one extreme, the affine network model assumes that the cross-links are firmly fixed in space and are displaced affinely with the whole network under strain, as if they were permanently attached to some elastic background. In real networks, however, the cross-links can wiggle considerably around their average positions, and this wiggle room will reduce the cumulative stretching of the polymer strands and lower the network modulus [8]. This model of zero fluctuations thus overestimates the network elasticity.

On the other extreme of the phantom network model, the cross-links are not fixed but free to fluctuate around their average positions and macroscopic deformation is transmitted to the bulk of the network through the chains attached to its surface. The fluctuation magnitudes determine how strongly the macroscopic deformation is coupled to the deformation of individual chains. This model, however, overestimates the fluctuation magnitudes as it ignores the exclusive volume of neighboring polymer chains (thus the name of phantom) that would significantly reduce the wiggle room. This model of maximum fluctuations unsurprisingly underpredicts the network elasticity. Between these two extremes are more advanced models such as the constrained junction model that imposes an additional harmonic potential on the cross-links of the phantom network and the diffused constraint model that further allow the harmonic potential to "slide" along the polymer chains [9].

4.3.1 Dynamics Probed by Neutron Scattering

Although assumptions of cross-link fluctuations constitute the cornerstones for a variety of theoretical models for a few decades, experimental progress has yet been largely limited by a number of practical challenges to direct testify predictions on such fluctuations. (i) The cross-link points must be selectively labeled, (ii) the fluctuations of cross-links need to be followed with enough spatial and temporal resolutions, and (iii) the spatial heterogeneity (if any) of such fluctuations has to be properly accounted for [28, 29]. Richter et al. utilized neutron spin echo (NSE) measurements to extract cross-link dynamics in a tetrafunctional model network of polydimethysiloxane (PDMS) melt [28]. The network was deuterated while the cross-links remain protonated, such that only the latter are "visible" to neutron scattering. The NSE technique provides excellent temporal and spatial resolutions on the order of ~1 ns and ~0.1 nm, respectively. They discovered that the time scale of motion for the cross-links was comparable to that of segmental diffusion within the chains attached to the junctions (rouse parameter 0.54×10^9 vs. $1.73 \times 10^9\ \mathrm{nm^{-4}\ s^{-1}}$), clearly disfavoring network models that assume highly restricted cross-link mobility both in space and time. Each cross-link performs Gaussian spatial fluctuations arounds its average position with a magnitude rather extended (2.45 nm, Figure 4.9), very close to the radius of gyration ($R_g = 2.5$ nm) but smaller than the phantom network prediction (3.5 nm) [8]. This is an important

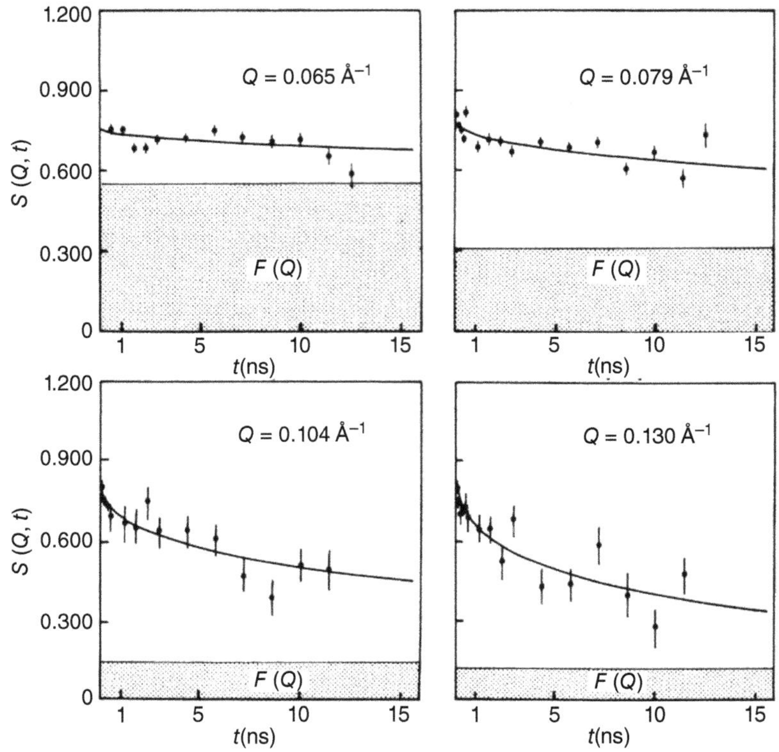

Figure 4.9 Neutron spin echo (NSE) spectra for labeled cross-links in PDMS networks at four different Q values. Included is a result of a fit with a sum of a constant EISF contribution $F(Q)$ and a Rouse relaxation spectrum. The shaded area displays the time-independent EISF part in the spectra. Source: Reproduced from Ref. [28].

piece of experimental work confirming that the real fluctuation is somewhat between what affine and phantom models predicted, yet it is an ensemble-averaged, reciprocal space means lacking the power to detect spatial heterogeneity inherent to many polymer networks.

4.3.2 Dynamics Probed by Direct Imaging

To tackle this long-standing problem of cross-link dynamics, Jiang et al. developed a comprehensive methodology called single-cross-link microscopy to directly visualize cross-links on single-molecule level in real space and real time [30]. The basic ideas are straightforward: selectively labeling the cross-linking points in virtue of filament breakage and fusion, recording real-time videos of the thermally fluctuating cross-links, extracting trajectories and fluctuations of cross-links from the videos, and finally calculating local elasticity and cross-correlations according to the trajectories (Figure 4.10a).

They chose to work with a biopolymer network of actin filaments cross-linked by biotin–avidin linkages rather than a common synthetic network for the following

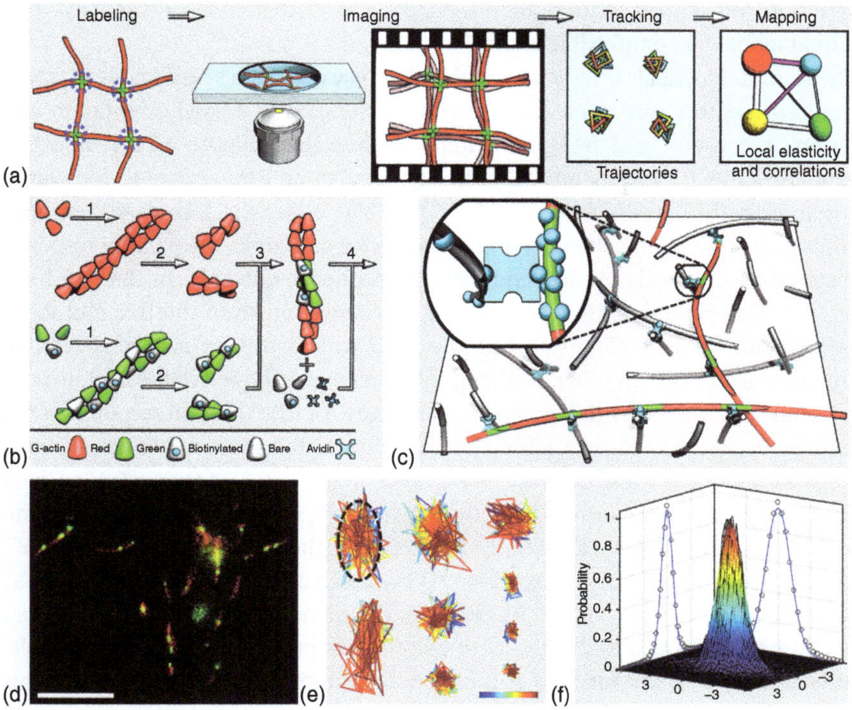

Figure 4.10 Single-cross-link microscopy in actin networks. (a) Four steps of single-cross-link microscopy. (b) Four steps to produce the network are (1) pre-polymerization to form red, biotin-free F-actin and green, biotin-loaded F-actin, (2) fragmentation to obtain ~100 nm segments by vigorous shear, (3) fusion to anneal the segments into long filaments with red, biotin-free backbones and sparsely distributed green segments with abundant biotin, and (4) network formation to obtain a cross-linked network of unlabeled and labeled filaments (ratio ~ 1000 : 1). (c) Schematic representation of the cross-linked network on the microscope focal plane, showing a few labeled filaments with red contour and cross-linkable, green segments in a background of unlabeled filaments. Inset highlights a biotin–avidin linkage in the orthogonal configuration. (d) Typical microscopy image of labeled filaments. The red contours are generally in the focal plane, and the green segments are traced as time elapses. Scale bar = 10 μm. (e) Color-coded cross-link trajectories highlighting anisotropicity and polydispersity. The color bar from blue to red covers 0–500 seconds with 0.5 seconds step. Length of color bar = 400 nm. (f) The ensemble probability distribution (black mesh) against displacements rescaled by (b) is fitted by a 2D Gaussian peak (rainbow color surface with color denoting height) with projections plotted on the side walls (circles and curves are data and fit, respectively). Source: Reproduced from Ref. [30].

practical reasons (Figure 4.10b,c). The mesh size of an actin network is significantly larger than that of a typical synthetic network (300 vs. 2 nm), giving marked cross-link fluctuations that can be easily tracked by an optical microscope. And the cross-links can be easily labeled with multiple dye molecules such that the cross-links are bright and resilient to photobleaching, ideal for longer observation. Fluctuations of each labeled cross-links were tracked, and the probability distributions were fitted by 2D Gaussian functions (Figure 4.10d–f). The anisotropic fluctuation was related to local strand lengths and further to local elasticity via a minimal model for semiflexible chains.

The minimal model is briefly recapitulated here. Consider a semiflexible filament of contour length ℓ and persistence length ℓ_p clamped to a fixed orientation and position at one end and free to fluctuate at the other (Figure 4.11a). Effective spring constants $k_\perp$ and $k_\parallel$ are, respectively, parallel and transverse to the clamp orientation with the ratio $k_\parallel/k_\perp = 2\ell_p/\ell$ [19, 20]. While the transverse response (bending) is of purely mechanical origin and hence enthalpic, the parallel response is dictated by thermally excited undulations and hence entropic. In the stiff limit of $\ell_p/\ell \gg 1$, the highly anisotropic, spring constants constrain the free end to an oblate spheroidal bulb (Figure 4.11a, the cyan pancake). One can consider two filaments normal to each other, clamped at four ends and joined in the middle with four equal clamp-to-cross-link distances ℓ_c (Figure 4.11b). Thermal response of the cross-link is determined by a parallel connection of the four individual effective springs with $k_\perp{}^c = 4k_\perp$ and $k_\parallel{}^c = 2k_\perp + 2k_\parallel$, such that $k_\parallel{}^c/k_\perp{}^c = (1 + \ell_p/\ell_c)/2$. Given $\ell_p = 16\,\mu\text{m}$ and $\ell_c \approx 1\,\mu\text{m}$ for the actin networks, $k_\parallel{}^c/k_\perp{}^c \gg 1$ and the cross-link trajectory is a prolate spheroidal bulb (Figure 4.11b, the cyan spindle). For each ℓ_p and ℓ_c, the fluctuation bulb is a specific, 3D Gaussian cloud with widths $\Delta_\perp$ and $\Delta_\parallel$ (Figure 4.11c). Projection of the 3D cloud on the plane is a 2D elliptical Gaussian distribution of widths a and b and with ellipticity dependent on θ; as an example, simulated data is shown for $\ell_p = 16\,\mu\text{m}$, $\ell_c = 1\,\mu\text{m}$ and $\theta = 40°$ (Figure 4.11d). This prediction of ellipticity is in qualitative agreement with the observations.

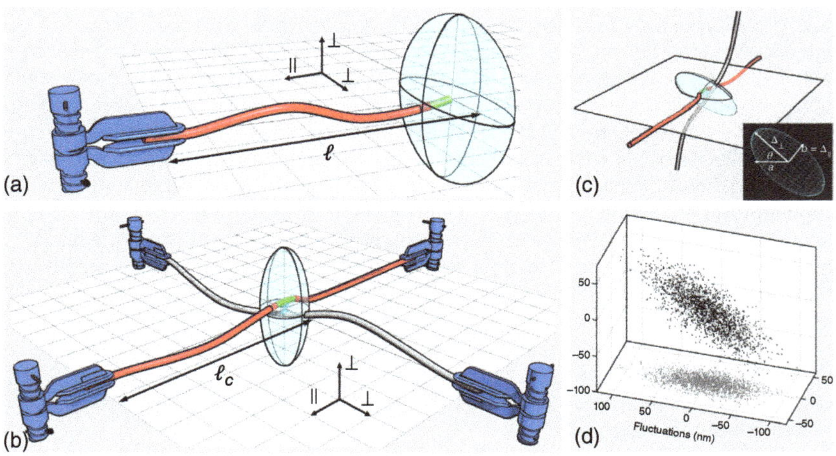

Figure 4.11 Theoretical model for single cross-links. (a) A filament (contour length ℓ) is clamped at one end and allowed to fluctuate thermally in a trajectory denoted by the pancake bulb at the other end. (b) Two filaments cross-linked at their middle such that the junction fluctuates thermally in a trajectory denoted by the spindle bulb. Here, ℓ_c is the clamp-to-junction distance (or cross-link distance). Parallel and vertical directions are denoted by $\parallel$ and $\perp$. (c) Projection of a 3D spindle on the focal plane is a 2D elliptical Gaussian cloud with long and short axes (a and b) experimentally measurable. (d) Given $\ell_p = 16\,\mu\text{m}$, $\ell_c = 1\,\mu\text{m}$, and $\theta = 40°$, a trajectory is simulated. Source: Reproduced from Ref. [30].

It is, however, impractical to directly visualize ℓ_c in real space as the cross-links are sparsely labeled. They took a sidestep to measure projected fluctuations of a target cross-link and to sequentially deduce its ℓ_c and local elasticity. A diagram of workflow demonstrates mutual dependences of the key variables, including two measurables (a and b), three parameters defining the spindle ($\Delta_\perp$, $\Delta_\parallel$, and θ), a central quantity (ℓ_c), local elasticities parallel and transverse to the filaments ($E_\parallel$ and $E_\perp$), and two input parameters ($\ell_p = 16\,\mu m$ for actin filaments and mesh size $\xi = 0.7\,\mu m$ for the current actin concentration of $0.2\,mg\,ml^{-1}$). Notably, at the heart of this diagram is ℓ_c that determines the spindle geometry and local elasticity. Seeking to test the model, the experimentally measured a' (binned by b values, gray points) are compared with the theoretically predicted a' (averaged over θ, gray line). The overall consistency is satisfactory, while the slight, systematic overestimation may suggest secondary contributions from unequal ℓ_c or off-orthogonal configurations. Finally, a real-space map was reconstructed with each spindle corresponding to a single cross-link, spindle sizes to fluctuation magnitudes, and spindle color to local elasticity (Figure 4.12). Large spatial heterogeneity is obvious for fluctuations

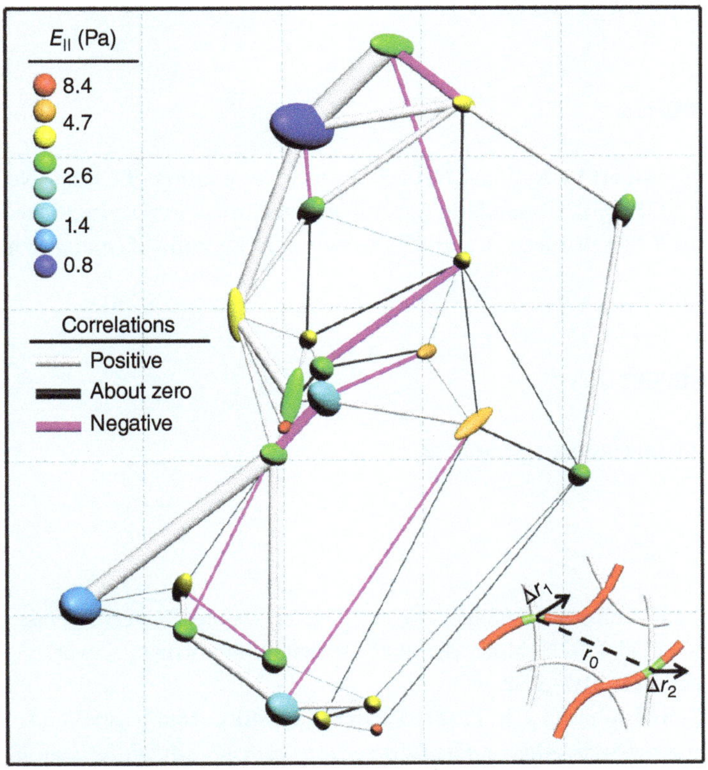

Figure 4.12 A spatial map of fluctuations, local elasticity, and correlations between cross-links. In this $42 \times 45\,\mu m^2$ view, the fluctuation bulbs are sized according to b (enlarged 50 times) and color coded by $E_\parallel$. The connected cross-links are not necessarily neighbors, and there could be many unlabeled cross-links in between them. Massive heterogeneity is noteworthy. Source: Reproduced from Ref. [30].

and local elasticity with the latter spanning a factor of 10 in this single view. Stiff and soft regions coexist with no apparent correlation between them.

4.4 Conclusion

Seeing is believing – this simple intuition promotes researchers to directly visualize polymer dynamics in entangled solutions or cross-linked networks, and the developments in optical imaging with enough spatial and temporal resolutions have recently enabled us to do so in real space and on real time. The reptative motion of polymer chains has been captured with great details, which on the one hand confirmed many theoretical predictions (such as the scaling laws of tube width and relaxation times) and on the other hand, arguably more importantly, presented deviations and complements to the existing models (such as the dependence of tube width on positions and shear and the interpolymer interactions). Although direct observation on the cross-link dynamics is still in its infancy, it may eventually settle debates about the transmission of force from macroscopic to molecular levels and to unveil hidden principles that govern the networks.

Acknowledgments

L. J. acknowledges support by National Natural Science Foundation of China (No. 22122201 and No. 22272060), Guangdong Natural Science Funds for Distinguished Young Scholar (No. 2018B030306011), and The Recruitment Program of Guangdong (No. 2016ZT06C322).

Conflict of Interest

The authors declare no competing interests.

References

1 Rubinstein, M. and Colby, R.H. (2003). *Polymer Physics*. Oxford University Press.
2 Danielsen, S.P.O. et al. (2021). Molecular characterization of polymer networks. *Chemical Reviews* 121: 5042–5092.
3 Gu, Y., Zhao, J., and Johnson, J.A. (2020). Polymer networks: from plastics and gels to porous frameworks. *Angewandte Chemie (International Edition in English)* 59: 5022–5049.
4 Wang R, Olsen B, Johnson J. *Quantifying the Impact of Cyclic Defects on Polymer Network Elasticity. American Physical Society*, (2017).
5 de Gennes, P.G. (1971). Reptation of a polymer chain in the presence of fixed obstacles. *The Journal of Chemical Physics* 55: 572–579.

6 Doi, M. and Edwards, S.F. (1986). *The Theory of Polymer Dynamics*. Oxford University Press.

7 Treloar, L.R.G. (1949). *The Physics of Rubber Elasticity*. Oxford University Press.

8 Flory, P.J., Gordon, M., and Mccrum, N.G. (1976). Statistical thermodynamics of random networks [and discussion]. *Proceedings of the Royal Society A: Mathematical, Physical and Engineering Sciences* 351: 351–380.

9 Rubinstein, M. and Panyukov, S. (2002). Elasticity of polymer networks. *Macromolecules* 35: 6670–6686.

10 Dai, X., Zhang, X., Gao, L. et al. (2022). Topology mediates transport of nanoparticles in macromolecular networks. *Nature Communications* 13: 4094.

11 Herrmann, A. et al. (2012). Mean square displacement and reorientational correlation function in entangled polymer melts revealed by field cycling 1H and 2H NMR relaxometry. *Macromolecules* 45: 6516–6526.

12 Brown, W. and Pu, Z. (1991). Solution properties of polyisobutylene investigated by using dynamic and static light scattering and pulsed field gradient NMR. *Macromolecules* 24: 5151–5157.

13 Monkenbusch, M. and Richter, D. (2007). High resolution neutron spectroscopy—a tool for the investigation of dynamics of polymers and soft matter. *Comptes Rendus Physique* 8: 845–864.

14 Schleger, P., Farago, B., Lartigue, C. et al. (1998). Clear evidence of reptation in polyethylene from neutron spin-echo spectroscopy. *Physical Review Letters* 81: 124–127.

15 Pajevic, S., Bansil, R., and Konak, C. (1995). Dynamic light scattering study of linear polyelectrolyte diffusion in gels. *Macromolecules* 28: 7536–7542.

16 Zettl, H., Zettl, U., Krausch, G. et al. (2007). Direct observation of single molecule mobility in semidilute polymer solutions. *Physical Review E* 75: 061804.

17 Käs, J., Strey, H., and Sackmann, E. (1994). Direct imaging of reptation for semiflexible actin filaments. *Nature* 368: 226–229.

18 Keshavarz, M. et al. (2016). Nanoscale study of polymer dynamics. *ACS Nano* 10: 1434–1441.

19 Hndler, T., Tutmarc, C., Glaser, M. et al. (2021). Measuring structural parameters of crosslinked and entangled semiflexible polymer networks with single-filament tracing. *Physical Review E* 103: 062501.

20 Betzig, E. et al. (2006). Imaging intracellular fluorescent proteins at nanometer resolution. *Science* 313: 1642–1645.

21 Huang, B., Wang, W., Bates, M., and Zhuang, X. (2008). Three-dimensional super-resolution imaging by stochastic optical reconstruction microscopy. *Science* 319: 810–813.

22 Abadi, M., Serag, M.F., and Habuchi, S. (2018). Entangled polymer dynamics beyond reptation. *Nature Communications* 9: 5098.

23 Yang, S. and Jiang, L. (2020). Biomimetic self-assembly of subcellular structures. *Chemical Communications (Cambridge)* 56: 8342–8354.

24 Tsang, B., Dell, Z.E., Jiang, L. et al. (2017). Dynamic cross-correlations between entangled biofilaments as they diffuse. *Proceedings of the National Academy of Sciences of the United States of America* 114: 3322–3327.

25 Wang, B., Guan, J., Anthony, S.M. et al. (2010). Confining potential when a biopolymer filament reptates. *Physical Review Letters* 104: 118301.

26 Glaser, J. et al. (2010). Tube width fluctuations in F-actin solutions. *Physical Review Letters* 105: 037801.

27 Zheng, K., Zhang, Z., Cao, B., and Granick, S. (2022). Biopolymer filament entanglement softens then hardens with shear. *Physical Review Letters* 129: 147801.

28 Oeser, R., Ewen, B., Richter, D., and Farago, B. (1988). Dynamic fluctuations of crosslinks in a rubber: a neutron-spin-echo study. *Physical Review Letters* 60: 1041–1044.

29 Hiroi, T., Ohl, M., Sakai, T., and Shibayama, M. (2014). Multiscale dynamics of inhomogeneity-free polymer gels. *Macromolecules* 47: 763–770.

30 Jiang, L., Xie, Q., Tsang, B., and Granick, S. (2019). Single-crosslink microscopy in a biopolymer network dissects local elasticity from molecular fluctuations. *Nature Communications* 10: 3314.

5

Recent Progress of Hydrogels in Fabrication of Meniscus Scaffolds

Chuanchuan Fan, Ziyang Xu, and Wenguang Liu

School of Material Science and Engineering, Tianjin Key Laboratory of Composite and Functional Materials, Tianjin University, Tianjin 300350, China

5.1 Introduction

The meniscus is a semilunar fibrocartilaginous disk located between the tibial plateau and femoral condyle that plays a pivotal role in distributing load, stability, lubrication, and nutrition [1]. Unfortunately, meniscal injuries often occur in active and aging populations, leading to restricted action ability due to severe pain [2]. Even worse, the injured meniscus has a limited intrinsic healing capability due to the absence of blood vessels (especially for the inner region of the meniscus) [3]. If timely treatment is not received, osteoarthritis (OA) would be developed due to the disordered joint biomechanical environment, formation of an inflammatory microenvironment, or cartilage wear [4].

Clinically, different treatments are adopted according to the location and severity of the injuries. Meniscus suture surgery is usually performed to repair circumferential tears in the peripheral vascularized portion but is not applicable for repairing the avascular zone of the meniscus with poor self-healing capability [5]. Meniscectomy is one of the most common procedures to treat larger sized and poorly healed meniscal injuries in the clinic [6]. However, the loss of the meniscus permanently changes the knee biomechanics and homeostasis, resulting in the degeneration of articular cartilage and the development of symptomatic OA [7]. Allografts are expected to replace the injured meniscus and restore its function in the knee joint [8]. However, limited donor and disease transfers have become major obstacles to expanding meniscus allografts [9]. In addition, the longevity of allografts and their effectiveness for arthritis pain relief should also be considered. Therefore, it is necessary to construct a meniscal scaffold to restore biomechanical functions or promote the regeneration of native tissues.

Recently, several polymer materials, such as polytetrafluoroethylene fiber woven mesh [10, 11], polyethylene glycol terephthalate felt [12], porous polyurethanes [13], porous collagen [14], porous silk fibroin [15], and porous extracellular matrix

C.C. Fan and Z.Y. Xu contributed equally to this work.

Dynamics and Transport in Macromolecular Networks: Theory, Modeling, and Experiments, First Edition. Edited by Li-Tang Yan.

(ECM) [16], have been employed for the construction of meniscus scaffolds. Nevertheless, some issues limit their practical application, such as poor wear resistance, toxic degradation products, poor diffusion of nutrients, and inferior protective effects of articular cartilage.

Hydrogels have attracted extensive attention as soft and wet materials with excellent biocompatibility because of their adjustable mechanical properties, high water content, and adapted rheological behavior [17]. Therefore, several hydrogels are widely used as cell delivery systems (or the niche for cell growth) to fabricate meniscal tissue-engineering scaffolds to promote meniscal regeneration [18]. What's more, in the past decades, several high-strength hydrogel systems have also been developed, which are expected to be used for fabricating the meniscus substitutes to partly or completely replace the native meniscus and perform the load-bearing functions. More importantly, 3D-printing technology has unlock new possibilities for personalized therapy by mimicking the complexity and personality of the meniscus microstructure [19, 20]. With the development of this technology, a series of high-strength hydrogel-based 3D printing inks and several 3D printing hydrogel-based meniscus substitutes have been developed in recent years [21].

In this chapter, we first introduce the microstructure and biomechanical functions of the meniscus and comprehensively review studies on hydrogel-based meniscus scaffolds from two aspects of "providing the microenvironment for cell growth" as well as "providing the load-bearing capability". Based on the research status of meniscus scaffolds, the challenges and key scientific issues in the design and construction of next-generation hydrogel-based meniscus scaffolds are further summarized, and the necessity of mimicking the anisotropic and hierarchical microstructures of menisci is highlighted.

5.2 Microstructure and Mechanical Properties of Meniscus

5.2.1 Meniscus Anatomy, Biochemical Content, and Cells

The meniscus is a semilunar and wedge-shaped cartilage tissue between the femoral condyle and tibial plateau [22]. Two pieces of menisci (medial and lateral meniscus) in each knee joint have distinct anatomy, with anterior and posterior horns attach to the tibial surface *via* the ligament (Figure 5.1a) [23, 24]. Based on whether it contains vessels and nerves, the meniscus can be divided into three distinct regions: red-red zone, white-white zone, and red-white zone. The vascularized "red-red zone" is only located in peripheral around 10-25% regions of the mature meniscus. In contrast, the avascular articular cartilage-like "white-white zone" refers to the inner third of the meniscus, which is nourished mainly relies on interstitial fluid (Figure 5.1b) [23]. However, the self-healing capability of each region is directly related to the degree of vascularization. Thus, white-white zone with low vascularization is difficult to repair spontaneously [22].

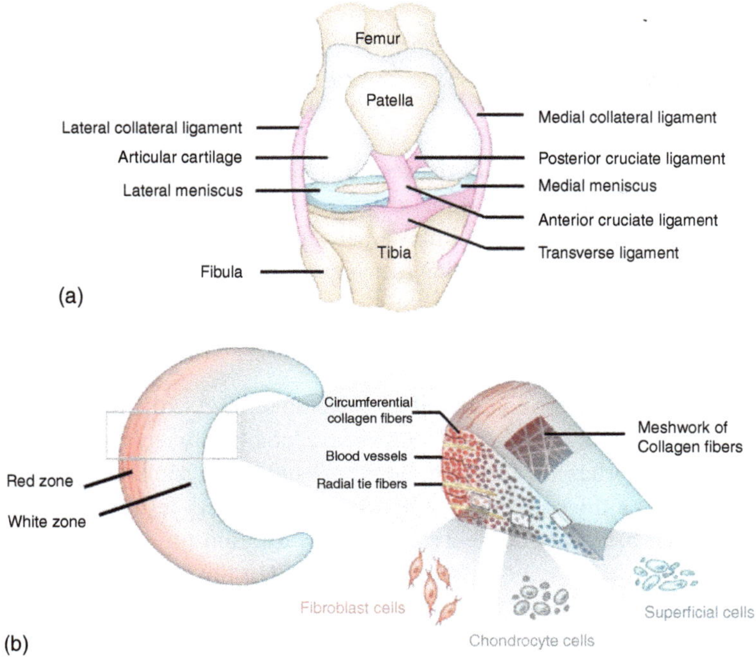

Figure 5.1 Diagram of meniscal anatomical features (a), collagen fiber microstructure, vascularization, and cell distribution of cell population (b) [23].

Similar to the vascularization, mature menisci cells vary in terms of number and topographic localization, which also differ in phenotypically and morphologically [24].

In particular, fibrochondrocytes or chondrocyte-like cells distributed in center of zone, while elongated fibroblast-like cells distributed in the peripheral zone. Moreover, the superficial zone of the meniscus exists a third cell population, which may be able to synthesize and secrete surface proteins. As reported, these cells are possibly specific progenitor cells accompanied by therapeutic and regenerative capabilities (Figure 5.1b) [23].

In addition to various cells, the meniscus is a complex tissue comprised of water and ECM with region-specific distribution features [25]. In general, the meniscus contains approximately 72% water (providing lubrication and nutrient transfer function) and 28% organic matter (abundant cells + meniscal ECM). Therein, the meniscal ECM contains collagen of 75%, glycosaminoglycans (GAGs) of 17%, and others composed of 2% DNA, less than 1% of elastin, and adhesive glycoproteins [1]. Collagen, as the most abundant organic matter in meniscus ECM, is the major component of meniscus skeleton microstructure [26]. In the peripheral area, collagen is dominated by type I collagen (approximately 80% of dry weight), and in the medial area, collagen accounts for 70% of dry weight (type II: 60%; type I: 40%) [27]. In addition, collagen fibers exhibit anisotropic structural characteristics and constitute a collagen skeleton structure composed of circumferential and radially oriented fibers in the ECM, providing specific load-bearing functions for the meniscus [28]. Proteoglycans are heavily glycosylated molecules composed of a core protein decorated with GAGs (main types: dermatan sulfate of 20–30%, chondroitin-4-sulfate of 10–20%, and keratin sulfate of 15%, in the normal human meniscus). Proteoglycan also has region-specific distribution characteristics, and the proportion of proteoglycan in the inner two-thirds of the region is higher than that in the outer one-third, which plays a role in absorbing water and bearing compression loads [29]. Adhesive glycoproteins serve as a linker to attach cells to components of the ECM, which contributes to cellular shaping recovery after loading [30].

5.2.2 Biomechanical Properties of the Meniscus

The unique microstructure of the collagen fibers skeleton infused with proteoglycan is a pivotal factor for resisting tensile and compressive loading of meniscus [31]. In general, the tensile properties are dominant by the oriented collagen fibers, and the different arrangement of the collagenous fibers in distinct layers results in highly heterogeneous tensile behaviors of meniscus, which vary from the surface to the interior sites and from the anterior to posterior regions. On the meniscus surface, the random-oriented collagen fibers give the nearly isotropic tensile properties. By comparison, in the deeper zones of the meniscus, the skeleton is consisted of the circumferentially and radially oriented collagen fibrils. During the load-bearing process, a big part of circumferentially direct forces and related tensile hoop stress occurs in the meniscus, and the circumferentially aligned collagenous fibers would resist hoop stresses to restrain displacement of the meniscal tissue from the tibial plateau [30, 32]. Therefore, the menisci of mammals usually have a high circumferential tensile modulus (bovine: ~139.0 MPa; canine: ~99.46 MPa; ovine ~217 MPa) [33], although various circumferential tensile moduli of the meniscus may be acquired based on different measurement methods [33–35].

In terms of the compressive properties, the proteoglycan component that governs the hydraulic permeability of meniscus is served as the major influence factor. As reported, the proteoglycan would absorb a large amount of water and maintain the interstitial pressure for a long time within the meniscus, resulting in efficient energy dissipation [32]. In contrast to the ultrahigh tensile modulus, the compressive modulus is significantly lower. According to Hull et al. [36], the axial and radial compressive moduli (at 12% strain) are 83.4 and 76.1 kPa at equilibrium state, which are lower than those at physiological strain rate (718 and 605 kPa, respectively). What is more, the compressive behaviors also exhibit the region-specific feature. Compared with the posterior region, the anterior region possessed a greater modulus under the same strain condition (anterior region: 1048 kPa; posterior region: 329 kPa, at 12% strains).

The meniscus shows strongly nonlinear behavior due to significant differences in the tensile and compressive moduli [36], and their complex biochemical compositions as well as anisotropic mechanical properties further cause difficulties in the fabrication of meniscal scaffolds similar to the natural tissue. To achieve this, material choice, architecture manufacturing, and mechanical matching should be comprehensively considered.

5.3 Biomaterial Requirements for Constructing Meniscal Scaffolds

Considering the complexity of the meniscal microstructure and its vital role in transferring forces, a meniscus scaffold would ideally need to simultaneously satisfy the following basic criteria (Figure 5.2). First, the meniscus scaffold should have similar semilunar and wedge-shaped geometrical characteristics to the native tissue, which is essential for transferring forces. Second, the anisotropic microstructure (e.g. circumferentially and radially oriented collagen fibers) and mechanical properties (e.g. high tensile modulus and relatively low compressive modulus) of the meniscus scaffold should match the native meniscus to restore the load-bearing function. Third, considering the common fixation method of the meniscus scaffold *in vivo* is suturing, the engineering biomaterials for fabricating the scaffold should exhibit strong suture resistance. Fourth and importantly, as an implant, good biocompatibility is essential.

In addition to these basic criteria, additional requirements must be considered for specific applications. For example, as a permanent meniscal substitute, the scaffold should also demonstrate antifatigue performance and superior swelling stability in physiological environments to maintain sustained and stable mechanical performance and shape during long-term implantation. On the other hand, meniscus tissue-engineering scaffolds require a microenvironment conducive to cell growth and bioactive mechanisms to promote meniscus regeneration. Moreover, the degradation rate matching the ingrowth rate of the new tissue should also be considered to achieve full meniscus regeneration.

In recent decades, many efforts have been made to find ideal materials for fabricating meniscal scaffolds to meet these requirements. Some synthetic polymers have been used to construct permanent meniscal replacements. However, limited processability, unmatched mechanical properties, poor tribological properties, and poor compatibility are the main limitations for its application in meniscal tissue engineering. In addition, to achieve regeneration of the meniscus, some resorbable polymers have also been fabricated into porous materials to support new tissue ingrowth. For example, porous scaffolds consisting of collagen type I obtained

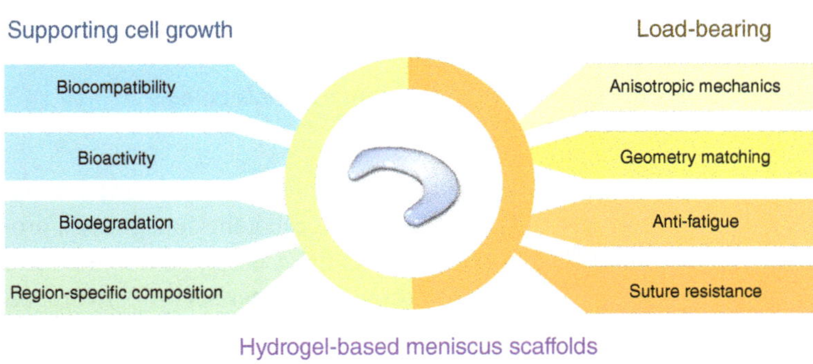

Figure 5.2 Window for general requirements of desired hydrogel meniscus scaffolds.

from bovine Achilles tendons have promoted meniscal repair [37]. Silk, another type of natural resorbable polymer, is also used in meniscus tissue engineering due to its predominant mechanical properties and versatile processability [38]. Although above two natural resorbable polymer meniscus scaffolds represent excellent biocompatibility, reabsorbability, and feasibility, relatively poor mechanical properties and nonadapted resorbable rate may cause the failure of meniscal transplant surgery. What is more, resorbable synthetic polymers have also been used to fabricate porous meniscus scaffolds. In 2006, Tienen et al. constructed a porous polymer material composed of aromatic polyurethane (Estane) and further implanted it into the knee joint of a dog to replace the resected lateral meniscus to evaluate its applicability to complete meniscus replacement [13]. Although tissue ingrowth shows regional specificity similar to the natural meniscus after six months of implantation, the protective effect of the Estane scaffold on articular cartilage is limited. In addition, toxic and carcinogenic methylene diphenylamine is formed after degradation, limiting its application as a meniscus scaffold. Furthermore, Buma et al. fabricated a porous scaffold composed of aliphatic polyurethane. However, it still have a poor protective effect on articular cartilage due to the weaker mechanical properties and slower biodegradation rate [39].

Although many biomaterials have been used to fabricate meniscus scaffolds, and their effectiveness has been widely investigated *in vitro* and *in vivo*, few studies have demonstrated their chondroprotective effect in humans. Therefore, exploring a biomaterial suitable for fabricating meniscal scaffold still remains a challenge. Among most biomaterials, hydrogels have a similar microstructure (a 3D network structure with high water content) to the ECM of native tissues and are acknowledged to be ideal candidates for constructing tissue scaffolds to replicate the mechanical features or microenvironments of native tissues. In fact, hydrogels are one of the most common biomaterials used to fabricate meniscal scaffolds. Depending on the purpose, the role of hydrogel in meniscus scaffolds can be divided into the following two aspects: (i) providing a suitable microenvironment for cell growth and (ii) providing load-bearing capacity. In the next part, we will introduce the application of hydrogels in meniscus scaffold construction in detail.

5.4 Hydrogel-Based Meniscus Scaffolds

5.4.1 Providing Matrix for Cell Growth and Biomacromolecules Delivery

One of the aims of meniscus tissue engineering is to promote meniscal regeneration. Therefore, an appropriate microenvironment to support cell adhesion, growth, proliferation, differentiation, and ECM secretion is essential. Compared to other solid materials (such as porous materials [40] and electrospun materials [41]), hydrogels are ideal candidates for supporting cell growth because of their specific 3D microenvironment and excellent ability to transport nutrients and wastes [42]. In the past decades, several natural polymer-based hydrogels (such as alginate [43], gelatin [44], etc.) have been used to encapsulate different cells, and the influence of materials on cell behaviors (such as proteoglycan synthesis and cell phenotype) has been investigated to evaluate their potential for meniscus repair *in vitro*. On this basis, many hydrogel-based meniscal tissue-engineering scaffolds have been fabricated, and their applicability for promoting meniscal regeneration has been demonstrated *in vivo*. According to the different forms, hydrogel-based tissue-engineering scaffolds can be divided into three categories: (i) injectable hydrogel-based meniscus tissue-engineering scaffolds, (ii) high-strength and biodegradable hydrogel-based meniscus tissue-engineering scaffolds, and (iii) 3D-printed polymer/hydrogel composite-based tissue-engineering scaffolds.

5.4.1.1 Injectable Hydrogel-Based Meniscus Tissue-Engineering Scaffolds

Injectable hydrogels are one of the most widely studied biomaterials, which can be administered through a syringe and rapidly form gel in the body [45]. In recent years, several injectable hydrogels based on stimuli response (e.g. pH [46], temperature [47], ionic concentration [48]) or chemical reactions (e.g. Michael addition [49], Schiff's base [50], and disulfide bond formation [51]) have been reported. Owing to their high moldability, injectable hydrogels can be used to load and deliver cells (or bioactive factors) *in vivo* in minimally invasive ways to provide local biological and mechanical cues for promoting tissue regeneration [52]. Therefore, injectable hydrogels have various applications in biomedical and pharmaceutical fields, such as vehicles for drug/gene/cell delivery [53], and tissue engineering [54].

The decellularized meniscus extracellular matrix (mECM) has a biochemical composition similar to that of the native tissue and has been widely used to repair meniscus defects [55]. Hong et al. prepared an injectable hydrogel consisting of the porcine meniscus-derived ECM. According to the *in vitro* and *in vivo* results, the hydrogel demonstrated a fibrous microstructure and the capability to promote the growth and infiltration of cells (e.g. bovine chondrocytes), indicating its potential for meniscal repair [56]. Yu et al. developed an mECM injectable hydrogel by modified decellularization and enzymatic digestion and investigated its capability to deliver rat bone marrow mesenchymal stem cells (BMSCs) *in situ* and promote the repair of meniscal defects in a SD rat model (Figure 5.3) [57]. The mECM injectable hydrogel significantly upregulated fibrochondrogenic marker expression by BMSCs, while preventing osteophyte formation, joint space narrowing, and OA development, suggesting its potential to deliver BMSCs and repair meniscus defects owing to the retention of the essential proteoglycans and collagens.

In addition to mECM, natural polymer hydrogels have also been used for fabricating injectable hydrogel-based meniscal tissue-engineering scaffolds owing to their similar components and characteristics to native tissues. Hwang et al. fabricated a tissue- adhesive injectable hydrogel based on tyrosinase (TYR)-mediated cross-linking between tyramine (TA)-conjugated hyaluronic acid (TA-HA) and gelatin [58]. According to the results, the TYR-cross-linked hydrogels exhibited controllable mechanical properties, degradation, gelation time, and good biocompatibility. More importantly, fibrochondrocytes encapsulated in this hydrogel

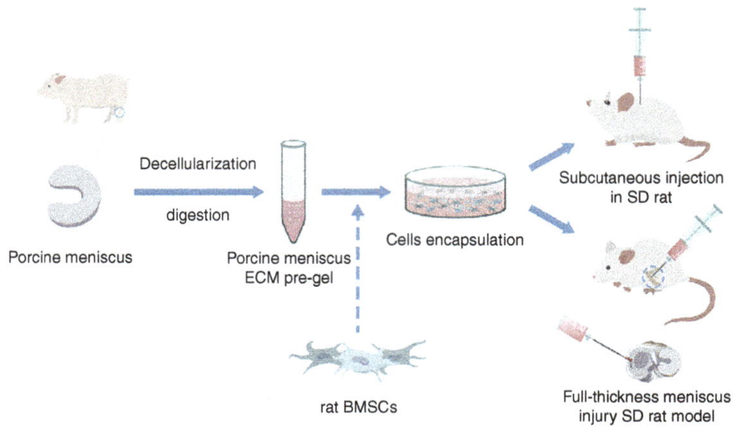

Figure 5.3 Schematic diagram of preparing mECM injectable hydrogel for meniscus defect repair [57].

showed cartilage-specific gene expression and matrix synthesis enhancement, representing its potential in meniscus tissue engineering. Joseph et al. developed an injectable hydrogel with a tunable gelation time and good cytocompatibility *via* borax complexation of oxidized alginate, followed by a self-cross-linking reaction with gelatin through Schiff's base reaction [59]. The injectable hydrogel exhibited good integration with the host meniscus tissue in the pig meniscal tear *ex vivo* model, suggesting that the hydrogel system can offer a niche for the growth of fibroblasts. In addition, injectable hydrogel systems have also been used to repair larger sized meniscus defects. An et al. developed an injectable semi-interpenetrated network (semi-IPN) hydrogel by adding Pluronic F127 (F127) and polymethyl methacrylate (PMMA) microbeads into fibrin (Fb) hydrogels [60]. According to the results, the above system can be injected and gelated within 50 seconds through a dual-channel injection device, and the introduction of F127 and PMMA can improve the mechanical properties and delay the rate of enzyme degradation. Furthermore, after injecting the Fb/F127/PMMA hydrogel into a segmental meniscus defect in rabbits for 12 weeks, meniscus repair was accelerated, indicating its potential as a meniscus tissue-engineering scaffold. Despite this, the biomechanical strength of regenerated tissues (3.59 ± 0.89 MPa) is still inferior to that of healthy tissue (6.63 ± 1.12 MPa).

Although the meniscal regeneration efficiency of injectable hydrogel systems has been confirmed, there are still some deficiencies: (i) almost all injectable hydrogels exhibit poor mechanical properties, which makes the injectable hydrogel scaffolds fail to provide appropriate load-bearing capability after implantation and (ii) scaffolds directly fabricated by injectable hydrogels are uniform, which makes them fail to regenerate meniscus tissue with region-specific distribution features. Therefore, injectable hydrogels can only be used to repair partial meniscal injury but not for regenerating whole meniscus tissue. A meniscus tissue-engineering scaffold should simultaneously provide load-bearing functions and a microenvironment for cellular growth to achieve total meniscus regeneration.

5.4.1.2 High Strength and Biodegradable Hydrogel-Based Meniscus Scaffolds

In order to meet the requirements of load-bearing and supporting ingrowth of new tissue simultaneously, researchers have been trying to develop the high-strength and biodegradable hydrogels meniscus scaffolds. For instance, Yin et al. prepared an amphiphilic polymer by grafting hydrophobic PCL with short chains onto the synthetic hydrophilic PLGA [61]. Then, PLGA-g-PCL was cross-linked with triethylene glycol (TEG) to form the chemically cross-linked network (CCN). After the solvent exchange process, the spindle-like stable hydrophobic PCL cluster was formed inside of the CCN, increasing its mechanical strength and elasticity. According to the results of mechanical measurement, the PLGA-g-PCL hydrogel demonstrated excellent resilience, and the maximum compressive strength was higher than 0.6 MPa (at a maximum strain of ~70%). What is more, the short chains of PCL accelerated the degradation of the PLGA-g-PCL hydrogel *in vivo* (within three months), which makes it suitable to be used as the tissue-engineering scaffolds. *In vivo* outcome demonstrated that the adipose-derived stem cells-loaded hydrogel promoted the regeneration of the meniscus-like tissue effectively and protected the articular cartilage from degeneration over a 16 weeks period. Despite this, the mechanical properties of this scaffold are still far from the native meniscus.

5.4.1.3 3D-Printed Polymer/Hydrogel Composite Tissue-Engineering Scaffolds

With the development of 3D printing technology, many 3D-printed polymer skeletons, such as polylactic acid (PLA) [62], polycaprolactone (PCL) [63], and polyurethane (PU) [64], have been fabricated to provide load-bearing functions. Based on this, to address these issues of injectable hydrogel systems, a series of 3D-printed polymer/hydrogel composite meniscus tissue-engineering scaffolds have been constructed.

Decellularized ECM-based hydrogels are ideal candidates for delivering cells (or bioactive factors) and providing a microenvironment to support cellular growth considering their excellent capability to accelerate meniscus regeneration. Therefore, many efforts have been made to develop 3D polymer/ECM gel composite meniscal tissue-engineering scaffolds. Asgarpour et al. fabricated a PCL/hydrogel scaffold by 3D printing a PCL scaffold with an optimized strand space (0.2 mm) followed by coating it with chitosan/ECM hydrogels [65]. On the one hand, the 3D-printed PCL skeleton is expected to provide a template and structural support for meniscus regeneration. On the other hand, chitosan/ECM hydrogel can recapitulate the structure and composition of native tissue. According to the results, biomimetic coating of chitosan/ECM hydrogel on scaffolds would not only promote the metabolic activity of cultured human adipose-derived stem cells (hASCs) but also induce the formation of fibrocartilage phenotype, suggesting the 3D-printed scaffolds are expected to provide a template for meeting the structural requirements of the meniscus scaffold.

Guo et al. prepared a mECM-based hydrogel by dissolving MECM particles into an alginate-Ca^{2+} system and investigated the effect of MECM particle concentration on the capability to promote cellular behaviors (e.g. proliferation, matrix forming) of meniscal fibrochondrocytes (MFCs) (Figure 5.4) [66]. According to the results,

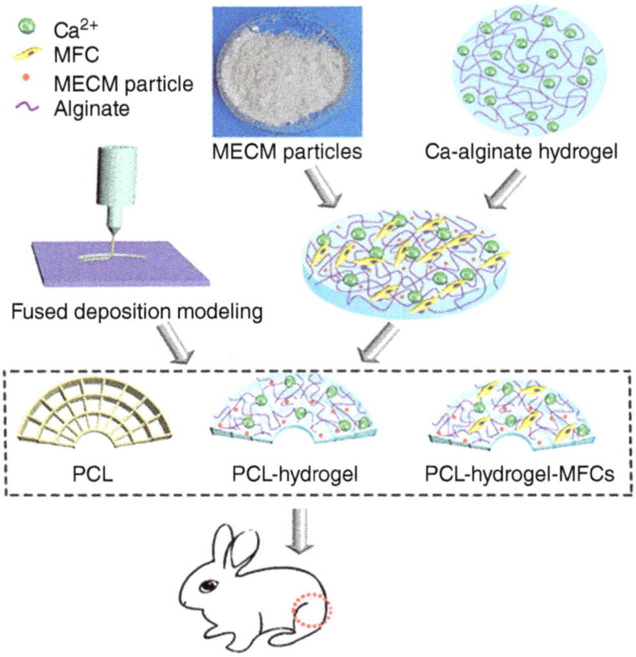

Figure 5.4 Schematic diagram of the PCL-hydrogel-MFCs scaffold fabrication [66].

a hydrogel with 2% MECM strongly enhanced chondrogenic marker mRNA expression and cell proliferation. On this basis, a hybrid meniscus scaffold with biomechanical properties similar to native tissues was further fabricated using extrusion-based 3D printing PCL as the skeleton and the infilled optimized MECM-based hydrogel as the cell delivery system. Furthermore, MFC-loaded PCL-MECM scaffolds were implanted into the knee joints of New Zealand rabbits after total medial meniscectomy to evaluate their effect on promoting meniscus regeneration *in vivo*. Six months after surgery, the PCL-hydrogel-MFC group exhibited good cartilage protection capability, and the histological structures, biochemical contents, and biomechanical properties of the regenerative tissue were similar to those of the native menisci in the sham operation group, demonstrating their potential for meniscus regeneration.

Although a method that combines printing with infusion has been used to mimic the unique anatomical structure and complex stress environment of the meniscus, the process remains complicated to some extent. To address this issue, Guo et al. first prepared a gelatin methacrylate (GelMA)/MECM-based bioink with good printability and cytocompatibility, which can be used for cell printing to construct a biomimetic meniscus scaffold using 3D bioprinting. Based on this, a customized 3D bioprinting system featuring dual nozzle + multitemperature printing was used to integrate the advantages of polycaprolactone (PCL) and MFCs-laden GelMA/MECM bioink into the biomimetic meniscal scaffold with biomimetic morphology and component features (Figure 5.5) [67]. According to

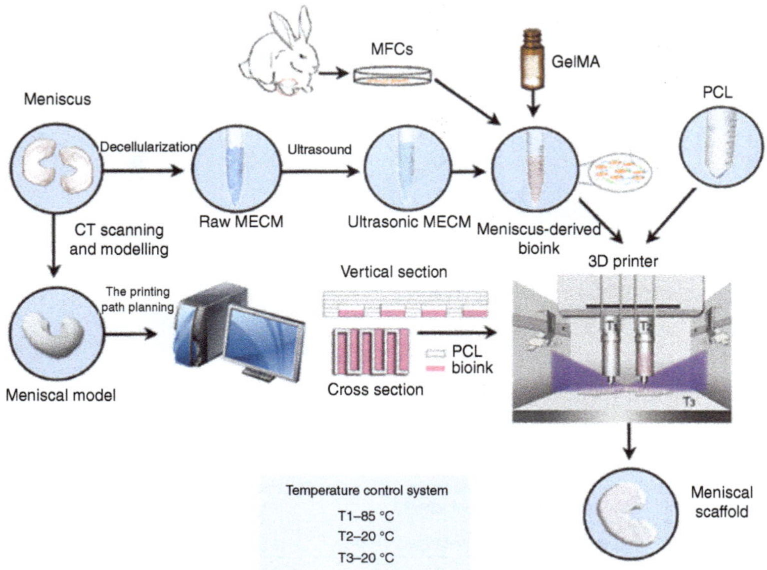

Figure 5.5 Process of fabricating the biomimetic PCL +MFCs-laden GelMA/MECM meniscal scaffold [67].

the cell viability, mechanics, biodegradation, and tissue formation results, the bioprinted PCL + hydrogel + MFCs scaffold exhibited sufficient feasibility as a meniscus tissue-engineering scaffold to promote meniscus regeneration.

In addition to providing a microenvironment for cell growth, hydrogels can also be used to load bioactive components to improve meniscus regeneration efficiency due to their interconnected microporous structures. To further improve the bioactivity of the scaffold, Guo et al. fabricated a composite meniscus scaffold composed of 3D-printed PCL skeleton (providing load-bearing capability) and a mECM gel containing a kartogenin (KGN)-loaded poly(lactic-*co*-glycolic) acid (PLGA) microsphere (μS) infill (as a drug delivery system) (Figure 5.6) [68]. Owing to the synergistic effect of MECM and sustained release of KGN, the meniscus scaffold exhibited excellent cell affinity and chondrogenic activity. More importantly, after implantation for 3 months, the regenerated meniscus of the PCL/MECM-KGN μS group was histologically more similar to the native meniscus than that of the other groups, indicating its potential for promoting meniscus regeneration.

In another study, Guo et al. first prepared bioactive-component-loaded bioinks by incorporating platelet-derived growth factor-BB (PDGF-BB) and small chondroinductive molecules (kartogenin, KGN) in MECM-based GelMA/HAMA hydrogels. Based on this, a cell-free dual drug-releasing biomimetic meniscus scaffold was fabricated using 3D bioprinting technology (Figure 5.7) [69]. Therein, incorporated PDGF-BB is expected to promote endogenous stem cell homing, and KGN in PLGA microspheres is employed to target the chondrogenesis of resident mesenchymal stem cells (MSCs). Subsequently, the bio-fabricated scaffolds were implanted in a

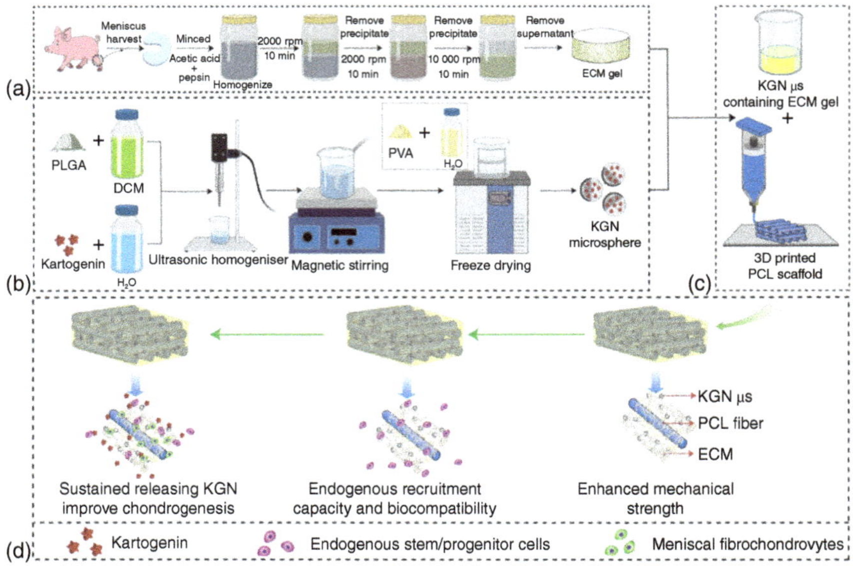

Figure 5.6 Diagram of the fabrication process of the scaffolds, including the preparation of (a) MECM gel, (b) KGN-containing PLGA microspheres as well as (c) PCL/MECM-KGN μs scaffold; (d) possible mechanism of meniscus regeneration. Enhanced mechanical strength, endogenous stem cells, and sustained releasing KGN contributed to meniscus regeneration in these experiments [68].

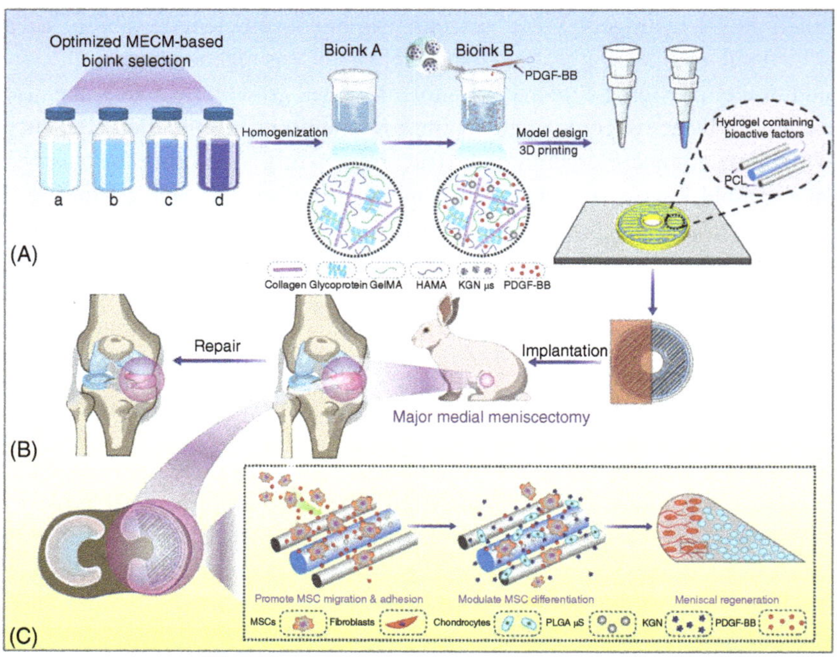

Figure 5.7 Schematic illustration of the scaffold construction and study design. (A) Optimized MECM-based bioink selection and fabrication of 3D-printed meniscal scaffolds; (B) schematic representation of 3D-printed meniscal scaffolds for *in situ* meniscus regeneration in rabbits; (C) schematic diagram of the potential meniscal regenerative process within the 3D-printed scaffolds [69].

rabbit model treated with critical-sized medial meniscectomy to evaluate the applicability of the biomimetic scaffold for promoting meniscus regeneration. According to the results, the dual drug-loaded scaffolds can enhance cell migration and promote MSC chondrogenic differentiation and neo-meniscal regeneration *in vivo* six months post-implantation, indicating their potential off-the-shelf products for the clinical treatment of meniscal injury and related degenerative joint diseases.

Although ECM-based hydrogels provide a similar biochemical composition to native tissue, most reported works fail to replicate region-specific distribution features, which is essential for providing biomechanical functions. Hence, different hydrogels have been incorporated into the outer and inner regions of the meniscus skeleton to construct meniscus scaffolds mimicking the microstructure and biochemical features of native tissue. Hasirci et al. designed and fabricated a PCL/hydrogel scaffold (Figure 5.8) [70]. On the one hand, 3D-printed PCL scaffolds with a shifted design and circumferential strands are expected to mimic the collagen organization in native tissue and lead to elongation and alignment of the human fibrochondrocytes. On the other hand, hydrogels are expected to enhance the ECM production both at mRNA and protein levels and to protect the articular cartilage from degeneration resulting from friction between the stiff PCL scaffold and the hyaline cartilage. Based on the results that GelMA is fibrogenic (induces expression of COL I), while GelMA-Ag is chondrogenic (induces expression of aggrecan and yields a high ratio of COL II/COL I), the PCL/hydrogel scaffold was constructed by impregnating the outer/inner region of the meniscus-shaped PCL scaffolds with GelMA and GelMA-Ag, respectively. After 6 weeks of culture, an increment in COL

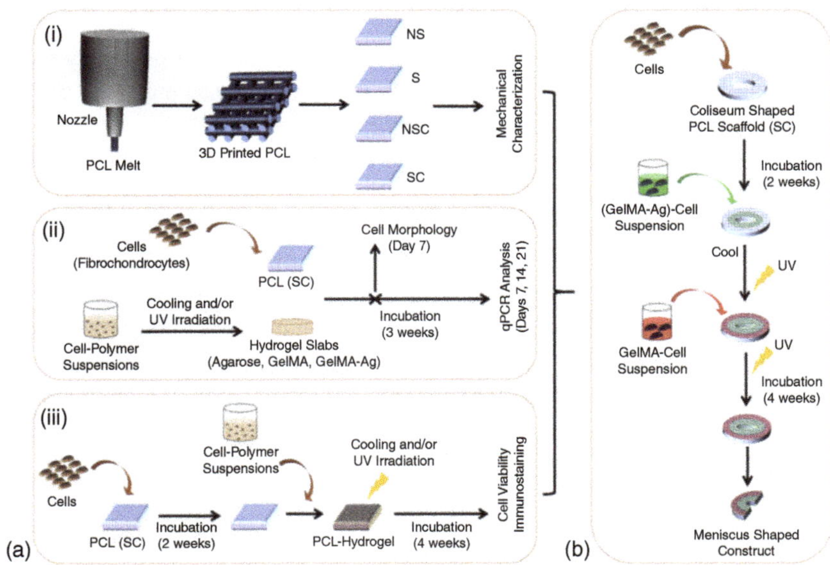

Figure 5.8 (a) Optimization of the scaffold fabrication conditions: (i) 3D printing of PCL scaffolds with different structural features and characterizing their mechanical properties; (ii) seeding the fibrochondrocytes on the 3D-printed PCL and Ag, GelMA, and GelMA-Ag hydrogels and characterizing the cell morphology and mRNA expression; (iii) constructing the cell-loaded PCL-hydrogel scaffolds and testing for cell viability as well as collagen production; (b) schematic diagram of meniscus-shaped scaffolds fabrication [70]. Bahcecioglua et al., 2019/With permission from ELSEVIER.

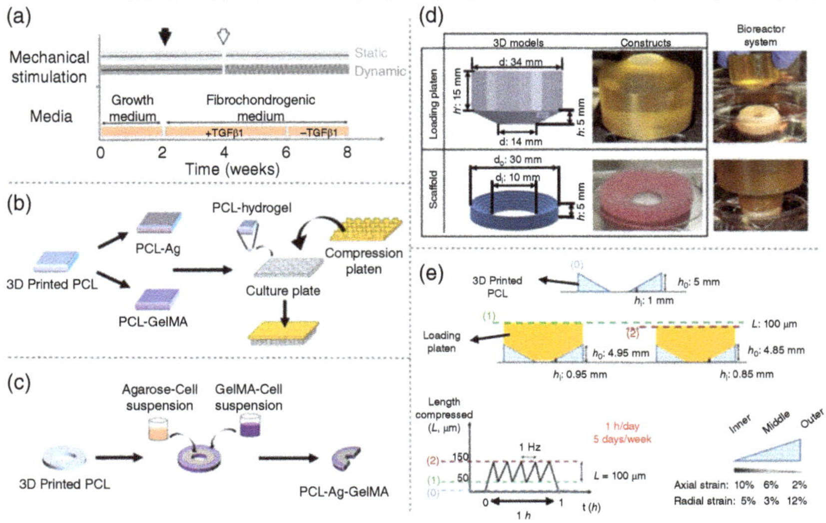

Figure 5.9 (a) Time plan showing the culture conditions (full arrowheads: hydrogel incorporation; empty arrowheads: dynamic stimulation). (b) Preparation of square prism constructs. Constructs were subjected to dynamic compression using a polysulfone-based compression platen that fits in a 24-well plate. (c) Preparation of the meniscus-shaped constructs (Inner region: GelMA; outer region: agarose); (d) 3D models and constructs of the meniscus-shaped constructs and the loading platen used for dynamic stimulation. (e) Meniscus-shaped constructs were subjected to biaxial dynamic loading. The magnitude of the strain increased gradually from the outer region toward the inner region [71]. Gokhan Bahcecioglu et al. 2019/Reproduced from IOP Publishing Ltd.

I expression and a decrease in COL II expression toward the outer periphery were observed, indicating the successful construction of the anatomical meniscus with a cartilage-like inner region and fibrocartilage-like outer region.

Hasirci et al. also constructed a meniscus scaffold mimicking the biochemical distribution of native tissue by 3D printing a meniscus-shaped PCL scaffold and then impregnating it with agarose (Ag) and gelatin methacrylate (GelMA) hydrogels in the inner and outer regions, respectively (Figure 5.9) [71]. Subsequently, the cell-loaded PCL/hydrogel scaffolds were cultured under dynamic stimulation at strain levels gradually increasing from the outer region (2% of the initial thickness) toward the inner region (10%) to mimic the physiological loading environment. According to the results, dynamic stimulation would increase COL II in the inner and COL I in the outer regions, suggesting that the scaffold holds potential for meniscus regeneration.

5.4.2 Providing Load-Bearing Capability

As a meniscus scaffold, the load-bearing capability is nonnegligible to restore the biomechanical microenvironment in the joint. However, the mechanical properties of conventional hydrogels are too poor to resist strong loads. In recent decades, the development of high-strength hydrogels has been expected to address this issue, and several high-strength hydrogel systems have been used to fabricate meniscus replacements to provide the load-bearing capability [72].

5.4.2.1 Polyvinyl Alcohol (PVA) Hydrogel-Based Meniscus Scaffolds

Polyvinyl alcohol (PVA) is a typical synthetic polymer that can form a series of physically or chemically cross-linked hydrogels with swelling stability in aqueous media [73]. PVA hydrogels are commonly used as load-bearing scaffolds owing to their good biocompatibility, suitable mechanical strength, high water content, and processability [74]. In 2003, Kobatashi et al. demonstrated that PVA hydrogels with viscoelastic characteristics resembling human meniscal tissue could be obtained by regulating the water content, which is expected to be used as an artificial meniscus *in vivo* [75]. To confirm this, a PVA hydrogel-based meniscus replacement was fabricated by processing a cylindrical rod PVA hydrogel with a water content of 90% into the meniscal shape. Subsequently, the replacement was implanted into the knees of female rabbits to replace the lateral meniscus, and the state of the knee joint and the degree of cartilage injury were observed at certain periods (4-, 6-months and 1-, 1.5-, 2-years) [76–78]. According to the results, in the control group (treated by lateral meniscectomy), the erosion and degeneration of articular cartilage worsened over time, and OA-like changes were even observed after one year of operation. Moreover, although partial and incomplete regenerative fibrocartilage-like tissue was formed 1.5–2 years post-operation, progressive OA change was still not remitted. In contrast, after treatment with the PVA meniscus substitute, although slight changes (luster loss and color change in some parts) and erosion were found on the articular cartilage surface after short-term (four and six months) implantation, no obvious deterioration was observed in terms of cartilage degeneration one to two years post-implantation. In addition, the implanted PVA artificial meniscus remained attached to the peripheral tissue and demonstrated no breakage, deformation, or dislocation after being transplanted for 1.5–2 years. The above results indicate that the PVA hydrogel meniscal scaffolds with high water content and robust mechanical properties exhibit durability *in vivo* for two years and can be fixed stably on the tibia platform using a sutural method to provide load-bearing functions. Subsequently, to further verify the feasibility of PVA hydrogel as a meniscus replacement in large animals, Kelly et al. implanted a PVA hydrogel-based scaffold into a mature sheep knee joint treated with lateral meniscectomy and evaluated the protective effect of the scaffold on articular cartilage at 2-, 4-, and 12 months post-implantation [79]. In short-term implantation, compared with the control group (treated by lateral meniscectomy), the implanted scaffold significantly relieved articular cartilage degeneration, and the efficiency was comparable to that of the normal group as well as the allograft group. However, after one year, more severe articular cartilage degeneration than in the positive groups and radial tear of the replacement was observed. According to the authors, graft failure may be attributed to size mismatch, inadequate peripheral fixation, and lack of an anisotropic structure.

To further improve the strength of PVA hydrogel scaffolds, Palmese et al. [80] developed fiber-reinforced PVA hydrogels by introducing ultrahigh-molecular-weight polyethylene (UHMWPE) or polypropylene fibers into the PVA matrix. By regulating the volume percentage of fibers, the UHMWPE-PVA (258.1 MPa) and PP-PVA (8.2 MPa) composite hydrogels exhibited tensile moduli similar to that of the native meniscus in the circumferential (94–295 MPa) and radial (2–23 MPa) directions, respectively. However, this work did not achieve the anisotropic mechanical structure resembling a human meniscus. What is more, further studies have not confirmed its applicability as a meniscus substitute *in vitro* or *in vivo*. Thus, further research requires investigating whether PVA hydrogels with robust mechanical properties reinforced by fibers can be used as a substitute to achieve meniscus function recovery. The swelling behaviors of hydrogels, i.e. whether the hydrogels can retain their initial shape and size after transplantation *in vivo*, are significant to construct an adequate model. Neutral phosphate-buffered saline (PBS) is commonly chosen as a swelling medium *in vitro* to study the swelling behaviors of hydrogels, but it does not resemble the swelling pressure of load-bearing joints, which would result in a distinct swelling ratio between those located *in vitro* and *in vivo*. Palmese et al. [81, 82] compared the *in vitro* swelling behavior of PVA hydrogels in an osmotic solution with swelling pressure in the knee to that immersed in PBS, which indicated that the swelling ratio decreased in osmotic pressure solutions compared to that in PBS. The long-term stability of the hydrogel is a key factor in achieving artificial meniscal replacement *in vivo*. Kobayashi et al. evaluated the mechanical stability of a PVA hydrogel artificial meniscus after *in vivo* transplantation for two years. However, we cannot help but wonder whether the volume, mass, and mechanical properties of hydrogels keep stable after maintaining a physiological environment for a longer time. In 2014, Awadallah-F [83] reported a series of novel polyvinyl alcohol/polyvinyl pyrrolidone/polyacrylic acid (PVA/PVP/PAAc) hydrogels achieved by polymerization of different concentrations and ratios of monomers *via* irradiation of γ-rays, which are considered to be used as artificial menisci. In this study, the mass stability, yield stress, and stress-relaxation change of the hydrogel with a PVA: PVP molar ratio of 95/5 and AAc concentration of 9 *v/v*% were measured after immersion in PBS at 37 °C for different times and thereafter for up to five years. The results showed that almost no mass loss occurred after five years, the value of yield stress continued to increase until it reached equilibrium two years later and remained stable for the next three years, the marked stress-relaxation was observed at nine months, and its value at three years resembled that at five years. This work proceeded for five years to evaluate the possibility of using synthesized hydrogels as artificial meniscus substitutes. However, the research was performed *in vitro*, and the physiological environment *in vivo* is more complicated than that *in vitro*. Thus, further tests should be performed to evaluate the biocompatibility, immunocompatibility, and other *in vivo* properties of hydrogel-based meniscal scaffolds.

Although several PVA-based hydrogels have been reported for fabricating meniscus substitutes to provide load-bearing functions, most of the reported PVA hydrogel-based meniscus replacements are fabricated using the casting method, which fails to duplicate the unique anisotropic microstructure (e.g. the circumferential orientation of collagenous fibers) and mechanical features of the native meniscus. With the development of 3D printing technology, several inks have been prepared for 3D printing of high-strength hydrogels. Therefore, 3D-printed high-strength hydrogels are expected to be used for constructing tissue substitutes with complex structures that are more similar to the native meniscus.

5.4.2.2 Poly(*N*-acryloyl glycinamide) (PNAGA) Hydrogel-Based Meniscus Scaffolds

PNAGA is a synthetic polymer whose structural unit contains double amide separated by an inactive C atom in the side chain [84]. PNAGA systems have been studied for a long time as UCST-type phase separation [85–90] or thermoreversible systems [91] until 2015 when our groups obtained hydrogen-bonding cross-linked nonswelling and high-strength PNAGA supramolecular polymer hydrogels by increasing the initial monomer concentration [92]. Based on this, several PNAGA-based hydrogels have been used to construct load-bearing tissue scaffolds. Gao et al. prepared a supramolecular copolymer hydrogel *via* one-step copolymerization between NAGA and *N*-[tris(hydroxymethyl)methyl] acrylamide (THMMA). Owing to the weakening effect of THMMA on the hydrogen bond cross-linking density of PNAGA, the obtained hydrogel exhibited a thermoreversible gel ⇔ sol transition and printability. On this basis, a high-strength biohybrid gradient hydrogel scaffold was 3D-printed *via* extrusion-based printing technology, and its repair effect for osteochondral defects was confirmed [93]. Despite this, copolymerization would sacrifice the swelling and mechanical properties when obtaining extrudability, which may reduce the "processing window" of PNAGA-based hydrogels. To address this issue, inspired by the concentration-dependent hydrogen-bonding reinforcement mechanism of PNAGA systems, Xu et al. proposed a novel self-thickening and self-strengthening strategy for 3D printing high-strength and antiswelling supramolecular polymer hydrogels (Figure 5.10) [94]. On the one hand, the thermoreversible PNAGA soft gel is used to thicken the concentrated NAGA monomer solution, affording an appropriate viscosity for thermal-assisted extrusion 3D printing; on the other hand, the post compensatory NAGA monomers can be further polymerized under the UV irradiation to significantly enhance the hydrogen bonding density in the 3D-printed constructs, which eventually generate the high-strength and antiswelling hydrogels. Thus, high-strength PNAGA hydrogels can be extrusion-based printed without sacrificing mechanical strength, circumventing the trade-off between printability and mechanical properties. Based on this, a high-strength and antiswelling PNAGA hydrogel-based meniscus scaffold was 3D printed and implanted in the rabbit knee to replace the medial meniscus. According to the *in vivo* outcome, the PNAGA hydrogel-based meniscus substitute demonstrated an appealing ability to alleviate cartilage wear efficiently during short-term implantation.

Recently, to develop a meniscus substitute with anisotropic microstructure as well as mechanical performance for providing long-term load-bearing functions, Zhang et al. fabricated a high-strength supramolecular polymer hydrogel-cushioned biomimetic structured meniscus replacement by infusing the PNAGA hydrogel into a 3D-printed poly(*e*-caprolactone) (PCL) fiber framework (Figure 5.11) [95]. On the one hand, the radially and circumferentially oriented PCL filament can

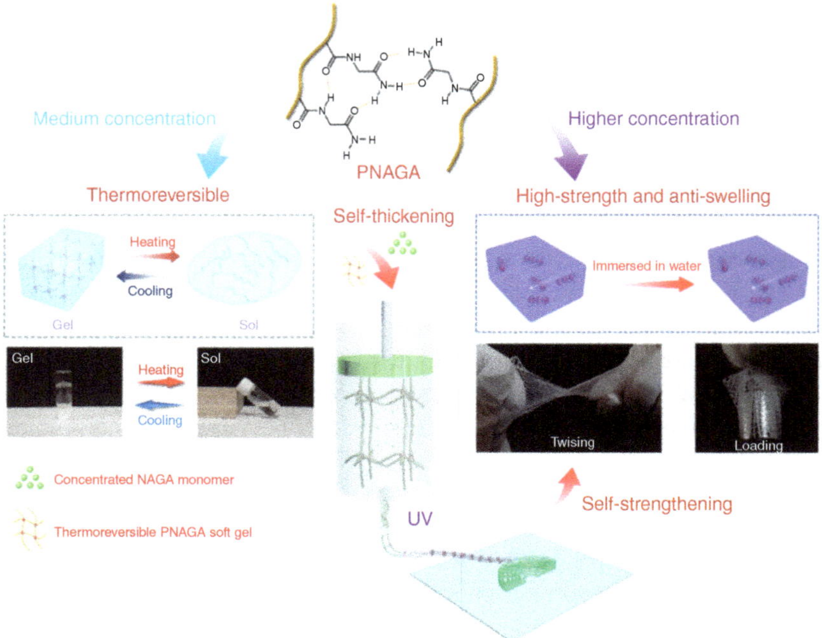

Figure 5.10 Schematic illustration of the self-thickening and self-strengthening strategy for 3D printing the high-strength and antiswelling PNAGA supramolecular polymer hydrogel-based meniscus substitute [94]. Ziyang Xu et al. 2021/Reproduced from JOHN WILEY & SONS, INC.

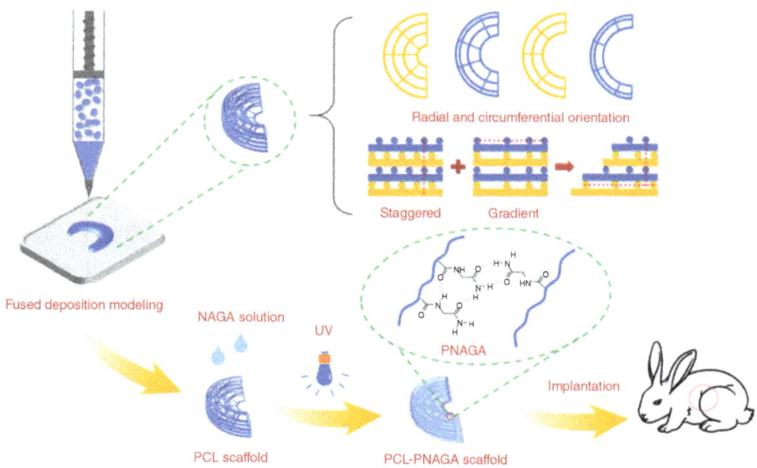

Figure 5.11 Schematic illustration of 3D-printed high-strength supramolecular polymer hydrogel-cushioned radially and circumferentially oriented meniscus substitute [95].

be employed to mimic the anisotropic collagen fibers in native tissue and provide a higher tensile modulus; on the other hand, the infilled PNAGA hydrogel is expected to replicate the function of proteoglycan in resisting axial compressive loads and provide energy absorption as well as cushion functions. According to the results, the PNAGA-cushioned PCL construct exhibits excellent mechanical properties (Young's modulus: 20.15 ± 1.37 MPa in a circumferential direction; compressive modulus: 1.11 ± 0.14 MPa; tearing energy of 17.00 ± 2.07 kJ/m^2). Furthermore, its applicability as a meniscal substitute was evaluated *in vivo*. Twelve weeks after surgery, the implanted 3D-printed PCL-PNAGA meniscus scaffold demonstrated efficient protection against cartilage wear, which ameliorated the development of OA.

5.4.2.3 Poly(*N*-acryloylsemicarbazide) (PNASC) Hydrogel-Based Meniscus Scaffold

Although many high-strength hydrogels have been reported, few exhibit a high tensile modulus similar to that of the native meniscus. In 2020, Fan et al. designed and synthesized an unprecedented feature monomer, *N*-acryloylsemicarbazide (NASC). Compared to the NAGA monomer, one extra NH is contained in the chemical structure of NASC, which causes a higher H-bonding density. Thus, the PNASC hydrogel exhibited an ultrahigh tensile modulus of approximately 100 MPa (close to the tensile modulus of the native meniscus) and excellent swelling stability in a physiological environment, which is expected to be fabricated into the meniscus scaffold (Figure 5.12) [96].

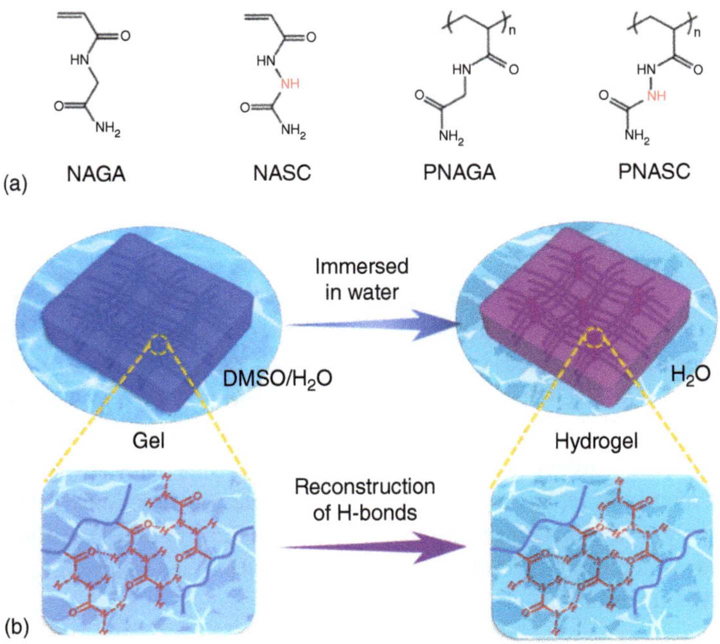

Figure 5.12 (a) Molecular structure of NAGA, NASC, PNAGA, and PNASC. (b) Schematic illustration of the reconstruction of hydrogen bonding of amide and urea in PNASC gels after completed solvent exchanges in aqueous [96].

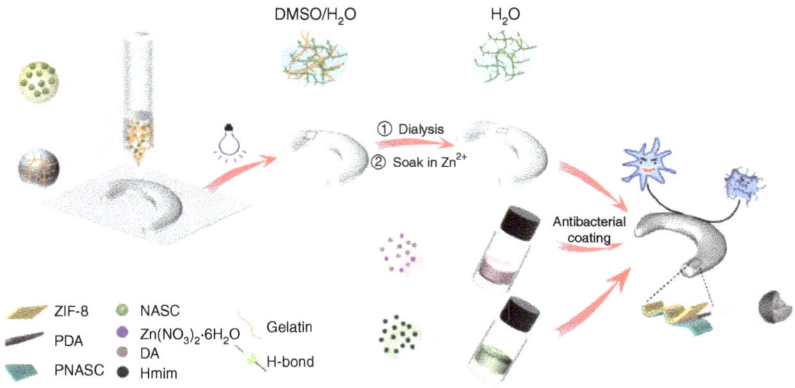

Figure 5.13 Schematic illustration of the sacrificial material strategy used for 3D printing high-strength PNASC hydrogel and subsequent one-pot synthesis of antibacterial coating on the surface of hydrogel [97].

Subsequently, the PNASC-based hydrogel meniscus scaffolds were constructed. For example, inspired by fugitive ink generally adopted in supporting bath printing, Yang et al. prepared extrudable inks by introducing gelatin into the NASC monomer solution and fabricated a 3D-printed stiff hydrogel-based construct (Tensile modulus: 26.25 MPa; compressive modulus: 4.22 MPa) *via* extrusion-based 3D printing technology followed by UV irradiation, solvent exchange, and high-temperature dialysis. According to the femoral wear simulation test results, the above-printed hydrogels did not exhibit noticeable wear to cartilage even after 10^4 cycles under a pressure of 28 N, demonstrating its excellent chondroprotective effect. In addition, through superficial modification with polydopamine coated-ZIF-8, the printed PNASC/gelatin constructs exhibited antibacterial properties without changing the mechanical performance and wear property, which is expected to be used for fabricating antibacterial and stiff hydrogel-based meniscus replacements to provide load-bearing functions while preventing implant failure caused by bacterial infection (Figure 5.13) [97].

In addition, Fan et al. prepared stiff and tough copolymer hydrogels by copolymerizing NASC monomers with the zwitterion monomer carboxybetaine acrylamide (CBAA) and fabricated a 3D-printed stiff and tough PNASC–PCBAA hydrogel based on the self-thickening and self-strengthening strategy (Figure 5.14)[98]. Because of the dense hydrogen bonding reconstruction during the solvent exchange, the above construct exhibits a higher tensile modulus (~ 10.98 MPa), which is expected to be used as a meniscus replacement to provide load-bearing functions. More importantly, the polyzwitterion component increased the lubricity of the 3D-printed meniscus scaffold, which can efficiently ameliorate cartilage surface wear within a set 12-week *in vivo* implantation.

5.4.2.4 Other Systems

In addition to the above high-strength hydrogels, some other systems also have the potential to construct meniscus substitutes to provide load-bearing functions. For example, Wang et al. [99] designed a highly stiff hydrogel *via* copolymerization

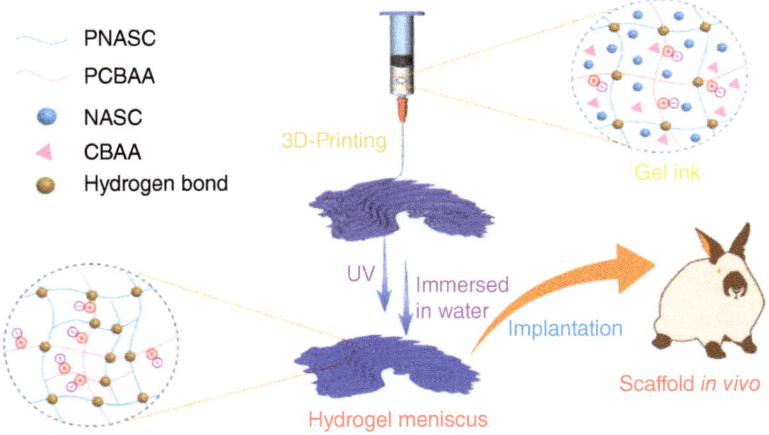

Figure 5.14 Schematic illustration of the 3D printing of the PNASC–PCBAA-6-40 hydrogel meniscal scaffold and its application as a meniscus substitute [98].

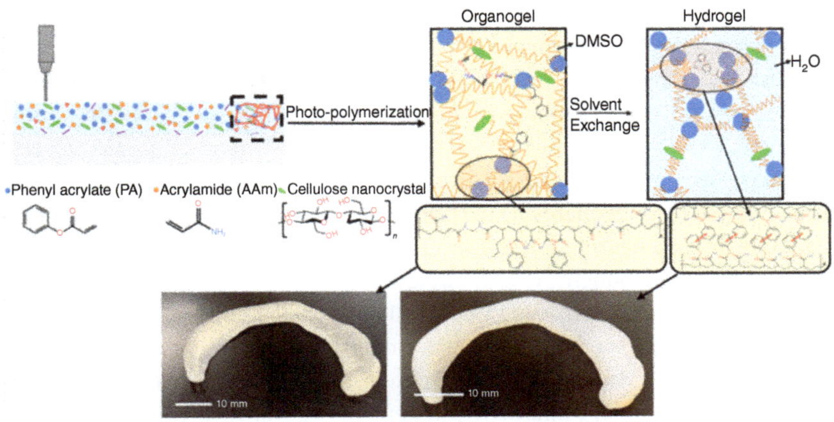

Figure 5.15 Illustration of the 3D printing process and the photo of printed meniscus scaffolds [99]. Zimeng Zhang et al. 2019/Reproduced from American Chemical Society.

between AAm (hydrophilic segment) and phenyl acrylate (PA, hydrophobic segment), followed by a solvent exchange in deionized water. Furthermore, cellulose nanocrystals were employed as the thickener and introduced into the precursor solution to fabricate the 3D printing construct with excellent tensile performance (Figure 5.15). In particular, Young's modulus of printed hydrogels is up to 232.4 MPa, which is much higher than those of reported printed hydrogels and meniscus circumferential tensile modulus. Finally, an artificial meniscus with clear features and a geometrical shape similar to that of a natural meniscus was fabricated *via* extrusion-based 3D printing technology, which has the potential for meniscus substitution. Regrettably, although the modulus of printed hydrogels has improved significantly, no animal experiments were performed in this study, and the *in vivo* feasibility was unknown.

5.5 Mimicking Microstructure: The Key to Constructing the Next-Generation Meniscus Scaffolds

Various hydrogel-based meniscus scaffolds have been constructed with the unremitting efforts of researchers, and their functions in load bearing or providing a microenvironment for cell growth have been investigated. Despite this, many issues should be considered to promote the development of new-generation meniscal scaffolds.

First, the basic task of meniscus substitutes is to mimic the biomechanical behaviors of the native meniscus, particularly the high tensile modulus and low compressive modulus. Currently, some ultrastiff hydrogels (such as enzyme-catalyzed mineralized hydrogels [100], hydrogen-bonding cluster-reinforced hydrogels [101], and anisotropic hydrogels [102]) have been reported, which are expected to mimic the circumferential tensile properties of collagen fibers. Despite this, homogenous high-modulus hydrogels cannot obtain a higher tensile modulus and lower compressive modulus simultaneously. Fabrication of a high-modulus hydrogel into a meniscus skeleton consisting of radial and circumferential hydrogel filaments should be considered to address this issue.

Second, considering that permanent meniscus replacement would continuously suffer repeated loading in the joint, excellent antifatigue performance is essential for the replacement to provide a long-term load-bearing function. Unfortunately, most of the reported studies did not investigate antifatigue performance. Recently, several strategies have been proposed to construct fatigue-resistant hydrogels, such as introducing a low hysteresis mechanism [103] and high-energy phases [104]. The latter is more suitable for load-bearing soft tissue replacement. Thus, to fabricate antifatigue meniscus scaffolds, introducing a high-energy phase with anisotropic features similar to those of the native meniscus ECM should be considered.

In addition, unlike permanent replacements, meniscal tissue-engineering scaffolds need to provide load-bearing function at the initial stage of implantation and provide a suitable microenvironment for cell growth during meniscus regeneration. To achieve these goals, the commonly used method to construct meniscus tissue-engineering scaffolds is to take 3D-printed polymers as the skeleton to provide the load-bearing function, followed by filling natural polymer (or ECM) hydrogels to provide a microenvironment for promoting tissue regeneration. Although several studies have demonstrated that these scaffolds exhibit good function for promoting meniscus regeneration, the degradation rate of 3D-printed skeleton is slow and cannot match the ingrowth rate of regenerated meniscus tissue, so it is difficult to achieve complete meniscus repair. Therefore, the design and construction of natural polymer hydrogels with excellent mechanical properties (close to the natural meniscus) and their use as a skeleton material for meniscal tissue-engineering scaffolds are expected to be a breakthrough to solve the above problems. A biomimetic approach is one of the most promising options to achieve this. On this basis, the construction of natural polymer hydrogels with anisotropic and hierarchical microstructures remains an issue to be considered.

It is worth noting that almost all above issues can be solved by mimicking the microstructure of the meniscus. As we all know, anisotropic and hierarchical collagen fiber microstructure is common in almost all the load-bearing soft tissues (e.g. tendon [105], ligament [106], and muscle [107]), which provides special biomechanical properties. In addition, unlike other tissues whose collagen fibers are uniaxially oriented, the anisotropic microstructure of the meniscus is more complex, and its extracellular matrix skeleton is mainly composed of circumferential and radial collagen fibers [108]. Although extrusion-based 3D-printed thermosensitive polymer (such as PCL and PLA) skeletons have been fabricated to mimic the anisotropic structure of the meniscus, the replication of its microstructure has not been realized (only likeness in shape but not in spirit). Therefore, mimicking the anisotropic and hierarchical microstructure of the meniscus should be a key consideration for designing and constructing next-generation meniscus scaffolds. Currently, several strategies to construct anisotropic hydrogels, such as prestretching orientation [102], directional freezing [109], and the introduction of nanofillers with a high aspect ratio [110], have been proposed. However, most of these strategies are not applicable for mimicking the radial and circumferential anisotropic hierarchical microstructures of the menisci. On the one hand, the method of directional freezing and the introduction of nanofillers fail to construct a hierarchical structure. On the other hand, the prestretching method is the most common and efficient strategy to introduce the hierarchical structure; however, most of the reported works are uniaxial. Therefore, an upgraded preloading method should be developed to construct an anisotropic and hierarchical structure similar to a meniscus.

5.6 Conclusion

The anisotropic microstructure and region-specific distribution of the components endow the meniscus with specific biomechanical behaviors. To restore the load-bearing functions of native tissues after meniscectomy, the construction of a meniscus scaffold with a microstructure and composition analogous to that of the native meniscus is essential. As the closest material to natural tissues, hydrogels have been widely used to construct meniscus scaffolds to provide load-bearing functions or a microenvironment for promoting cell growth. Nevertheless, the currently reported hydrogel-based meniscus scaffolds are far from ideal. To address the issues of meniscus scaffolds for replacement and regeneration (e.g. how to mimic the anisotropic mechanical properties, develop the antifatigue meniscus scaffold, and construct the natural polymer high-strength hydrogel-based meniscus skeleton), we presented corresponding suggestions and possible ways to solve these problems. Unexpectedly, we found that all of the above issues can be addressed by mimicking anisotropic and hierarchical microstructures. In other words, mimicking the microstructure of the meniscus is key to constructing a next-generation meniscus scaffold that is more similar to the native meniscus, and the achievement of this goal is dependent on the cooperation and development of different fields. It can be expected that with the efforts of researchers in materials science, biology, medicine, and other fields, the fabrication of meniscus scaffolds similar to native tissues will be achieved in the future.

References

1 Bilgen, B., Jayasuriya, C.T., and Owens, B.D. (2018). Current concepts in meniscus tissue engineering and repair. *Advanced Healthcare Materials* 7: 1701407.

2 Gee, S.M., Tennent, D.J., Cameron, K.L., and Posner, M.A. (2020). The burden of meniscus injury in young and physically active populations. *Clinics in Sports Medicine* 39: 13–27.

3 Haddad, B., Pakravan, A.H., Konan, S. et al. (2013). A systematic review of tissue engineered meniscus: cell-based preclinical models. *Current Stem Cell Research & Therapy* 8: 222–231.

4 Chang, P.S. and Brophy, R.H. (2020). As goes the meniscus goes the knee: early, intermediate, and late evidence for the detrimental effect of meniscus tears. *Clinics in Sports Medicine* 39: 29–36.

5 Yoon, K.H. and Park, K.H. (2014). Meniscal repair. *Knee Surgery & Related Research* 26: 68–76.

6 Pillai, M.M., Gopinathan, J., Selvakumar, R., and Bhattacharyya, A. (2018). Human knee meniscus regeneration strategies: a review on recent advances. *Current Osteoporosis Reports* 16: 224–235.

7 Veronesi, F., Vandenbulcke, F., Ashmore, K. et al. (2020). Meniscectomy-induced osteoarthritis in the sheep model for the investigation of therapeutic strategies: a systematic review. *International Orthopaedics* 44: 779–793.

8 Young, J., Tudor, F., Mahmoud, A., and Myers, P. (2017). Meniscal transplantation: procedures, outcomes, and rehabilitation. *Orthopedic Research and Reviews* 9: 35–43.

9 Twomey, K.J. and Jayasuriya, C.T. (2020). Meniscus repair and regeneration: a systematic review from a basic and translational science perspective. *Clinics in Sports Medicine* 39: 125–163.

10 Messner, K. (1994). The concept of a permanent sythetic meniscus prothesis: a critical discussion after 5 years of experimental investigation using Decron and Teflon implants. *Biomaterials* 15: 243–250.

11 Messner, K., Lohmander, L.S., and Gillquist, J. (1992). Cartilage mechanics and morphology, synovitis and proteoglycan fragments in rabbit joint fluid after prosthetic meniscal substitution. *Biomaterials* 14: 163–168.

12 Sommerlath, K., Gallino, M., and Gillquist, J. (1992). Biomechanical characteristics of different artificial substitutes for rabbit medial meniscus and effect of prosthesis size on knee cartilage. *Clinical biomechanics* 7: 97–103.

13 Tienen, T.G., Heijkants, R.G., de Groot, J.H. et al. (2006). Replacement of the knee meniscus by a porous polymer implant: a study in dogs. *The American Journal of Sports Medicine* 34: 64–71.

14 Moran, C.J., Withers, D.P., Kurzweil, P.R., and Verdonk, P.C. (2015). Clinical application of scaffolds for partial meniscus replacement. *Sports Medicine and Arthroscopy Review* 23: 156–161.

15 Mandal, B.B., Park, S.H., Gil, E.S., and Kaplan, D.L. (2011). Multilayered silk scaffolds for meniscus tissue engineering. *Biomaterials* 32: 639–651.

16 Gao, S., Guo, W.M., Chen, M.X. et al. (2017). Fabrication and characterization of electrospun nanofibers composed of decellularized meniscus extracellular matrix and polycaprolactone for meniscus tissue engineering. *Journal of Materials Chemistry B* 5: 2273–2285.

17 Wang, W., Zhang, Y.Y., and Liu, W.G. (2017). Bioinspired fabrication of high strength hydrogels from non-covalent interactions. *Progress in Polymer Science* 71: 1–25.

18 Koh, R.H., Jin, Y.J., Kang, B.J., and Hwang, N.S. (2017). Chondrogenically primed tonsil-derived mesenchymal stem cells encapsulated in riboflavin-induced photocrosslinking collagen-hyaluronic acid hydrogel for meniscus tissue repairs. *Acta Biomaterialia* 53: 318–328.

19 McNulty, A.L. and Guilak, F. (2015). Mechanobiology of the meniscus. *Journal of Biomechanics* 48: 1469–1478.

20 Gao, F., Ruan, C.S., and Liu, W.G. (2019). High-strength hydrogel-based bioinks. *Materials Chemistry Frontiers* 3: 1736–1746.

21 Li, J., Wu, C., Chu, P.K., and Gelinsky, M. (2020). 3D printing of hydrogels: rational design strategies and emerging biomedical applications. *Materials Science & Engineering R: Reports* 140: 100543.

22 Xia, B.A., Kim, D.H., Bansal, S. et al. (2021). Development of a decellularized meniscus matrix-based nanofibrous scaffold for meniscus tissue engineering. *Acta Biomaterialia* 128: 175–185.

23 Murphy, C.A., Garg, A.K., Silva, C.J. et al. (2019). The meniscus in normal and osteoarthritic tissues: facing the structure property challenges and current treatment trends. *Annual Review of Biomedical Engineeing* 21: 495–521.

24 Markes, A.R., Hodax, J.D., and Ma, C.B. (2020). Meniscus form and function. *Clinical in Sports Medicine* 39: 1–12.

25 Herwig, J., Egner, E., and Buddecke, E. (1984). Chemical changes of human knee joint menisci in various stages of degeneration. *Annals of the Rheumatic Diseases* 43: 635–640.

26 Tienen, T.G., Hannink, G., and Buma, P. (2009). Meniscus replacement using synthetic materials. *Clinical in Sports Medicine* 28: 143–156.

27 Cheung, H.S. (1987). Distribution of type I, II, III and V in the pepsin solubilized collagens in bovine menisci. *Connective Tissue Research* 16: 343–356.

28 Li, Q., Qu, F.N., Han, B. et al. (2017). Micromechanical anisotropy and heterogeneity of the meniscus extracellular matrix. *Acta Biomaterialia* 54: 356–366.

29 Danso, E.K., Oinas, J.M.T., Saarakkala, S. et al. (2017). Structure-function relationships of human meniscus. *Journal of the Mechanical Behavior of Biomedical* 67: 51–60.

30 Mcdevitt, C.A. and Webber, R.J. (1990). The ultrastructure and biochemistry of meniscal cartilage. *Clinical Orthopaedics and Related Research* 252: 8–18.

31 Tissakht, M. and Ahmed, A.M. (1995). Tensile stress-strain characteristics of the human meniscal. *Journal of Biomechanics* 28: 411–422.

32 Setton, L.A., Guilak, F., Hsu, E.W., and Vail, T.P. (1999). Biomechanical factors in tissue engineered meniscal repair. *Clinical Orthopaedics and Related Research* 367: S254–S272.

33 Sweigart, M.A. and Athanasiou, K.A. (2005). Tensile and compressive properties of the medial rabbit meniscus. *Proceedings of the Institution of Mechanical Engineers. Part H, Journal of Engineering in Medicine* 219: 337–347.

34 Fithian, D.C., Kelly, M.A., and Mow, V.C. (1990). Material properties and structure-function relationships in the menisci. *Clinical Orthopaedics and Related Research* 252: 19–31.

35 Lechner, K., Hull, M.L., and Howell, S.M. (2000). Is the circumferential tensile modulus within a human medial meniscus affected by the test sample location and cross-sectional area? *Journal of Orthopaedic Research* 18: 945–951.

36 Chia, H.N. and Hull, M.L. (2008). Compressive moduli of the human medial meniscus in the axial and radial directions at equilibrium and at a physiological strain rate. *Journal of Orthopaedic Research* 26: 951–956.

37 Buma, P., van Tienen, T., and Veth, R.P. (2007). The collagen meniscus implant. *Expert Review of Medical Devices* 4: 507–516.

38 Yan, L.P., Oliveira, J.M., Oliveira, A.L. et al. (2012). Macro/microporous silk fibroin scaffolds with potential for articular cartilage and meniscus tissue engineering applications. *Acta Biomaterialia* 8: 289–301.

39 Welsing, R.T.C., Tienen, T.G., and Buma, P. (2008). Effect on tissue differentiation and articular cartilage degradation of a polymer meniscus implant: a 2-year follow-up study in dogs. *The American Journal of Sports Medicine* 36: 1978–1989.

40 Yan, R.J., Chen, Y.W., Gu, Y.J. et al. (2019). A collagen-coated sponge silk scaffold for functional meniscus regeneration. *Journal of Tissue Engineering and Regenerative Medicine* 13: 156–173.

41 Gao, S., Chen, M.X., Wang, P. et al. (2018). An electrospun fiber reinforced scaffold promotes total meniscus regeneration in rabbit meniscectomy model. *Acta Biomaterialia* 73: 127–140.

42 Vernerey, F.J., Sridhar, S.L., Muralidharan, A., and Bryant, S.J. (2021). Mechanics of 3D cell-hydrogel interactions: experiments, models, and mechanisms. *Chemical Reviews* 121: 11085–11148.

43 Narayanan, L.K., Huebner, P., Fisher, M.B. et al. (2016). 3D-bioprinting of polylactic acid (PLA) nanofiber-alginate hydrogel bioink containing human adipose-derived stem cells. *ACS Biomaterials Science & Engineering* 2: 1732–1742.

44 Ishida, K., Kuroda, R., Miwa, M. et al. (2007). The regenerative effects of platelet-rich plasma on meniscal cells in vitro and its in vivo application with biodegradable gelatin hydrogel. *Tissue Engineering* 13: 1103–1112.

45 Li, Y.L., Rodrigues, J., and Tomás, H. (2012). Injectable and biodegradable hydrogels: gelation, biodegradation and biomedical applications. *Chemical Society Reviews* 41: 2193–2221.

46 Cimen, Z., Babadag, S., Odabas, S. et al. (2021). Injectable and self-healable pH-responsive gelatin–PEG/Laponite hybrid hydrogels as long-acting implants for local cancer treatment. *ACS Applied Polymer Materials* 3: 3504–3518.

47 Heffernan, J.M., Overstreet, D.J., Vernon, B.L. et al. (2022). In vivo evaluation of temperature-responsive antimicrobial-loaded PNIPAAm hydrogels for prevention of surgical site infection. *Journal of Biomedical Materials Research Part B, Applied biomaterials* 110: 103–114.

48 Amini, A.A. and Nair, L.S. (2012). Injectable hydrogels for bone and cartilage repair. *Biomedical Materials* 7: 024105.

49 Pupkaite, J., Rosenquist, J., Hilborn, J., and Samanta, A. (2019). Injectable shape-holding collagen hydrogel for cell encapsulation and delivery cross-linked using thiol-michael addition click reaction. *Biomacromolecules* 20: 3475–3484.

50 Yang, X., Li, P.L., Tang, W.T. et al. (2021). A facile injectable carbon dot/oxidative polysaccharide hydrogel with potent self-healing and high antibacterial activity. *Carbohydrate Polymers* 251: 117040.

51 Xu, K.K., Yao, H., Fan, D. et al. (2021). Hyaluronic acid thiol modified injectable hydrogel: synthesis, characterization, drug release, cellular drug uptake and anticancer activity. *Carbohydrate Polymers* 254: 117286.

52 Ding, H.C., Li, B.Q., Liu, Z.L. et al. (2021). Nonswelling injectable chitosan hydrogel via UV crosslinking induced hydrophobic effect for minimally invasive tissue engineering. *Carbohydrate Polymers* 252: 117143.

53 Zhao, Z.Y., Wang, Z., Li, G. et al. (2021). Microfluidic hydrogel microspheres: injectable microfluidic hydrogel microspheres for cell and drug delivery. *Advanced Functional Materials* 31: 2103339.

54 Ghaderinejad, P., Najmoddin, N., Bagher, Z. et al. (2021). An injectable anisotropic alginate hydrogel containing oriented fibers for nerve tissue engineering. *Chemical Engineering Journal* 420: 130465.

55 Guo, W.M., Liu, S.Y., Zhu, Y. et al. (2015). Advances and prospects in tissue-engineered meniscal scaffolds for meniscus regeneration. *Stem Cells International* 2015: 517520.

56 Wu, J.L., Ding, Q., Dutta, A. et al. (2015). An injectable extracellular matrix derived hydrogel for meniscus repair and regeneration. *Acta Biomaterialia* 16: 49–59.

57 Zhong, G., Yao, J., Huang, X. et al. (2020). Injectable ECM hydrogel for delivery of BMSCs enabled full-thickness meniscus repair in an orthotopic rat model. *Bioactive Materials* 5: 871–879.

58 Kim, S.H., An, Y.H., Kim, H.D. et al. (2018). Enzyme-mediated tissue adhesive hydrogels for meniscus repair. *International Journal of Biological Macromolecules* 110: 479–487.

59 Resmi, R., Parvathy, J., John, A., and Joseph, R. (2020). Injectable self-crosslinking hydrogels for meniscal repair: a study with oxidized alginate and gelatin. *Carbohydrate Polymers* 234: 115902.

60 An, Y.H., Kim, J.A., Yim, H.G. et al. (2021). Meniscus regeneration with injectable Pluronic/PMMA-reinforced fibrin hydrogels in a rabbit segmental meniscectomy model. *Journal of Tissue Engineering* 12: 204173142110501.

61 Zhang, K.X., Wu, J., Zhang, W.J. et al. (2018). In situ formation of hydrophobic clusters to enhance mechanical performance of biodegradable poly(l-glutamic acid)/poly(ε-caprolactone) hydrogel towards meniscus tissue engineering. *Journal of Materials Chemistry B* 6: 7822–7833.
62 Gupta, S., Sharma, A., Kumar, J.V. et al. (2020). Meniscal tissue engineering via 3D printed PLA monolith with carbohydrate based self-healing interpenetrating network hydrogel. *International Journal of Biological Macromolecules* 162: 1358–1371.
63 Olubamiji, A.D., Izadifar, Z., Si, J.L. et al. (2016). Modulating mechanical behaviour of 3D-printed cartilage-mimetic PCL scaffolds: influence of molecular weight and pore geometry. *Biofabrication* 8: 025020.
64 Ding, G.C., Li, X.P., Sun, M.Y. et al. (2022). Meniscal transplantation and regeneration using functionalized polyurethane bionic scaffold and digital light processing 3D printing. *Chemical Engineering Journal* 431: 133861.
65 Asgarpour, R., Masaeli, E., and Kermani, S. (2021). Development of meniscus-inspired 3D-printed PCL scaffolds engineered with chitosan/extracellular matrix hydrogel. *Polymers for Advanced Technologies* 32: 4721–4732.
66 Chen, M.X., Feng, Z.X., Guo, W.M. et al. (2019). PCL-MECM-based hydrogel hybrid scaffolds and meniscal fibrochondrocytes promote whole meniscus regeneration in a rabbit meniscectomy model. *ACS Applied Materials & Interfaces* 11: 41626–41639.
67 Zhou, J., Tian, Z., Tian, Q.Y. et al. (2021). 3D bioprinting of a biomimetic meniscal scaffold for application in tissue engineering. *Bioactive Materials* 6: 1711–1726.
68 Li, H., Liao, Z.Y., Yang, Z. et al. (2021). 3D printed poly(ε-caprolactone)/meniscus extracellular matrix composite scaffold functionalized with kartogenin-releasing PLGA microspheres for meniscus tissue engineering. *Frontiers in Bioengineering and Biotechnology* 9: 662381.
69 Liu, H., Zhao, T.Y., Yang, Z. et al. (2022). Biofabrication of cell-free dual drug-releasing biomimetic scaffolds for meniscal regeneration. *Biofabrication* 14: 015001.
70 Bahcecioglua, G., Bilgen, B., Hasirci, N., and Hasirci, V. (2019). Anatomical meniscus construct with zone specific biochemical composition and structural organization. *Biomaterials* 218: 119361.
71 Bahcecioglu, G., Hasirci, N., Bilgen, B., and Hasirci, V. (2019). A 3D printed PCL/hydrogel construct with zone-specific biochemical composition mimicking that of the meniscus. *Biofabrication* 11: 025002.
72 Zhao, X.H.C. and X.Y., Yuk H., Lin S.T., Liu X.Y., Parada G. (2021). Soft materials by design: unconventional polymer networks give extreme properties. *Chemical Reviews* 121: 4309–4372.
73 Kamoun, E.A., Loutfy, S.A., Hussein, Y., and Kenawy, E.S. (2021). Recent advances in PVA-polysaccharide based hydrogels and electrospun nanofibers in biomedical applications: a review. *International Journal of Biological Macromolecules* 187: 755–768.

74 Kumar, A. and Han, S.S. (2016). PVA-based hydrogels for tissue engineering: a review. *International Journal of Polymeric Materials and Polymeric Biomaterials* 66: 159–182.

75 Kobayashi, M., Toguchida, J., and Oka, M. (2003). Development of an artificial meniscus using polyvinyl alcohol-hydrogel for early return to, and continuance of, athletic life in sportspersons with severe meniscus injury. I: mechanical evaluation. *The Knee* 10: 47–51.

76 Kobayashi, M., Toguchida, J., and Oka, M. (2003). Preliminary study of polyvinyl alcohol-hydrogel (PVA-H) artificial meniscus. *Biomaterials* 24: 639–647.

77 Kobayashi, M. (2004). A study of polyvinyl alcohol-hydrogel (PVA-H) artificial meniscus in vivo. *Bio-medical Materials and Engineering* 14: 505–515.

78 Kobayashi, M., Chang, Y.S., and Oka, M. (2005). A two year in vivo study of polyvinyl alcohol-hydrogel (PVA-H) artificial meniscus. *Biomaterials* 26: 3243–3248.

79 Kelly, B.T., Robertson, W., Potter, H.G. et al. (2007). Hydrogel meniscal replacement in the sheep knee: preliminary evaluation of chondroprotective effects. *The American Journal of Sports Medicine* 35: 43–52.

80 Holloway, J.L., Lowman, A.M., and Palmese, G.R. (2010). Mechanical evaluation of poly(vinyl alcohol)-based fibrous composites as biomaterials for meniscal tissue replacement. *Acta Biomaterialia* 6: 4716–4724.

81 Holloway, J.L., Spiller, K.L., Lowman, A.M., and Palmese, G.R. (2011). Analysis of the in vitro swelling behavior of poly(vinyl alcohol) hydrogels in osmotic pressure solution for soft tissue replacement. *Acta Biomaterialia* 7: 2477–2482.

82 Holloway, J.L., Lowman, A.M., VanLandingham, M.R., and Palmese, G.R. (2014). Interfacial optimization of fiber-reinforced hydrogel composites for soft fibrous tissue applications. *Acta Biomaterialia* 10: 3581–3589.

83 Awadallah, F.A. (2014). Five years in vitro study of (poly vinyl alcohol/poly vinyl pyrrolidone/poly acrylic acid) hydrogel to mimic the knee joint meniscus. *Polymers for Advanced Technologies* 25: 581–587.

84 Xu, Z.Y. and Liu, W.G. (2018). Poly(N-acryloyl glycinamide): a fascinating polymer that exhibits a range of properties from UCST to high-strength hydrogels. *Chemical Communications* 54: 10540–10553.

85 Seuring, J. and Agarwal, S. (2010). Non-ionic homo- and copolymers with H-donor and H-acceptor units with an UCST in water. *Macromolecular Chemistry and Physics* 211: 2109–2117.

86 Seuring, J., Bayer, F.M., Huber, K., and Agarwal, S. (2012). Upper critical solution temperature of poly(N-acryloyl glycinamide) in water: a concealed property. *Macromolecules* 45: 374–384.

87 Liu, F.Y., Seuring, J., and Agarwal, S. (2013). Atom transfer radical polymerization as a tool for making poly(N-acryloylglycinamide) with molar mass independent UCST-type transitions in water and electrolytes. *Polymer Chemistry* 4: 3123–3131.

88 Liu, F.Y., Seuring, J., and Agarwal, S. (2012). Controlled radical polymerization of N-acryloylglycinamide and UCST-type phase transition of the polymers. *Journal of Polymer Science Part A: Polymer Chemistry* 50: 4920–4928.

89 Seuring, J. and Agarwal, S. (2012). First example of a universal and cost-effective approach: polymers with tunable upper critical solution temperature in water and electrolyte solution. *Macromolecules* 45: 3910–3918.

90 Seuring, J. and Agarwal, S. (2013). Polymers with upper critical solution temperature in aqueous solution: unexpected properties from known building blocks. *ACS Marco Letters* 2: 597–600.

91 Boustta, M., Colombo, P.E., Lenglet, S. et al. (2014). Versatile UCST-based thermoresponsive hydrogels for loco-regional sustained drug delivery. *Journal of Controlled Release* 174: 1–6.

92 Dai, X.Y., Zhang, Y.Y., Gao, L.N. et al. (2015). A mechanically strong, highly stable, thermoplastic, and self-healable supramolecular polymer hydrogel. *Advanced Materials* 27: 3566–3571.

93 Gao, F., Xu, Z.Y., Liang, Q.F. et al. (2018). Direct 3D printing of high strength biohybrid gradient hydrogel scaffolds for efficient repair of osteochondral defect. *Advanced Functional Materials* 28: 1706644.

94 Xu, Z.Y., Fan, C.C., Zhang, Q. et al. (2021). A self-thickening and self-strengthening strategy for 3D printing high-strength and antiswelling supramolecular polymer hydrogels as meniscus substitutes. *Advanced Functional Materials* 31: 2100462.

95 Zhang, Q., Xu, Z.Y., Zhang, X.P. et al. (2021). 3D printed high-strength supramolecular polymer hydrogel-cushioned radially and circumferentially oriented meniscus substitute. *Advanced Functional Materials* 32: 2200360.

96 Fan, C.C., Liu, B., Xu, Z.Y. et al. (2020). Polymerization of N-acryloylsemicarbazide: a facile and versatile strategy to tailor-make highly stiff and tough hydrogels. *Materials Horizons* 7: 1160–1170.

97 Yang, R., Fan, C.C., Dou, Y.M. et al. (2021). 3D printing stiff antibacterial hydrogels for meniscus replacement. *Applied Materials Today* 24: 101089.

98 Fan, C.C., Xu, Z.Y., Wu, T.L. et al. (2021). 3D printing of lubricative stiff supramolecular polymer hydrogels for meniscus replacement. *Biomaterials Science* 9: 5116–5126.

99 Zhang, Z.M., Liu, R.C., Zepeda, H. et al. (2019). 3D printing super strong hydrogel for artificial meniscus. *ACS Applied Polymer Materials* 1: 2023–2032.

100 Rauner, N., Meuris, M., Zoric, M., and Tiller, J.C. (2017). Enzymatic mineralization generates ultrastiff and tough hydrogels with tunable mechanics. *Nature* 543: 407–410.

101 Wang, Y.J., Zhang, X.N., Song, Y. et al. (2019). Ultrastiff and tough supramolecular hydrogels with a dense and robust hydrogen bond network. *Chemistry of Materials* 31: 1430–1440.

102 Mredha, T.I., Guo, Y.Z., Nonoyama, T. et al. (2018). A facile method to fabricate anisotropic hydrogels with perfectly aligned hierarchical fibrous structures. *Advanced Materials* 30: 1704937.

103 Lei, H., Dong, L., Li, Y. et al. (2020). Stretchable hydrogels with low hysteresis and anti-fatigue fracture based on polyprotein cross-linkers. *Nature Communications* 11: 4032.

104 Liang, X.Y., Chen, G.D., Lin, S.T. et al. (2021). Anisotropically fatigue-resistant hydrogels. *Advanced Materials* 33: 2102011.

105 Wang, J.H.C. (2006). Mechanobiology of tendon. *Journal of Biomechanics* 39: 1563–1582.

106 Puetzer, J.L., Ma, T.C., Sallent, I. et al. (2021). Driving hierarchical collagen fiber formation for functional tendon, ligament, and meniscus replacement. *Biomaterials* 269: 120527.

107 Jana, S., Levengood, S.K.L., and Zhang, M.Q. (2016). Anisotropic materials for skeletal-muscle-tissue engineering. *Advanced Materials* 28: 10588–10612.

108 Abbadessa, A., Campo, J.C., and Alonso, M.J. (2021). Engineering anisotropic meniscus: zonal functionality and spatiotemporal drug delivery. *Tissue Engineering. Part B-Reviews* 27: 133–154.

109 Feng, Y.B., Liu, H., Zhu, W.H. et al. (2021). Muscle-inspired MXene conductive hydrogels with anisotropy and low-temperature tolerance for wearable flexible sensors and arrays. *Advanced Functional Materials* 31: 2105264.

110 Prince, E., Alizadehgiashi, M., Campbell, M. et al. (2018). Patterning of structurally anisotropic composite hydrogel sheets. *Biomacromolecules* 19: 1276–1284.

6

Strong, Tough, and Fast-Recovery Hydrogels

Bin Xue[1,2] *and Yi Cao*[1,2,3]

[1] *Collaborative Innovation Center of Advanced Microstructures, National Laboratory of Solid State Microstructure, Department of Physics, Nanjing University, Nanjing 210093, China*
[2] *Jinan Microecological Biomedicine Shandong Laboratory, Department of Regenerative Medicine and Tissue Engineering, Jinan 250000, China*
[3] *Chemistry and Biomedicine innovation Center, Nanjing University, Nanjing 210093, China*

6.1 Current Progress on Strong and Tough Hydrogels

Numerous load-bearing tissues in nature exhibit remarkable mechanical stability and reliability, swiftly restoring their baseline properties and enduring repetitive mechanical cycles over extended periods. A prime example is our articular cartilage, which can withstand compression of over 40% strain and forces six times the weight of the human body, at a frequency of 0.5 Hz, without displaying noticeable mechanical fatigue [1–4]. These exceptional qualities observed in natural load-bearing tissues have inspired extensive research in the field of hydrogel engineering. The aim is to develop hydrogels that can withstand intense mechanical loads for various applications, including musculoskeletal repair, cartilage regeneration, and soft robot construction [5–7].

A crucial approach in enhancing the toughness of hydrogels is the implementation of sacrificial bonds or networks to dissipate energy [8, 9]. One notable contribution in this area was made by Gong et al., who introduced the concept of a second network formed through covalent cross-linking [10–12]. This innovation led to the development of the renowned double-network hydrogel (DN hydrogel) as depicted in Figure 6.1. DN hydrogels typically consist of two interconnected networks: one with short chains (1st network) and the other with longer chains (2nd network). The primary network with tight cross-linking enables energy dissipation, while the secondary network with looser cross-linking provides extension. The combination of these two networks significantly enhances the toughness of hydrogels by facilitating the breakage of covalent bonds in the rigid primary network, thereby dissipating energy effectively.

On the contrary, despite the considerable enhancements in strength and toughness achieved by the double-network structure in hydrogels, the rupture of the short-chain network leads to permanent damage [13]. Consequently, DN hydrogels

Dynamics and Transport in Macromolecular Networks: Theory, Modeling, and Experiments, First Edition. Edited by Li-Tang Yan.

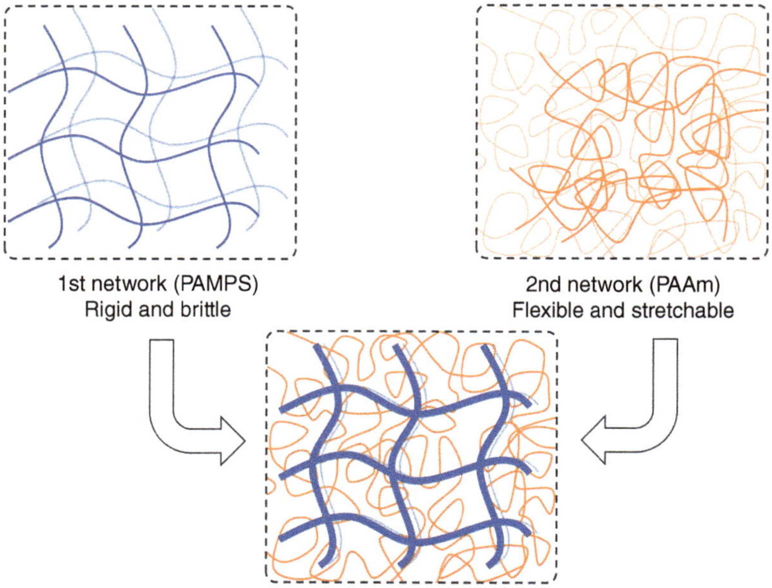

Figure 6.1 Schematic of double-network hydrogels presented by Gong et al., adapted from (Jian Ping Gong et al., 2014) [12]. A double-network (DN) hydrogel was prepared from the combination of two different polymer networks. The covalently cross-linked tight network was used as the sacrificial network.

are unable to recover from these damages following the initial loading, resulting in irreversible softening under cyclic mechanical loads. To address this issue, dynamic cross-linked networks have been employed as sacrificial networks, aiming to enhance the reversibility of DN hydrogels. These dynamic networks utilize various mechanisms such as hydrophobic interactions [14–16], ionic pairing [17–19], hydrogen bonding [20, 21], coordination interactions [22, 23], host–guest interactions [24], and microcrystals [25, 26]. Among these dynamic interactions, metal-coordinated interactions raised special attentions due to the relatively high mechanical strength as well as the fast recovery after the rupture of coordination (Figure 6.2) [27].

An exemplary study by Suo et al. demonstrated the development of highly stretchable and tough DN hydrogels that can be stretched beyond 20 times their original length, exhibiting a fracture energy of approximately $9\,\mathrm{kJ\,m^{-2}}$ [17]. In this case, the sacrificial network utilized was an ionically cross-linked alginate network, as illustrated in Figure 6.3. By integrating this dynamic cross-linked network, the reversibility of the DN hydrogels was significantly improved, addressing the issue of irreversibility associated with the rupture of the short-chain network.

Despite efforts to enhance the toughness of hydrogels through molecular engineering of dynamic cross-linking, there is often a trade-off between strength and toughness since these properties are mutually exclusive [28–30]. According to the Lake–Thomas model, which applies to single covalently bonded networks,

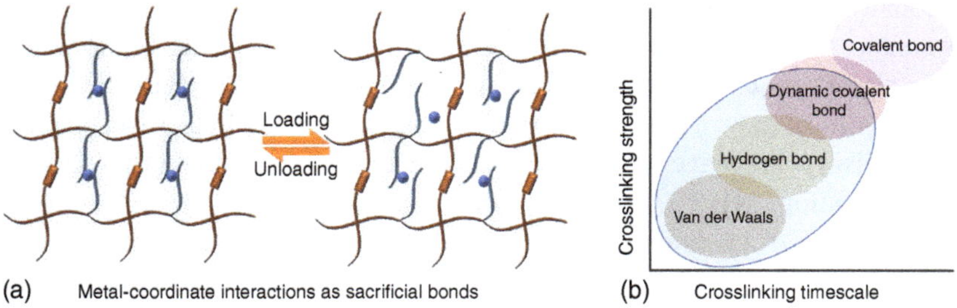

Figure 6.2 Energy dissipation and fast recovery enabled by metal-coordinated bonds, adapted from (Eesha Khare et al., 2021) [27]. (a) Sacrificial networks cross-linked by metal coordination bonds can rupture after loading to dissipate energy and recovery after unloading, thereby increasing fracture energy and preventing catastrophic material failure. (b) Metal-coordination bonds span a range of cross-linking strengths and timescales.

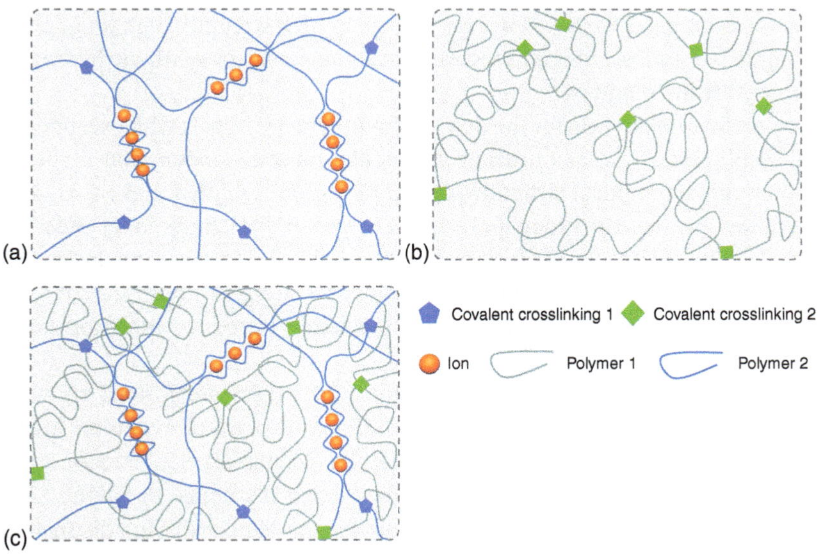

Figure 6.3 Schematics of highly stretchable and tough hydrogels with the dynamically cross-linked network acting as the sacrificial network, adapted from (Zhigang Suo et al., 2012) [17]. (a) Alginate hydrogel formed by ionically cross-linking through Ca^{2+} (red circles). (b) Polyacrylamide hydrogel formed by covalently cross-linking through *N*,*N*-methylenebisacrylamide (MBAA; green squares). (c) Alginate–polyacrylamide hybrid hydrogel comprised of alginate and polyacrylamide networks. The two types of polymer network are intertwined and joined by covalent cross-links between amine groups on polyacrylamide chains and carboxyl groups on alginate chains.

the toughness, Γ, displays a direct relationship with the cross-linking density, C. Specifically, Γ is inversely proportional to $C^{-1/2}$, indicating that as C increases, Γ decreases. On the other hand, the Flory's polymer model suggests that stiffness, E, is proportional to C. Obviously, it is impossible to concurrently increase the toughness and stiffness of a polymer network by solely varying its cross-linking density.

Additionally, although the weak interactions at the molecular level are dynamic and reversible, the hydrogels exhibit slow and limited recovery after deformation due to the significant entropy cost associated with reforming these dynamic bonds within the hydrogel networks [17, 31–33]. Consequently, the hydrogels lack an efficient mechanism to prevent crack propagation, rendering them vulnerable to fatigue under cyclic loading [34].

Recently, some advancements have partially addressed the trade-offs between mechanical properties through strategies such as strain-induced self-reinforcement [35], chain entanglement [36], or mechanochemically activable hidden length [37]. For instance, Liu et al. employed a damageless reinforcement approach by inducing strain-induced crystallization, resulting in hydrogels with high toughness and rapid recovery [35]. Kim et al. utilized dense entanglements, which outnumber cross-links in hydrogels, to achieve high toughness, strength, and fatigue resistance [36]. Furthermore, the inclusion of hidden length in fibril structures or folded proteins offers additional extensibility upon release, thereby enhancing the toughness of hydrogels [37–40]. However, achieving high strength, high toughness, robust fatigue resistance, and rapid recovery simultaneously in synthetic hydrogels remains a significant challenge.

Nature provides inspiration for the design of synthetic hydrogels with exceptional mechanical properties, as load-bearing tissues like muscle, tendon, and cartilage possess intricate hierarchical structures spanning various length scales [4, 41, 42]. Recent studies by the He group [43] and Liu group [44] have independently demonstrated the effectiveness of introducing hierarchical structures into hydrogels through freeze-casting and additional processing, resulting in improved strength and toughness. However, these approaches inevitably introduce inhomogeneity and anisotropy to the hydrogels.

In natural load-bearing materials, hierarchical structures are formed through complex self-assembly processes [45]. Over millions of years of evolution, these materials have developed nanoscale structures and molecular-level interactions that integrate high strength, toughness, rapid recovery, and resistance to fatigue [23]. Biological tissue networks, such as tendon, muscle, and bone, exhibit a distinctive combination of self-assembled fibrous nanostructures and abundant nonfibrillar organic matrix [46]. This combination efficiently mediates load distribution and energy dissipation, serving as a hallmark feature of these biological tissues.

6.2 Polymer-Supramolecular Double-Network Hydrogels

Drawing inspiration from the self-assembled fibrous nanostructures found in natural load-bearing tissues, we embarked on introducing a supramolecular network composed of self-assembled peptides as the second network within double network hydrogels. Our aim was to enhance the recovery rate of hydrogels. Emulating the unique polymer-supramolecular polymer double network architecture observed in natural articular cartilage [47], we devised the polymer-supramolecular double network hydrogel (PS-DN hydrogel, Figure 6.4) [48]. This novel hydrogel

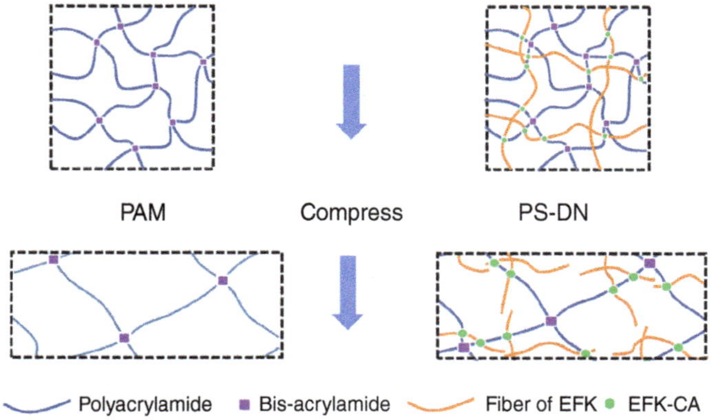

Figure 6.4 Schematic illustration of the PS-DN hydrogel networks made of PAM and EFK, adapted from (Wenxu Sun et al., 2016) [48]. Bis-acrylamide was used as the cross-linker for the PAM network, and EFK-CA was used as the anchoring points to connect the two networks. The schematic for the PAM gel was also shown for comparison.

consisted of a polymer network composed of polyacrylamide and a supramolecular network formed by self-assembling short ionic complementary octopeptides (N-Lys-Phe-Glu-Phe-Lys-Phe-Glu-Phe-C, referred to as EFK). The EFK peptide is renowned for its ability to autonomously self-assemble, forming a fibrous network within the hydrogel [49, 50]. Notably, the self-assembled EFK network exhibited rapid recovery of its mechanical properties following high-strain destruction (as depicted in right of Figure 6.4). Consequently, EFK proved to be an ideal candidate for the sacrificial supramolecular network in our PS-DN hydrogel system.

The disruption of the supramolecular network within the PS-DN hydrogel enables efficient energy dissipation, leading to enhanced mechanical strength. Additionally, the reassembly of the supramolecular network facilitates rapid recovery of the hydrogel's mechanical properties. As a result, the PS-DN gel exhibits notable mechanical characteristics, including high strength ranging from approximately 0.32–0.57 MPa, work of rupture in the range of 300–2670 $J\,m^{-3}$, compressibility of around 66% to 90%, and swift recovery within seconds. Remarkably, the PS-DN hydrogel also displays energy dissipation, strain stiffening, and stress relaxation behaviors akin to articular cartilage. Furthermore, the mechanical properties of the PS-DN hydrogel can be finely tuned by adjusting its chemical components. This flexibility in customization offers promising prospects for applications in cartilage regeneration and the repair of other load-bearing tissues. The unique combination of mechanical properties exhibited by the PS-DN hydrogel positions it as a compelling candidate for such therapeutic endeavors.

6.3 Hybrid Networks with Peptide-Metal Complexes

In addition to the nanoscale structures, the cooperative nature of weak interactions at the molecular level has played a crucial role in integrating high strength, toughness, rapid recovery, and antifatigue properties in natural load-bearing tissues [41, 51–55]. Motivated by this concept, our aim was to extend these principles to the design of synthetic hydrogels, enabling the integration of high mechanical strength, toughness, and fast recovery using simple building blocks. To realize this, we introduced hybrid network (HN) hydrogels, employing a specially designed peptide-metal complex as the physical cross-linker [23]. Metal coordination complexes are known for their rapid binding and unbinding kinetics, but they often fail to produce hydrogels with high mechanical strength due to their excessive dynamism [56, 57]. However, nature has provided a solution to this challenge in load-bearing structural proteins, where specifically evolved metal coordination sites contain multiple ligands within each binding site [53, 54]. These evolved binding sites exhibit considerably improved stability when binding metal ions. Furthermore, load-bearing proteins frequently feature multiple binding sites arranged in a tandem configuration, promoting cooperativity among different binding sites to achieve strong binding [41, 55]. By incorporating these concepts, we demonstrated the engineering of HN hydrogels with remarkable mechanical properties (Figure 6.5). The use of the peptide-metal complex allowed for fast binding and unbinding kinetics, while the presence of multiple ligands within each binding site and the cooperative arrangement of binding sites enhanced the stability and strength of the hydrogel. This approach harnessed the advantages of weak interactions and their cooperativity, enabling the synthesis of synthetic hydrogels with integrated high mechanical strength, toughness, and rapid recovery.

Through the strategic design of ligand cooperativity to create efficient metal-binding sites within a peptide sequence, we successfully developed hydrogels exhibiting exceptional mechanical stability, toughness, and rapid recovery [23]. Our approach involved the design of a histidine-rich decapeptide featuring two tandem zinc-binding motifs (Figure 6.5a). Leveraging allosteric structural changes that

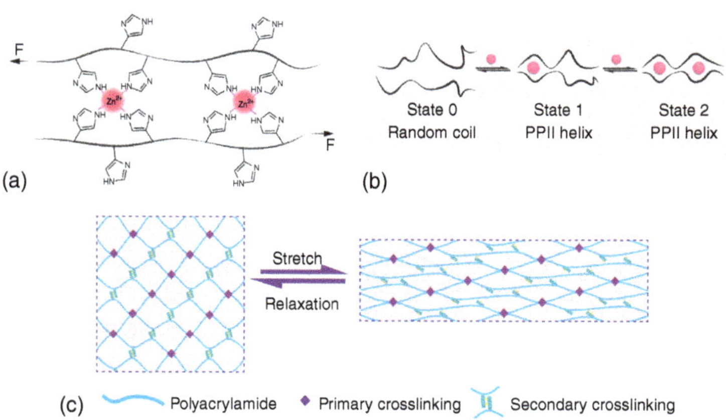

Figure 6.5 Molecular engineering of metal coordination interactions for strong, tough, and fast-recovery hydrogels, adapted from (Wenxu Sun et al., 2020) [23]. (a) Metal ion coordination complexes formed by two metal chelation sites in tandem (PH_6). (b) Schematic illustration of the cooperative Zn^{2+}-binding mechanism of PH_6 peptide. (c) Schematic illustration of the network structure of HN-PH_6 hydrogels in stretch-relaxation cycles.

induce cooperative binding, this decapeptide exhibited superior thermodynamic stability, enhanced binding strength, and accelerated binding kinetics compared to single-binding motifs or isolated ligands (Figure 6.5b).

The incorporation of the engineered peptide-zinc complex into HN hydrogels resulted in impressive mechanical properties (Figure 6.5c). These hydrogels demonstrated a break stress of approximately 3.0 MPa, toughness of approximately 4.0 MJ m^{-3}, and rapid recovery within seconds. We envision that these hydrogels can serve effectively as scaffolds for load-bearing tissue engineering and as fundamental building blocks for soft robotics applications. Furthermore, this research offers a general pathway for fine-tuning the mechanical and dynamic properties of hydrogels at the molecular level, opening up new avenues for tailoring their performance in diverse fields.

6.4 Hydrogels Cross-Linked with Hierarchically Assembled Peptide Structures

Taking inspiration from the integrated nanoscale structures and molecular-level interactions observed in natural load-bearing tissues, we embarked on engineering hydrogels with exceptional strength, toughness, fatigue resistance, and rapid mechanical recovery through the synergistic combination of self-assembly and metal coordination [58]. In conventional double-network hydrogels, the primary tightly cross-linked network dissipates energy, while the secondary loosely cross-linked network provides hidden length. Although these hydrogels theoretically regain their baseline mechanical properties upon load release, the restoration of their original shape relies solely on the recoil of the secondary network, while the reformation of the tight network is entropically unfavorable.

To overcome these challenges, we present a novel hydrogel network comprised of picot fibers that integrate both hidden length and energy dissipation capabilities (Figure 6.6a). These picot fibers consist of long, flexible polymers adorned with self-assembling peptides. The self-assembly of these peptides drives the zipping of peptide links into picots, which contribute to the hidden length. The picots possess varying lengths, and the peptides are randomly interconnected. By rupturing strong interpeptide interactions, such as hydrogen bonding within the picot fibers, energy can be efficiently dissipated, releasing the hidden length without compromising the network connectivity of the hydrogels (Figure 6.6b). Consequently, this design reconciles the traditional trade-off between strength and toughness in polymer networks. Furthermore, the reformation of individual picot fibers occurs locally and independently, ensuring the rapid recovery of the hydrogel's structure. Through cross-linking these fibers, the resulting hydrogels exhibit an extraordinary combination of high mechanical strength (fracture stress ~4.1 MPa), ultrahigh toughness (fracture energy ~25.3 kJ m^{-2}), excellent fatigue resistance (fatigue threshold ~424 J m^{-2}), and nearly 100% mechanical recovery within one second. This innovative design approach offers a versatile pathway for integrating typically

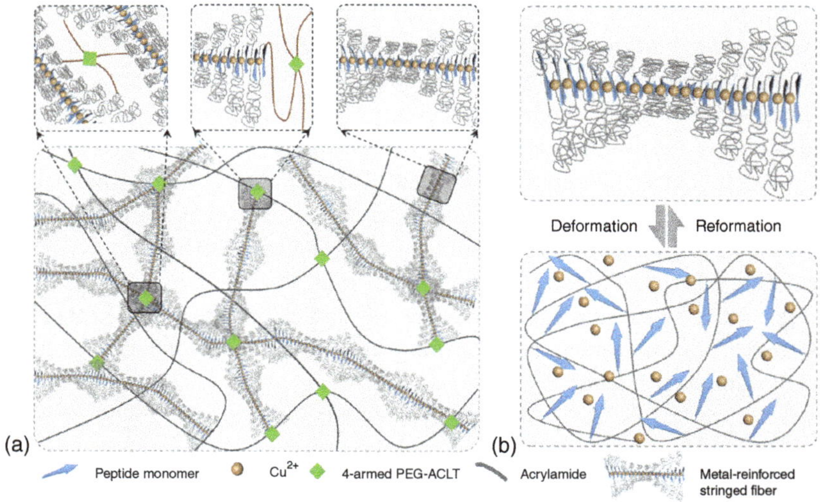

Figure 6.6 Hydrogel design based on metal ion-clad picot fibers, adapted from (Bin Xue et al., 2023) [58]. (a) Schematic of the hydrogel constructed by picot fibers made of self-assembling peptide strands with zipped flexible hidden lengths. Upon deformation of the hydrogel, the picot fibers are extended to dissipate energy efficiently and release the polyacrylamide hidden length without reducing the network connectivity of the hydrogels. (b) Schematic of rupture and recovery of the metal ion-clad picot fibers in hydrogels during loading and unloading.

incompatible mechanical properties within synthetic hydrogels, opening up new possibilities for advanced material applications.

6.5 Outlook

High strength, toughness, and rapid recovery are traits inherent in many natural materials, such as spider silk [59–61] or nacre [62, 63]. These materials inspire the quest for synthetic counterparts, particularly hydrogels, which have broad potential applications ranging from biomedical devices to soft robotics [64–70]. Combining these properties in hydrogels, however, remains a significant challenge due to their unique network structures and distinct mechanical responses.

Experimental, theoretical, and computational approaches are indispensable tools in this pursuit. Experimentally, sophisticated fabrication methods such as double-network formation or dynamic cross-linking can be used to increase strength and toughness [10, 71, 72]. Furthermore, high-throughput experimentation can help in screening various conditions and compositions [73–75]. Simulations and modelling [76–78], on the other hand, provide a lens to visualize and predict the microstructural evolution and mechanical behavior of these hydrogels. Finite element analysis [79–81] and molecular dynamics simulations [82–84] can be particularly useful in understanding the interplay between various network features and mechanical properties. Artificial intelligence (AI), especially machine learning, brings another dimension [85–88]. By training models on large experimental and simulation datasets, AI can predict optimal hydrogel compositions and fabrication conditions that yield desired properties, thereby significantly speeding up the development process. Engineering principles also come into play. Concepts like fracture mechanics [89–91] can guide the design of tough hydrogels, while principles from materials science, such as phase transformation [92–94] or strain hardening [95–98], can be used to enhance strength and enable rapid recovery.

The potential applications of these advanced hydrogels are vast. In medicine, they can be used for wound dressings [99–102], drug delivery systems [103–107], or tissue engineering scaffolds [5, 108–111]. In soft robotics, they can serve as actuators or sensors, thanks to their tunable mechanical properties. Moreover, their potential for self-healing can lead to durable, reusable materials in various industries.

In conclusion, by harnessing nature's inspiration, combining experiments, simulations, and AI, and applying engineering principles, we can engineer hydrogels that combine high strength, toughness, and fast recovery. The resulting materials promise to open new frontiers in science and technology.

References

1 Huey, D.J., Hu, J.C., and Athanasiou, K.A. (2012). Unlike bone, cartilage regeneration remains elusive. *Science* 338: 917–921.

2 Thambyah, A., Nather, A., and Goh, J. (2006). Mechanical properties of articular cartilage covered by the meniscus. *Osteoarthritis and Cartilage* 14: 580–588.

3 Robinson, D.L., Kersh, M.E., Walsh, N.C. et al. (2016). Mechanical properties of normal and osteoarthritic human articular cartilage. *Journal of the Mechanical Behavior of Biomedical Materials* 61: 96–109.

4 Burgin, L.V., Edelsten, L., and Aspden, R.M. (2014). The mechanical and material properties of elderly human articular cartilage subject to impact and slow loading. *Medical Engineering & Physics* 36: 226–232.

5 Drury, J.L. and Mooney, D.J. (2003). Hydrogels for tissue engineering: scaffold design variables and applications. *Biomaterials* 24: 4337–4351.

6 Bodugoz-Senturk, H., Macias, C.E., Kung, J.H., and Muratoglu, O.K. (2009). Poly(vinyl alcohol)–acrylamide hydrogels as load-bearing cartilage substitute. *Biomaterials* 30: 589–596.

7 Moutos, F.T., Freed, L.E., and Guilak, F. (2007). A biomimetic three-dimensional woven composite scaffold for functional tissue engineering of cartilage. *Nature Materials* 6: 162–167.

8 Creton, C. (2017). 50th anniversary perspective: networks and gels: soft but dynamic and tough. *Macromolecules* 50: 8297–8316.

9 Mayumi, K., Guo, J., Narita, T. et al. (2016). Fracture of dual crosslink gels with permanent and transient crosslinks. *Extreme Mechanics Letters* 6: 52–59.

10 Gong, J.P., Katsuyama, Y., Kurokawa, T., and Osada, Y. (2003). Double-network hydrogels with extremely high mechanical strength. *Advanced Materials* 15: 1155–1158.

11 Gong, J.P. (2010). Why are double network hydrogels so tough? *Soft Matter* 6: 2583–2590.

12 Gong, J.P. (2014). Materials both tough and soft. *Science* 344: 161–162.

13 Webber, R.E., Creton, C., Brown, H.R., and Gong, J.P. (2007). Large strain hysteresis and mullins effect of tough double-network hydrogels. *Macromolecules* 40: 2919–2927.

14 Zhang, H.J., Sun, T.L., Zhang, A.K. et al. (2016). Tough physical double-network hydrogels based on amphiphilic triblock copolymers. *Advanced Materials* 28: 4884–4890.

15 Gonzalez, M.A., Simon, J.R., Ghoorchian, A. et al. (2017). Strong, tough, stretchable, and self-adhesive hydrogels from intrinsically unstructured proteins. *Advanced Materials* 29: 1604743.

16 Wang, T. et al. (2021). Regulating mechanical properties of polymer-supramolecular double-network hydrogel by supramolecular self-assembling structures. *Chinese Journal of Chemistry* 39: 2711–2717.

17 Sun, J.-Y., Zhao, X., Illeperuma, W.R. et al. (2012). Highly stretchable and tough hydrogels. *Nature* 489: 133.

18 Yang, Y., Wang, X., Yang, F. et al. (2018). Highly elastic and ultratough hybrid ionic–covalent hydrogels with tunable structures and mechanics. *Advanced Materials* 30: 1707071.

19 Long, T., Li, Y., Fang, X., and Sun, J. (2018). Salt-mediated polyampholyte hydrogels with high mechanical strength, excellent self-healing property, and satisfactory electrical conductivity. *Advanced Functional Materials* 28: 1804416.

20 Chen, Q., Zhu, L., Zhao, C. et al. (2013). A robust, one-pot synthesis of highly mechanical and recoverable double network hydrogels using thermoreversible sol-gel polysaccharide. *Advanced Materials* 25: 4171–4176.

21 Hu, X., Zhou, J., Daniel, W.F. et al. (2017). Dynamics of dual networks: strain rate and temperature effects in hydrogels with reversible H-bonds. *Macromolecules* 50: 652–659.

22 Lin, P., Ma, S., Wang, X., and Zhou, F. (2015). Molecularly engineered dual-crosslinked hydrogel with ultrahigh mechanical strength, toughness, and good self-recovery. *Advanced Materials* 27: 2054–2059.

23 Sun, W., Xue, B., Fan, Q. et al. (2020). Molecular engineering of metal coordination interactions for strong, tough, and fast-recovery hydrogels. *Science Advances* 6: eaaz9531.

24 Liu, J., Tan, C.S.Y., Yu, Z. et al. (2017). Tough supramolecular polymer networks with extreme stretchability and fast room-temperature self-healing. *Advanced Materials* 29: 1605325.

25 Li, J., Suo, Z., and Vlassak, J.J. (2014). Stiff, strong, and tough hydrogels with good chemical stability. *Journal of Materials Chemistry B* 2: 6708–6713.

26 Wang, W., Zhang, Y., and Liu, W. (2017). Bioinspired fabrication of high strength hydrogels from non-covalent interactions. *Progress in Polymer Science* 71: 1–25.

27 Khare, E., Holten-Andersen, N., and Buehler, M.J. (2021). Transition-metal coordinate bonds for bioinspired macromolecules with tunable mechanical properties. *Nature Reviews Materials* 6: 421–436.

28 Ritchie, R.O. (2011). The conflicts between strength and toughness. *Nature Materials* 10: 817–822.

29 Zhu, H., Zhu, S., Jia, Z. et al. (2015). Anomalous scaling law of strength and toughness of cellulose nanopaper. *Proceedings of the National Academy of Sciences of the United States of America* 112: 8971–8976.

30 Liao, X., Dulle, M., de Souza e Silva, J.M. et al. (2019). High strength in combination with high toughness in robust and sustainable polymeric materials. *Science* 366: 1376–1379.

31 Yu, A.C., Lian, H., Kong, X. et al. (2021). Physical networks from entropy-driven non-covalent interactions. *Nature Communications* 12: 746.

32 Zhang, D., Yang, F., He, J. et al. (2020). Multiple physical bonds to realize highly tough and self-adhesive double-network hydrogels. *ACS Applied Polymer Materials* 2: 1031–1042.

33 Sun, T.L., Kurokawa, T., Kuroda, S. et al. (2013). Physical hydrogels composed of polyampholytes demonstrate high toughness and viscoelasticity. *Nature Materials* 12: 932–937.

34 Li, X. and Gong, J.P. (2022). Role of dynamic bonds on fatigue threshold of tough hydrogels. *Proceedings of the National Academy of Sciences of the United States of America* 119: e2200678119.

35 Liu, C., Morimoto, N., Jiang, L. et al. (2021). Tough hydrogels with rapid self-reinforcement. *Science* 372: 1078–1081.

36 Kim, J., Zhang, G., Shi, M., and Suo, Z. (2021). Fracture, fatigue, and friction of polymers in which entanglements greatly outnumber cross-links. *Science* 374: 212–216.

37 Wang, Z., Zheng, X., Ouchi, T. et al. (2021). Toughening hydrogels through force-triggered chemical reactions that lengthen polymer strands. *Science* 374: 193–196.

38 Smith, B.L., Schäffer, T.E., Viani, M. et al. (1999). Molecular mechanistic origin of the toughness of natural adhesives, fibres and composites. *Nature* 399: 761–763.

39 Fantner, G.E., Oroudjev, E., Schitter, G. et al. (2006). Sacrificial bonds and hidden length: unraveling molecular mesostructures in tough materials. *Biophysical Journal* 90: 1411–1418.

40 Fantner, G.E., Hassenkam, T., Kindt, J.H. et al. (2005). Sacrificial bonds and hidden length dissipate energy as mineralized fibrils separate during bone fracture. *Nature Materials* 4: 612–616.

41 Zhao, H. and Waite, J.H. (2006). Proteins in load-bearing junctions: the histidine-rich metal-binding protein of mussel byssus. *Biochemistry* 45: 14223–14231.

42 Boschetti, F. and Peretti, G.M. (2008). Mechanical properties of normal and osteoarthritic human articular cartilage. *Journal of Biomechanics* 41: S171.

43 Hua, M., Wu, S., Ma, Y. et al. (2021). Strong tough hydrogels via the synergy of freeze-casting and salting out. *Nature* 590: 594–599.
44 Liang, X., Chen, G., Lin, S. et al. (2021). Anisotropically fatigue-resistant hydrogels. *Advanced Materials* 33: 2102011.
45 Harrington, M.J. and Fratzl, P. (2021). Natural load-bearing protein materials. *Progress in Materials Science* 120: 100767.
46 Burla, F., Mulla, Y., Vos, B.E. et al. (2019). From mechanical resilience to active material properties in biopolymer networks. *Nature Reviews Physics* 1: 249–263.
47 Mow, V. and Lai, W. (1980). Recent developments in synovial joint biomechanics. *SIAM Review* 22: 275–317.
48 Sun, W., Xue, B., Li, Y. et al. (2016). Polymer-supramolecular polymer double-network hydrogel. *Advanced Functional Materials* 26: 9044–9052.
49 Marini, D.M., Hwang, W., Lauffenburger, D.A. et al. (2002). Left-handed helical ribbon intermediates in the self-assembly of a β-sheet peptide. *Nano Letters* 2: 295–299.
50 Measey, T.J., Markiewicz, B.N., and Gai, F. (2013). Amide I band and photoinduced disassembly of a peptide hydrogel. *Chemical Physics Letters* 580: 135–140.
51 Wu, J., Li, P., Dong, C. et al. (2018). Rationally designed synthetic protein hydrogels with predictable mechanical properties. *Nature Communications* 9: 620.
52 Lv, S., Dudek, D.M., Cao, Y. et al. (2010). Designed biomaterials to mimic the mechanical properties of muscles. *Nature* 465: 69.
53 Harrington, M.J., Masic, A., Holten-Andersen, N. et al. (2010). Iron-clad fibers: a metal-based biological strategy for hard flexible coatings. *Science* 328: 216.
54 Degtyar, E., Harrington, M.J., Politi, Y., and Fratzl, P. (2014). The mechanical role of metal ions in biogenic protein-based materials. *Angewandte Chemie. International Edition* 53: 12026–12044.
55 Lee, B.P., Messersmith, P.B., Israelachvili, J.N., and Waite, J.H. (2011). Mussel-inspired adhesives and coatings. *Annual Review of Materials Research* 41: 99–132.
56 Grindy, S.C., Learsch, R., Mozhdehi, D. et al. (2015). Control of hierarchical polymer mechanics with bioinspired metal-coordination dynamics. *Nature Materials* 14: 1210.
57 Fullenkamp, D.E., He, L., Barrett, D.G. et al. (2013). Mussel-inspired histidine-based transient network metal coordination hydrogels. *Macromolecules* 46: 1167–1174.
58 Xue, B., Bashir, Z., Guo, Y. et al. (2023). Strong, tough, rapid-recovery, and fatigue-resistant hydrogels made of picot peptide fibres. *Nature Communications* 14: 2583.
59 Porter, D., Guan, J., and Vollrath, F. (2013). Spider silk: super material or thin fibre? *Advanced Materials* 25: 1275–1279.
60 Linke, W.A. (2010). Spider strength and stretchability. *Nature Chemical Biology* 6: 702–703.

61 Lewis, R.V. (2006). Spider silk: ancient ideas for new biomaterials. *Chemical Reviews* 106: 3762–3774.

62 Katti, K.S. and Katti, D.R. (2006). Why is nacre so tough and strong? *Materials Science and Engineering: C* 26: 1317–1324.

63 Wegst, U.G.K., Bai, H., Saiz, E. et al. (2015). Bioinspired structural materials. *Nature Materials* 14: 23–36.

64 Wang, J., Cheng, Q., and Tang, Z. (2012). Layered nanocomposites inspired by the structure and mechanical properties of nacre. *Chemical Society Reviews* 41: 1111–1129.

65 Katiyar, N.K., Goel, G., Hawi, S., and Goel, S. (2021). Nature-inspired materials: Emerging trends and prospects. *NPG Asia Materials* 13: 56.

66 Chan, N.J.-A., Gu, D., Tan, S. et al. (2020). Spider-silk inspired polymeric networks by harnessing the mechanical potential of β-sheets through network guided assembly. *Nature Communications* 11: 1630.

67 Gu, L., Jiang, Y., and Hu, J. (2019). Scalable spider-silk-like supertough fibers using a pseudoprotein polymer. *Advanced Materials* 31: 1904311.

68 Dou, Y., Wang, Z.P., He, W. et al. (2019). Artificial spider silk from ion-doped and twisted core-sheath hydrogel fibres. *Nature Communications* 10: 5293.

69 Zhao, X., Chen, F., Li, Y. et al. (2018). Bioinspired ultra-stretchable and anti-freezing conductive hydrogel fibers with ordered and reversible polymer chain alignment. *Nature Communications* 9: 3579.

70 Fu, F., Chen, Z., Zhao, Z. et al. (2017). Bio-inspired self-healing structural color hydrogel. *Proceedings of the National Academy of Sciences of the United States of America* 114: 5900–5905.

71 Xiang, C., Wang, Z., Yang, C. et al. (2020). Stretchable and fatigue-resistant materials. *Materials Today* 34: 7–16.

72 Ducrot, E., Chen, Y., Bulters, M. et al. (2014). Toughening elastomers with sacrificial bonds and watching them break. *Science* 344: 186–189.

73 MacLeod, B.P., Parlane, F.G.L., Brown, A.K. et al. (2022). Flexible automation accelerates materials discovery. *Nature Materials* 21: 722–726.

74 Ludwig, A. (2019). Discovery of new materials using combinatorial synthesis and high-throughput characterization of thin-film materials libraries combined with computational methods. *npj Computational Materials* 5: 70.

75 Hedden, R.C. (2020). High-throughput screening of polymeric membranes for liquid mixture separation. *Current Opinion in Chemical Engineering* 28: 83–89.

76 Popa, I. (2023). *Multicomponent Hydrogels: Smart Materials for Biomedical Applications* (ed. J.M. Dodda, K. Deshmukh, and D. Bezuidenhout). The Royal Society of Chemistry.

77 Zheng, S., Huang, R., Lin, R., and Liu, Z. (2022). A phase field solution for modelling hyperelastic material and hydrogel fracture in ABAQUS. *Engineering Fracture Mechanics* 276: 108894.

78 Zhang, M., Zhang, D., Chen, H. et al. (2021). A multiscale polymerization framework towards network structure and fracture of double-network hydrogels. *npj Computational Materials* 7: 39.

79 Liu, H., Du, C., Liao, L. et al. (2022). Approaching intrinsic dynamics of MXenes hybrid hydrogel for 3D printed multimodal intelligent devices with ultrahigh superelasticity and temperature sensitivity. *Nature Communications* 13: 3420.

80 Castro, A.P.G., Yao, J., Battisti, T., and Lacroix, D. (2018). Poroelastic modeling of highly hydrated collagen hydrogels: experimental results vs. numerical simulation with custom and commercial finite element solvers. *Frontiers in Bioengineering and Biotechnology* 6: 142.

81 Xue, B., Sheng, H., Li, Y. et al. (2021). Stretchable and self-healable hydrogel artificial skin. *National Science Review* 9.

82 Roy, J.K., Pinto, H.P., and Leszczynski, J. (2021). Interaction of epoxy-based hydrogels and water: a molecular dynamics simulation study. *Journal of Molecular Graphics & Modelling* 106: 107915.

83 Angelerou, M.G.F., Frederix, P.W., Wallace, M. et al. (2018). Supramolecular nucleoside-based gel: molecular dynamics simulation and characterization of its nanoarchitecture and self-assembly mechanism. *Langmuir* 34: 6912–6921.

84 Ou, X., Xue, B., Lao, Y. et al. (2020). Structure and sequence features of mussel adhesive protein lead to its salt-tolerant adhesion ability. *Science Advances* 6.

85 Guo, K., Yang, Z., Yu, C.-H., and Buehler, M.J. (2021). Artificial intelligence and machine learning in design of mechanical materials. *Materials Horizons* 8: 1153–1172.

86 Sha, W., Guo, Y., Yuan, Q. et al. (2020). Artificial intelligence to power the future of materials science and engineering. *Advanced Intelligent Systems* 2: 1900143.

87 Audus, D.J. et al. (2023). *Artificial Intelligence for Science*, 413–430. World Scientific.

88 Huang, J.S., Liew, J.X., Ademiloye, A.S., and Liew, K.M. (2021). Artificial intelligence in materials modeling and design. *Archives of Computational Methods in Engineering* 28: 3399–3413.

89 Zheng, S.Y., Yu, H.C., Yang, C. et al. (2020). Fracture of tough and stiff metallosupramolecular hydrogels. *Materials Today Physics* 13: 100202.

90 Long, R. and Hui, C.-Y. (2016). Fracture toughness of hydrogels: measurement and interpretation. *Soft Matter* 12: 8069–8086.

91 Zhang, W., Hu, J., Tang, J. et al. (2019). Fracture toughness and fatigue threshold of tough hydrogels. *ACS Macro Letters* 8: 17–23.

92 Zhou, Y. and Jin, L. (2023). Mechanics underpinning phase separation of hydrogels. *Macromolecules* 56: 426–439.

93 Chen, F., Li, X., Yu, Y. et al. (2023). Phase-separation facilitated one-step fabrication of multiscale heterogeneous two-aqueous-phase gel. *Nature Communications* 14: 2793.

94 Wang, M., Zhang, P., Shamsi, M. et al. (2022). Tough and stretchable ionogels by in situ phase separation. *Nature Materials* 21: 359–365.

95 Jaspers, M., Dennison, M., Mabesoone, M.F. et al. (2014). Ultra-responsive soft matter from strain-stiffening hydrogels. *Nature Communications* 5: 5808.

96 Myung, D., Koch, W., Ko, J. et al. (2007). Biomimetic strain hardening in interpenetrating polymer network hydrogels. *Polymer* 48: 5376–5387.
97 Huang, G., Zhang, H., Liu, Y. et al. (2017). Strain hardening behavior of poly(vinyl alcohol)/borate hydrogels. *Macromolecules* 50: 2124–2135.
98 Cui, Y., Tan, M., Zhu, A., and Guo, M. (2014). Strain hardening and highly resilient hydrogels crosslinked by chain-extended reactive pseudo-polyrotaxane. *RSC Advances* 4: 56791–56797.
99 Xue, B., Gu, J., Li, L. et al. (2021). Hydrogel tapes for fault-tolerant strong wet adhesion. *Nature Communications* 12: 7156.
100 Liang, Y., He, J., and Guo, B. (2021). Functional hydrogels as wound dressing to enhance wound healing. *ACS Nano* 15: 12687–12722.
101 Tavakoli, S. and Klar, A.S. (2020). Advanced hydrogels as wound dressings. *Biomolecules* 10.
102 Maleki, A., He, J., Bochani, S. et al. (2021). Multifunctional photoactive hydrogels for wound healing acceleration. *ACS Nano* 15: 18895–18930.
103 Li, J. and Mooney, D.J. (2016). Designing hydrogels for controlled drug delivery. *Nature Reviews Materials* 1: 16071.
104 Dreiss, C.A. (2020). Hydrogel design strategies for drug delivery. *Current Opinion in Colloid and Interface Science* 48: 1–17.
105 Hoare, T.R. and Kohane, D.S. (2008). Hydrogels in drug delivery: progress and challenges. *Polymer* 49: 1993–2007.
106 Ashley, G.W., Henise, J., Reid, R., and Santi, D.V. (2013). Hydrogel drug delivery system with predictable and tunable drug release and degradation rates. *Proceedings of the National Academy of Sciences of the United States of America* 110: 2318–2323.
107 Tibbitt, M.W., Dahlman, J.E., and Langer, R. (2016). Emerging frontiers in drug delivery. *Journal of the American Chemical Society* 138: 704–717.
108 Lee, K.Y. and Mooney, D.J. (2001). Hydrogels for tissue engineering. *Chemical Reviews* 101: 1869–1880.
109 Xue, B., Tang, D., Wu, X. et al. (2021). Engineering hydrogels with homogeneous mechanical properties for controlling stem cell lineage specification. *Proceedings of the National Academy of Sciences of the United States of America* 118: e2110961118.
110 Zhang, Y.S. and Khademhosseini, A. (2017). Advances in engineering hydrogels. *Science* 356: eaaf3627.
111 Annabi, N., Tamayol, A., Uquillas, J.A. et al. (2014). 25th anniversary article: rational design and applications of hydrogels in regenerative medicine. *Advanced Materials* 26: 85–124.

7

Diffusio-Mechanical Theory of Polymer Network Swelling

Zhaoyu Ding[1], *Peihan Lyu*[1], *and Xingkun Man*[1,2]

[1]*Beihang University, Center of Soft Matter Physics and its Applications, School of Physics, Xueyuan Road, Beijing 100191, China*
[2]*Beihang University, Peng Huanwu Collaborative Center for Research and Education, Xueyuan Road, Beijing 100191, China*

7.1 Introduction

Polymer networks are percolating structures made by chemically or physically bonded macro-molecules. A dried polymer network or gel placed in a solvent absorbs the solvent from the surroundings, resulting in the gel swelling. Swelling of polymer gels is abundant and common in nature (Figure 7.1A,B), such as the plant tissues regulating internal flow [1, 2], expansion or shrinking of soil [3], and swelling of the brain after traumatic injury [4]. Gel swelling is also one of the classical challenges in polymer science and technology. Many important ramifications for technical applications, ranging from simulating biological tissues [5] to controlled drug release [6] to gel actuators [7–9], are directly related to gel swelling, as shown in Figure 7.1C–E. Despite a century of active research about gel swelling, the theory of these dynamics remains an outstanding challenge. In this chapter, we are going to give a brief description of the development of swelling theory of polymer networks. We shall show detailed derivations of the diffusio-mechanical models for gel swelling based on the Onsager variational principle (OVP).

As gel swelling is caused by absorbing solvents into gels, one may understand the swelling process as a diffusion process of the solvent in a polymer network. Although this picture captures an essential aspect of the phenomena, the diffusion in gels is different from the diffusion in solutions. The difference arises from the fact that gel is elastic material, and therefore, the solvent diffusion and the deformation of the polymer network are taking place simultaneously, and they are coupled: solvent diffusion causes polymer network deformation, and polymer network deformation causes solvent diffusion. Such coupling is generally called diffusio-mechanical coupling [10, 11].

From a thermodynamic point of view, the swelling of a gel is determined by the competition between the elastic energy costed by the deformation of the polymer

Dynamics and Transport in Macromolecular Networks: Theory, Modeling, and Experiments, First Edition. Edited by Li-Tang Yan.

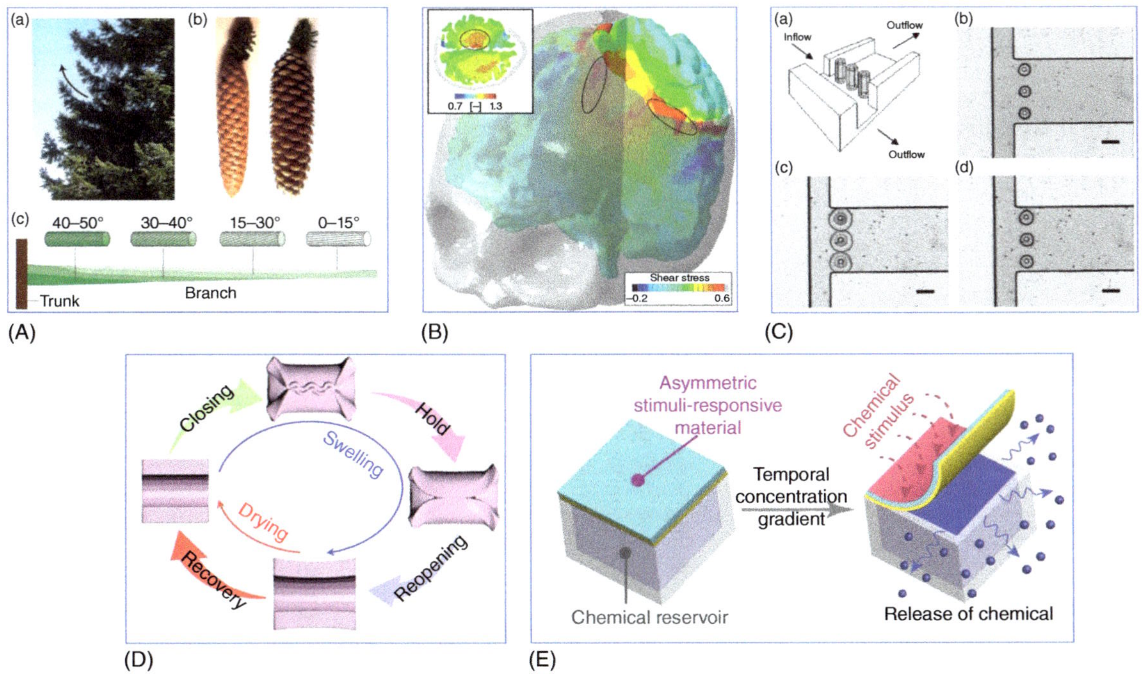

Figure 7.1 Examples of gel swelling in nature and industry. (A) Spruce branches and cones change their shape depending on the ambient humidity. Source: [2]. Peter Fratzl et al. 2009/Reproduced with permission from Springer Nature; (B) Swelling of the brain after traumatic injury. Source: Reproduced with permission from Ref. [4]. ©2016 by the American Physical Society; (C) Swelling gel as flow controlling device. Source: Reproduced with permission from Ref. [9]. ©2000 by the Springer Nature; (D) Swelling gel as actuator. Source: Reproduced with permission from Ref. [7]. ©2021 by the American Chemical Society/CC BY 4.0; (E) Swelling gel as drug release controller. Source: Reproduced with permission from Ref. [6]. ©2021 by the authors and American Association for the Advancement of Science.

network and the attraction energy and entropy gained by the mixture of polymer and solvent. The widely used Flory–Rehner's model provides a basic free energy that accounts for such a mechanism, where the elastic energy is the Neo–Hookean model derived from the Gaussian network theory and the mixing energy is the Flory–Huggins solution theory [11]. Later on, additional energy terms were taken into account in the Flory–Rehner free energy to study even more complicated gel systems. For example, the model has been used to study polyelectrolyte gels by introducing extra ionization and polarization energy [12, 13]. Other extensions include the photo-excitation energy of photo-excitation for photo-thermal gels [14], the magnetization energy for gels infused with magnetic particles exposed to a magnetic field [15], the phase transition energy for gels [16], and so on. Besides the Gaussian network theory, non-Gaussian elastic energy has also been developed for gels that undergo large deformation. Among many works, Li et al. used Gent model, an empirical two-constant constitutive model, to discuss the non-Gaussian chain effects in the large swelling behavior of the polyelectrolyte gel [17]; Chester and Anand used Arruda–Boyce model, a classical non-Gaussian statistical mechanics models of rubber elasticity, to describe the coupled fluid permeation and large viscoelastic deformation of polymeric gels [18, 19].

On the other hand, from a dynamic point of view the theory of poroelasticity has been widely used in the study of gel swelling. This is a continuum mechanics model, in which the gel is treated as a soft porous material with fluid occupying the pore space [20]. Though the poroelasticity theoretical framework was proposed by Biot as early as in 1941, the first theory of gel swelling was only given by Tanaka and Fillmore [21]. They proposed the Tanaka–Filmore equation (TF equation) that accounts for the effect of elasticity by two material constants, including the osmotic bulk modulus K and the shear modulus G. They analytically solved the TF equation for a spherical gel. One critical assumption of this model is that there is no motion of solvents during swelling for the case of G/K being negligibly small. In other words, the stress only drives the motion (or deformation) of polymer network. Then, a force balance equation between the viscous force and potential force is given by

$$\xi \dot{\boldsymbol{u}} = \nabla \cdot \boldsymbol{\sigma}, \tag{7.1}$$

where ξ is the friction constant per unit volume for the polymer motion relative to the solvent, $\boldsymbol{u}$ is the displacement vector of polymer, $\dot{\boldsymbol{u}}$ is the time derivative of $\boldsymbol{u}$, and $\boldsymbol{\sigma}$ is the stress acting on the polymer network that arises from the free energy of the gel, called the osmotic stress. More exactly, the osmotic stress $\boldsymbol{\sigma}$ contains two stresses, the network stress $\boldsymbol{\sigma}^{(n)}$ and the osmotic pressure Π, i.e. $\boldsymbol{\sigma} = \boldsymbol{\sigma}^{(n)} - \Pi$, where $\Pi(\phi) = \phi f'(\phi) - f$ with f being the density of the mixing energy, and ϕ is the fractional volume of the gel network. In the TF model, the stress is taken as the expression of small deformation in elasticity.

Later on, the TF equation was solved by Peters and Candau for nonnegligible values of G/K [22], and for cylinder and disk-shaped gels [23]. However, in 1990, Li and Tanaka [24] pointed out a fundamental deficit of the TF equation: the swelling time of a cylindrical gel in a radial direction and in an axial direction is different by a factor of the square of the aspect ratio (the ratio of the length and the radius). To remove the deficit, they realized that the relaxation of a polymer network needs to be accounted. Therefore, they proposed an approximate way to solve this problem. They found that the insufficiency of TF model is mainly due to the existence of the shear modulus of the network. When the shear modulus is non-ignorable, the relative motion between the solvents and polymer network has to be taken into account in the model. In other words, the motion of solvent becomes important in determining the swelling dynamics of gels. Their approximation is to assume that the solvent diffusion and the deformation of the polymer network are not taking place simultaneously. Instead, they divide the diffusio-mechanical coupling problem into two steps problem: first, a diffusion process of the solvent in a polymer network, then the relaxation process of the deformed polymer network. In order to deal with the later relaxation process, they added a governing equation for minimizing the shear energy

$$\delta A_{\rm sh} = 0, \tag{7.2}$$

where $A_{\rm sh}$ is the shear energy of a gel of arbitrary shape. Actually, this is equivalent to using a new force balance equation instead of Eq. (7.1)

$$\xi(\dot{\boldsymbol{u}} - \boldsymbol{V}_{\rm s}) = \nabla \cdot \boldsymbol{\sigma} \tag{7.3}$$

where $\boldsymbol{V}_{\rm s}$ is the velocity of solvent.

A similar improved model was made by Wang et al. [25], but this model involves an approximation that is valid for dilute polymer concentration, i.e. ϕ is very small. The incompressibility condition for small deformation is $\nabla \cdot [\phi \boldsymbol{V}_{\rm p} + (1-\phi)\boldsymbol{V}_{\rm s}] = 0$, which was reduced to $\nabla \cdot \boldsymbol{V}_{\rm s} = 0$ to calculate the solvent motion for the case of small ϕ.

A complete set of equations describing the diffusio-mechanical coupling was given in 1992 [26]. In contrast, Doi and Onuki used the variational method to give the following force balance equation

$$\frac{\xi}{1-\phi}(\dot{\boldsymbol{u}} - \boldsymbol{V}_{\rm s}) = \nabla \cdot \boldsymbol{\sigma}, \tag{7.4}$$

We can see that the motion of polymers relative to the solvent is driven by the gradient of osmotic stress. This model was used to study the swelling of a cylindrical gel by Yamaue and Doi [27].

All models mentioned so far are based on a set of nonlinear PDEs. Though these equations can be solved analytically for spherical, cylindrical, and disk-shape gels, they are usually difficult to solve, making it difficult to have an overview of or physical insight into the gel swelling dynamics. Recently, Man and Doi developed the diffusio-mechanical theory based on OVP, and also provide a new strategy to solve the diffusio-mechanical problem [28]. Although the model is consistent with previous calculations based on mechanical analysis, the calculation based on OVP is straightforward and much simple. We show the detailed derivations of the diffusio-mechanical models based on OVP calculations of gel swelling in the following.

7.2 Swelling Model

We consider a state of the gel in equilibrium when all external forces are removed, and call this the reference state. Let $\boldsymbol{u}(\boldsymbol{r}, t)$ be the displacement vector of the point at $\boldsymbol{r}$ in the reference state, i.e.

$$\boldsymbol{u}(\boldsymbol{r}, t) = \tilde{\boldsymbol{r}}(\boldsymbol{r}, t) - \boldsymbol{r}, \tag{7.5}$$

where $\boldsymbol{r}$ is the Lagrangian (material) coordinate, and $\tilde{\boldsymbol{r}}(\boldsymbol{r}, t)$ is the current position of the point that is initially at $\boldsymbol{r}$. We take three types of gels (spherical, cylindrical, and disk-shape) as examples:

(a) For a spherical gel with radius R_0, we take a spherical coordinate system (r, θ, ϕ) with origin at the center of the sphere. The deformation $(r, \theta, \phi) \to (\tilde{r}(r, t), \theta, \phi) = (r + u(r, t), \theta, \phi)$ is spherically symmetric, i.e.

$$\boldsymbol{u}(\boldsymbol{r}, t) = u(r, t)\boldsymbol{e}_r. \tag{7.6}$$

(b) For a cylindrical gel with length L_0 and radius R_0 ($R_0 \ll L_0$), we take a cylindrical coordinate (r, θ, z), the z-axis is taken to be parallel to the cylinder axis. The deformation is $(r, \theta, z) \to (\tilde{r}(r, t), \theta, \tilde{z}(z, t)) = (r + u(r, t), \theta, z + \varepsilon(t)z)$. The displacement vector can be written as

$$\boldsymbol{u}(\boldsymbol{r}, t) = u_r\boldsymbol{e}_r + u_z\boldsymbol{e}_z = u(r, t)\boldsymbol{e}_r + \varepsilon(t)z\boldsymbol{e}_z. \tag{7.7}$$

If the length of the gel is infinite, the swelling in the z-direction should be independent of z, i.e. $u_z = \varepsilon(t)z$. Here, we assume that the displacement in the z-direction is also uniform in the case of $R_0 \ll L_0$.

(c) For a disk-shaped gel that was initially at an equilibrium reference state with radius R_0 and half thickness L_0, we choose the cylindrical coordinate system such that the top and bottom surfaces of the disk are at $\pm L_0$, respectively. The origin $z = 0$ corresponds to the plane of symmetry of the disk. When such a gel is placed in a solution, the gel absorbs the solvent and starts to swell toward a new equilibrium state. The deformation is uniform in the radial direction but is nonuniform in the axial direction, i.e. $(r, \theta, z) \to (\tilde{r}(r, t), \theta, \tilde{z}(z, t)) = (r + \varepsilon(t)r, \theta, z + u(z, t))$. Therefore, the displacement vector is given by

$$\boldsymbol{u}(\boldsymbol{r}, t) = u_r\boldsymbol{e}_r + u_z\boldsymbol{e}_z = \varepsilon(t)r\boldsymbol{e}_r + u(z, t)\boldsymbol{e}_z. \tag{7.8}$$

The mapping $\boldsymbol{r} \to \tilde{\boldsymbol{r}}$ can be nonlinear for large deformation. However, the local deformation is always linear, i.e.

$$\mathrm{d}\tilde{\boldsymbol{r}} = \mathrm{d}\boldsymbol{r} \cdot \frac{\partial}{\partial \boldsymbol{r}}\tilde{\boldsymbol{r}}(\boldsymbol{r}), \tag{7.9}$$

or in terms of components

$$\mathrm{d}\tilde{r}_\alpha = F_{\alpha\beta}\mathrm{d}r_\beta, \tag{7.10}$$

where $\alpha, \beta = x, y, z$, and $F_{\alpha\beta}$ is defined by

$$F_{\alpha\beta}(\boldsymbol{r}) = \frac{\partial \tilde{r}_\alpha}{\partial r_\beta}, \tag{7.11}$$

and the Einstein summation convention is adopted. The volume V of the gel in the reference becomes $\tilde{V}$ after deformation, and for volume element $\mathrm{d}V$ becomes $\mathrm{d}\tilde{V} = J\mathrm{d}\tilde{V}, J = \det(F_{\alpha\beta})$.

The OVP states that the evolution of a system is determined by the minimum of the Rayleighian function which is defined by

$$\mathcal{R} = \dot{A} + \Phi, \tag{7.12}$$

where $\dot{A}$ is the time derivative of the free energy A, and Φ is the energy dissipation function of the system.

Let $A[\tilde{\boldsymbol{r}}]$ be the total free energy of the gel. It can be written as the integral of the local free energy,

$$A = \int \mathrm{d}\boldsymbol{r} f_{\mathrm{gel}}(F_{\alpha\beta}(\boldsymbol{r})), \tag{7.13}$$

where f_{gel} stands for the deformation free energy of the gel per unit volume in the reference state. The expression of the free energy density f has been discussed in the textbook of polymer physics [29]. For example, Flory–Rehner's model takes the following form

$$f_{\mathrm{gel}} = \frac{G}{2}\left(\lambda_1^2 + \lambda_2^2 + \lambda_3^2 - 3\right) + f_{\mathrm{mix}}(\phi), \tag{7.14}$$

where G is the shear modulus, λ_1, λ_2, and λ_3 are the principal stretch ratios, ϕ is the polymer volume fraction and $f_{\mathrm{mix}}(\phi)$ is the mixing free energy of polymer and solvent: ϕ can be written in terms of λ_i $(i = 1, 2, 3)$ and the volume fraction ϕ_0 at the reference state:

$$\phi = \frac{\phi_0}{\lambda_1\lambda_2\lambda_3} = \frac{\phi_0}{J}. \tag{7.15}$$

The energy dissipation is caused by the relative motion of polymer and solvent. Let $\mathbf{V}_{\mathrm{p}}(\boldsymbol{r})$ and $\mathbf{V}_{\mathrm{s}}(\boldsymbol{r})$ are the velocity vectors of a polymer and that of a solvent at point $\boldsymbol{r}$. The energy dissipation function is given by

$$\Phi = \int_{\tilde{V}} \frac{\xi(\phi)}{2}\left[\mathbf{V}_{\mathrm{s}}(\tilde{\boldsymbol{r}}) - \mathbf{V}_{\mathrm{p}}(\tilde{\boldsymbol{r}})\right]^2 \mathrm{d}\tilde{\boldsymbol{r}} = \int_V \frac{\xi(\phi)}{2}\left[\mathbf{V}_{\mathrm{s}}(\tilde{\boldsymbol{r}}(\boldsymbol{r})) - \mathbf{V}_{\mathrm{p}}(\tilde{\boldsymbol{r}}(\boldsymbol{r}))\right]^2 J\mathrm{d}\boldsymbol{r}, \tag{7.16}$$

where the region of the first integral is $\tilde{V}$, the region of the second integral is V, ξ is the friction constant per unit volume, $J = \det(\nabla_{\boldsymbol{r}}\tilde{\boldsymbol{r}}) = \det(\boldsymbol{I} + \nabla_{\boldsymbol{r}}\boldsymbol{u}(\boldsymbol{r}))$ is the Jacobian determinant and can be ignored as $J \approx 1$ for small deformation. The velocity of the polymer $\mathbf{V}_{\mathrm{p}}(\boldsymbol{r})$ is $\dot{\boldsymbol{u}}(\boldsymbol{r})$, and $\mathbf{V}_{\mathrm{s}}(\boldsymbol{r})$ can be determined by the incompressibility condition

$$\nabla \cdot [\phi \boldsymbol{V}_{\mathrm{p}} + (1 - \phi)\boldsymbol{V}_{\mathrm{s}}] = 0. \tag{7.17}$$

Here, we must point out that ∇ is the gradient operator in the real space (coordinate of the current state). In other words, ∇ should apply to real coordinates rather than Lagrangian coordinates, i.e. $\nabla = \nabla_{\tilde{\boldsymbol{r}}} \neq \nabla_{\boldsymbol{r}}$. According to the chain rule, we have

$$\frac{\partial}{\partial r_\beta} = F_{\alpha\beta}\frac{\partial}{\partial \tilde{r}_\alpha}. \tag{7.18}$$

For small deformation, ∇ can be replaced by ∇_r that makes the calculation much easier.

Then, we also give the three kinds of gels as examples to show the usage of the incompressibility condition to obtain the $\mathbf{V}_s(\mathbf{r})$ in terms of $\mathbf{V}_p(\mathbf{r})$:

(a) For a spherical gel, according to the spherical symmetry, the relative motion takes place in r-direction only. Then, the incompressibility condition (Eq. (7.17)) becomes

$$\frac{1}{\tilde{r}^2}\frac{\partial}{\partial \tilde{r}}\left[\tilde{r}^2\left(\phi\dot{u} + (1-\phi)V_{sr}\right)\right] = 0, \tag{7.19}$$

where ϕ is the volume fraction of polymer. Then V_{sr} is the radial component of $\mathbf{V}_s$, can be solved as

$$V_{sr} = -\frac{\phi}{1-\phi}\dot{u}. \tag{7.20}$$

(b) For a thin cylindrical gel, the relative motion between the polymer and solvent in the z-direction does not take place since such motion causes very large energy dissipation. Therefore we assume that $V_{sz} = V_{pz} = \dot{\varepsilon}z$. Then, the incompressibility condition gives

$$\frac{1}{\tilde{r}}\frac{\partial}{\partial \tilde{r}}\left[\tilde{r}\left(\phi\dot{u}_r + (1-\phi)V_{sr}\right)\right] + \frac{\partial}{\partial \tilde{z}}\left[\left(\phi\dot{u}_z + (1-\phi)V_{sz}\right)\right] = 0. \tag{7.21}$$

This equation is solved for V_{sr} as

$$V_{sr} = \frac{-\frac{1}{2}\frac{\dot{\varepsilon}}{1+\varepsilon}(r+u) - \phi\dot{u}}{(1-\phi)}. \tag{7.22}$$

Here, we have used the relations that $\tilde{r}(r,t) = r + u(r,t)$ and $\partial z/\partial\tilde{z} = 1/(1+\varepsilon)$.

(c) For a disk-shape gel, the radial and axial components of $\boldsymbol{V}_p$ are given by $V_{pr} = \dot{\varepsilon}r$ and $V_{pz} = \dot{u}$. We may assume $V_{pr} = V_{sr}$, and the relative motion takes place only in the z-direction. The incompressibility condition gives that

$$\frac{\partial}{\partial \tilde{z}}\left[\phi V_{pz} + (1-\phi)V_{sz}\right] + 2\frac{\partial r}{\partial \tilde{r}}\frac{\partial}{\partial r}(\dot{\varepsilon}r) = 0. \tag{7.23}$$

Then, the axial component of $\boldsymbol{V}_s$ is written as

$$V_{sz} = \frac{-2(z+u)\frac{\dot{\varepsilon}}{1+\varepsilon} - \phi\dot{u}}{1-\phi}. \tag{7.24}$$

With Eqs. (7.20), (7.22), and (7.24), the dissipation function Φ (Eq. (7.16)) can be written as a function of $\dot{\boldsymbol{u}}$, resulting in a Rayleighian $\mathcal{R}$ that is also a function of $\dot{\boldsymbol{u}}$. Then, the condition $\delta\mathcal{R}/\partial\dot{\boldsymbol{u}} = 0$ gives the evolution equations of $\dot{\boldsymbol{u}}$ that describe the swelling dynamics of gels. More precisely, we should minimize $\mathcal{R}$ respect to (a) $\dot{u}$ for spherical gels, (b) $\dot{u}$ and $\dot{\varepsilon}$ for cylindrical gels, and (c) $\dot{u}$ and $\dot{\varepsilon}$ for disk-shaped gels.

In the following, we shall first show the derivation of evolution equations of the swelling dynamics based on a general free energy. The free energy can be the Flory–Rehner's model and other model that is used to study large deformation (or nonlinear) cases. It is worth noticing that the Flory–Rehner free energy can be modified by adding extra terms such as Coulomb interaction energy to study even more complicated cases. We shall also show the detailed derivation of the swelling model for small deformation (or linear) cases, where a specific quadratic model of the free energy has been used. Moreover, we shall solve the diffusio-mechanical equation analytically for a spherical, a cylindrical, and a disk-shaped gel, separately. Analytical expressions are given for the shape evolution and swelling time, indicating that how these characteristic features of swelling dynamics of gels depend on material parameters such as shear modulus G and osmotic bulk modulus K.

7.2.1 General Theoretical Framework

7.2.1.1 Spherical Gel

For a spherical gel, the free energy is written as

$$A = \int_0^{R_0} 4\pi r^2 f_{\text{gel}}(u, u')\text{d}r, \tag{7.25}$$

where f_{gel} is a general free energy density of the gel, which can be different for different cases. In this stage, we use such a general form to show how we can obtain a general model for the gel swelling problems for various kinds of gel systems. Here, we use the notations that $u' = \partial u(r,t)/\partial r$ and $\dot{u} = \partial u(r,t)/\partial t$. We must point out that f_{gel} is the function of u and u', not just u'. We know that f_{gel} is the function of $F_{\alpha\beta}$ in the x–y–z-rectangular coordinates. If we use spherical coordinates, the coordinate transformation will result in a free energy expression with u in it. In the Section 7.2.2, we can see this more clearly in the theory of small deformation of spherical gels. The time derivative of the total free energy is

$$\dot{A} = \int_0^{R_0} 4\pi r^2 \left[\frac{\partial f_{\text{gel}}}{\partial u}\dot{u} + \frac{\partial f_{\text{gel}}}{\partial u'}\frac{\partial \dot{u}}{\partial r}\right] \text{d}r. \tag{7.26}$$

With Eqs. (7.16) and (7.20), the dissipation function can be written as

$$\Phi = \int_0^{R} \frac{\xi}{2}\frac{\dot{u}^2}{(1-\phi)^2}4\pi\tilde{r}^2\text{d}\tilde{r} = \int_0^{R_0} \frac{\xi}{2}\frac{\dot{u}^2}{(1-\phi)^2}4\pi\tilde{r}^2\tilde{r}'\text{d}r, \tag{7.27}$$

where $\tilde{r}' = \partial\tilde{r}(r,t)/\partial r$. Then, the time derivative of the free energy (Eq. (7.26)) and the dissipation function (Eq. (7.27)) construct a Rayleighian function $\mathcal{R}$ for a spherical gel. The condition $\delta\mathcal{R}/\delta\dot{u} = 0$ gives the Euler–Lagrange equation

$$\frac{\partial}{\partial r}\left(r^2\frac{\partial f_{\text{gel}}}{\partial u'}\right) = r^2\frac{\partial f_{\text{gel}}}{\partial u} + \frac{\xi\tilde{r}^2\tilde{r}'}{(1-\phi)^2}\dot{u}, \tag{7.28}$$

and the boundary condition at $r = R_0$

$$\left.\frac{\partial f_{\text{gel}}}{\partial u'}\right|_{r=R_0} = 0. \tag{7.29}$$

Finally, Eq. (7.28) together with the boundary condition, Eq. (7.29), are the diffusio-mechanical model of the swelling of a spherical gel. Once the form of f_{gel} is given, such set of equations can be solved numerically.

7.2.1.2 Cylindrical Gel

For a cylindrical gel, the free energy is

$$A = L_0 \int_0^{R_0} 2\pi r f_{\text{gel}}(u, u', \varepsilon) \mathrm{d}r, \tag{7.30}$$

where the appearance of u in f_{gel} is still due to the coordinate transformation. Then, the time derivative of the total free energy is

$$\dot{A} = L_0 \int_0^{R_0} 2\pi r \left[\frac{\partial f_{\text{gel}}}{\partial u}\dot{u} + \frac{\partial f_{\text{gel}}}{\partial u'}\frac{\partial \dot{u}}{\partial r} + \frac{\partial f_{\text{gel}}}{\partial \varepsilon}\dot{\varepsilon} \right] \mathrm{d}r. \tag{7.31}$$

Moreover, Eqs. (7.16) and (7.22) give the dissipation function of a cylindrical gel as

$$\Phi = (1+\varepsilon)L_0 \int_0^{R} \frac{\xi}{2} \frac{(\frac{1}{2}\frac{\dot{\varepsilon}}{1+\varepsilon}\tilde{r} + \dot{u})^2}{(1-\phi)^2} 2\pi\tilde{r}\mathrm{d}\tilde{r} = (1+\varepsilon)L_0 \int_0^{R_0} \xi \frac{(\frac{1}{2}\frac{\dot{\varepsilon}}{1+\varepsilon}\tilde{r} + \dot{u})^2}{(1-\phi)^2} \pi\tilde{r}\tilde{r}' \mathrm{d}r. \tag{7.32}$$

Then, we can obtain the Rayleighian function $\mathcal{R}$ of a cylindrical gel from Eqs. (7.31) and (7.32). Again, the condition $\delta\mathcal{R}/\delta\dot{u} = 0$ gives

$$\frac{\partial}{\partial r}\left(r\frac{\partial f_{\text{gel}}}{\partial u'} \right) = r\frac{\partial f_{\text{gel}}}{\partial u} + \frac{\tilde{r}\tilde{r}'\xi(1+\varepsilon)}{(1-\phi)^2}\left(\frac{1}{2}\frac{\dot{\varepsilon}}{1+\varepsilon}\tilde{r} + \dot{u} \right), \tag{7.33}$$

and the boundary condition

$$\left.\frac{\partial f_{\text{gel}}}{\partial u'}\right|_{r=R_0} = 0. \tag{7.34}$$

$\delta\mathcal{R}/\delta\dot{\varepsilon} = 0$ give

$$\int_0^{R_0} 2\pi r \frac{\partial f_{\text{gel}}}{\partial \varepsilon}\mathrm{d}r + \int_0^{R_0} \frac{\xi}{(1-\phi)^2}\left(\frac{1}{2}\frac{\dot{\varepsilon}}{1+\varepsilon}\tilde{r} + \dot{u} \right) \pi\tilde{r}^2\tilde{r}' \mathrm{d}r = 0. \tag{7.35}$$

Finally, Eqs. (7.33) and (7.35), together with the boundary condition, Eq. (7.34), are the set of diffusio-mechanical equations of the swelling of a cylindrical gel. Noticing that in the cylindrical case, the $\varepsilon(t)$ is unknown, and has to be solved simultaneously together with $u(t)$. This is a characteristic of the diffusio-mechanical coupling problem. Equations (7.33) and (7.35) indicate that the displacement in the axial direction is coupled with the displacement in the radial direction. Such coupling arises from the mechanical coupling of the polymer network in the gel.

7.2.1.3 Disk-Shaped Gel

For a disk-shaped gel, the free energy is

$$A = \pi R_0^2 \int_{-L_0}^{L_0} f_{\text{gel}}(u', \varepsilon) \mathrm{d}z, \tag{7.36}$$

where $u' = \partial u(z,t)/\partial z$. The time derivative of free energy is

$$\dot{A} = \pi R_0^2 \int_{-L_0}^{L_0} \left[\frac{\partial f_{\text{gel}}}{\partial u'}\frac{\partial \dot{u}}{\partial z} + \frac{\partial f_{\text{gel}}}{\partial \varepsilon}\dot{\varepsilon} \right] \mathrm{d}z. \tag{7.37}$$

From the definition of dissipation function (Eq. (7.16)) and the expression of solvent velocity for a disk-shaped gel (Eq. (7.24)), the dissipation function is given as

$$\Phi = \pi R^2 \int_{-L}^{L} \frac{\xi}{2} \frac{(2\tilde{z}\frac{\dot{\epsilon}}{1+\epsilon} + \dot{u})^2}{(1-\phi)^2} d\tilde{z} = \pi R_0^2 (1+\epsilon)^2 \int_{-L_0}^{L_0} \frac{\xi}{2} \frac{(2\tilde{z}\frac{\dot{\epsilon}}{1+\epsilon} + \dot{u})^2}{(1-\phi)^2} \tilde{z}' dz, \quad (7.38)$$

where $\tilde{z}' = \partial\tilde{z}(z,t)/\partial z$. Equations (7.37) and (7.38) construct a Rayleighian function for a disk-shaped gel. Then, the condition $\delta\mathcal{R}/\delta\dot{u} = 0$ gives

$$\frac{\partial}{\partial z}\left(\frac{\partial f_{\text{gel}}}{\partial u'}\right) = \frac{(1+\epsilon)^2 \xi \tilde{z}'}{(1-\phi)^2} \left(\frac{2\tilde{z}\dot{\epsilon}}{1+\epsilon} + \dot{u}\right), \quad (7.39)$$

and the boundary condition

$$\left.\frac{\partial f_{\text{gel}}}{\partial u'}\right|_{z=\pm L_0} = 0. \quad (7.40)$$

$\delta\mathcal{R}/\delta\dot{\epsilon} = 0$ give

$$\int_{-L_0}^{L_0} \frac{\partial f_{\text{gel}}}{\partial \epsilon} dz + \int_{-L_0}^{L_0} \frac{2\tilde{z}\tilde{z}'\xi}{(1-\phi)^2} [2\tilde{z}\dot{\epsilon} + (1+\epsilon)\dot{u}]\, dz = 0. \quad (7.41)$$

Finally, Eqs. (7.39) and (7.41) together with the boundary condition, Eq. (7.40), are the set of the diffusio-mechanical equations of the swelling of the disk-shaped gel. The same as the cylindrical case that $\epsilon(t)$ has to be solved simultaneously together with $u(t)$. In other words, the displacement in the z-direction is coupled with the displacement in the radial direction, which shows the diffusio-mechanical coupling characteristic feature of the swelling of gels again.

In principle, if f_{gel} is known, the set of diffusio-mechanical equations for three types of gel can be solved numerically. However, solving such set of equations is not trivial. In the following, we shall restrict ourselves to the case of small deformation and use OVP to derive the reduced evolution equations of spherical, cylindrical and disk-shaped gels, separately. We shall show that the reduced model for small deformation is linear and can be solved exactly.

7.2.2 Diffusio-Mechanical Model for Small Deformation

7.2.2.1 Spherical Gel

We consider a spherical gel, which was initially at an equilibrium reference state with radius R_0, as shown in Figure 7.2a. We assume that the swelling of a spherical gel has radial symmetry, and only consider the cases where the deformation from the reference state is small. We take a spherical coordinate system (r, θ, ϕ) with origin at the center of the sphere. For small deformation, the deformation-free energy density of the gel has a standard form

$$f(\boldsymbol{r}) = \frac{K}{2}(w - 3\epsilon_{\text{eq}})^2 + \frac{G}{4}\left(\epsilon_{\alpha\beta} + \epsilon_{\beta\alpha} - \frac{2}{3} w \delta_{\alpha\beta}\right)^2, \quad (7.42)$$

where $\epsilon_{\alpha\beta} = \partial u_\alpha / \partial r_\beta$, $w = \nabla \cdot \boldsymbol{u} = \epsilon_{\alpha\alpha}$, r_α are the components of the Cartesian coordinate system, $\alpha, \beta = x, y, z$, and the Einstein summation convention is adopted.

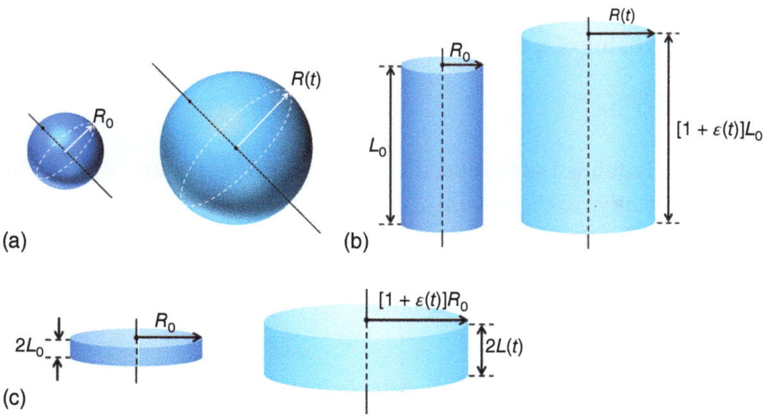

Figure 7.2 Schematic of the swelling of (a) spherical gel, (b) cylindrical gel, and (c) disk-shaped gel. The spherical gel with an initial radius R_0 swells by adsorbing solvent. The radius becomes $R(t)$ at time t for a swollen state. The cylindrical gel with length L_0 and radius R_0 ($R_0 \ll L_0$) swells to the state with length $[1 + \varepsilon(t)]L_0$ and radius $R(t)$ at time t. The disk-shaped gel with radius R_0 and half thickness L_0 ($L_0 \ll R_0$) swells to the state with half thickness $L(t)$ and radius $[1 + \varepsilon(t)]R_0$ by adsorbing solvents.

For uniaxial deformation, the total free energy A in the spherical coordinate is written as

$$A = \int f(\mathbf{r})\mathrm{d}\mathbf{r} = \int_0^{R_0} 4\pi r^2 \left[\frac{K}{2}\left(\frac{2u}{r} + \frac{\partial u}{\partial r} - 3\varepsilon_{\mathrm{eq}}\right)^2 + \frac{2G}{3}\left(\frac{u}{r} - \frac{\partial u}{\partial r}\right)^2\right] \mathrm{d}r, \tag{7.43}$$

where $\varepsilon_{\mathrm{eq}}$ is the equilibrium strain toward which $\partial u/\partial r$ change in time, and K and G are material parameters called osmotic bulk modulus and the shear modulus, respectively.

The time derivative of such free energy is easily obtained as

$$\dot{A} = \int_0^{R_0} 4\pi r^2 \left[K\left(\frac{2u}{r} + \frac{\partial u}{\partial r} - 3\varepsilon_{\mathrm{eq}}\right)\left(\frac{2\dot{u}}{r} + \frac{\partial \dot{u}}{\partial r}\right) + \frac{4G}{3}\left(\frac{u}{r} - \frac{\partial u}{\partial r}\right)\left(\frac{\dot{u}}{r} - \frac{\partial \dot{u}}{\partial r}\right)\right] \mathrm{d}r. \tag{7.44}$$

The energy dissipation of the gel is caused by the motion of the polymer network relative to that of the solvent. For small deformation, the energy dissipation function is given by

$$\Phi = \int \frac{\xi}{2}(\mathbf{V}_{\mathrm{s}} - \mathbf{V}_{\mathrm{p}})^2 \mathrm{d}\mathbf{r}. \tag{7.45}$$

According to the spherical symmetry, the relative motion takes place in r-direction only. The solvent velocity $\boldsymbol{V}_{\mathrm{s}}$ has to satisfy the incompressibility condition $\nabla \cdot [\phi \boldsymbol{V}_{\mathrm{p}} + (1 - \phi)\boldsymbol{V}_{\mathrm{s}}] = 0$, which is

$$\frac{1}{r^2}\frac{\partial}{\partial r}\left[r^2\left(\phi\dot{u} + (1 - \phi)V_{\mathrm{sr}}\right)\right] = 0. \tag{7.46}$$

Coincidentally, this equation gives the same V_{sr} as Eq. (7.20) for large deformation case. Inserting Eq. (7.20) into Eq. (7.45), the dissipation function becomes

$$\Phi = \int_0^{R_0} \tilde{\xi}\dot{u}^2 2\pi r^2 \mathrm{d}r, \tag{7.47}$$

where $\tilde{\xi} = \xi/(1-\phi)^2$ is defined as the effective friction constant per unit volume for the polymer motion relative to the solvent.

The Onsager principle states that the time evolution of $u(z, t)$ is determined by the condition that the functional derivative $\delta\mathcal{R}/\delta\dot{u}$ is equal to zero. By Eqs. (7.44) and (7.47), this gives the following evolution equation

$$\dot{u} = D_{\text{eff}}\frac{\partial}{\partial r}\left[\frac{1}{r^2}\frac{\partial}{\partial r}(r^2 u)\right], \tag{7.48}$$

where $D_{\text{eff}} = \left(K + \frac{4G}{3}\right)/\tilde{\xi}$, and the boundary condition at $r = R_0$

$$\left[K\left(2ru + r^2\frac{\partial u}{\partial r} - 3r^2\varepsilon_{\text{eq}}\right) - \frac{4G}{3}\left(ru - r^2\frac{\partial u}{\partial r}\right)\right]\Bigg|_{r=R_0} = 0. \tag{7.49}$$

Another boundary condition at $r = 0$ is

$$u(0, t) = 0. \tag{7.50}$$

This is due to the symmetry with respect to the spherical center that the position of the center is fixed for all times.

The evolution equation (7.48), boundary conditions Eqs. (7.49), (7.50) and the initial condition $u(r, 0) = 0$ are the set of the diffusio-mechanical equations of the swelling of the spherical gel, which determine the time evolution of $u(r, t)$. The set of equations is linear, and can be solved exactly.

Exact solution for a spherical gel. To solve the set of evolution equations, we should make the boundary condition Eq. (7.49) homogeneous. We define

$$g(r, t) = \frac{u(r, t) - r\varepsilon_{\text{eq}}}{1 + \varepsilon_{\text{eq}}}. \tag{7.51}$$

Then, Eq. (7.48), the boundary conditions and the initial condition become

$$\dot{g} = D_{\text{eff}}\frac{\partial}{\partial r}\left[\frac{1}{r^2}\frac{\partial}{\partial r}\left(r^2 g\right)\right], \tag{7.52}$$

$$2\left(K - \frac{2}{3}G\right)\frac{g(R_0, t)}{R_0} + \left(K + \frac{4G}{3}\right)\frac{\partial g}{\partial r}\bigg|_{r=R_0} = 0, \tag{7.53}$$

$$g(0, t) = 0, \tag{7.54}$$

$$g(r, 0) = -\frac{\varepsilon_{\text{eq}}}{1 + \varepsilon_{\text{eq}}}r. \tag{7.55}$$

This set of equations is exactly the same as the classical theory [21]. It is interesting to notice that Eq. (7.51) is the displacement vector (radial component) with the final state (rather than the initial state) as the reference state. This can be seen from the modified initial condition Eq. (7.55).

The above set of equations can be solved by the eigenfunction expansion method. With $g(r,t)$ assumed given by $g(r,t) = T(t)\rho(r)$, Eqs. (7.52)–(7.54) give

$$T'(t) + D_{\text{eff}}\frac{\lambda^2}{R_0^2}T(t) = 0 \quad (\lambda > 0), \tag{7.56}$$

$$\begin{cases} \rho(r) + \dfrac{2}{r}\rho'(r) + \left(\dfrac{\lambda^2}{R_0^2} - \dfrac{2}{r^2}\right)\rho(r) = 0 \\ \rho(0) = 0, 2\left(K - \dfrac{2}{3}G\right)\frac{\rho(R_0)}{R_0} + \left(K + \frac{4G}{3}\right)\left.\frac{\partial\rho}{\partial r}\right|_{r=R_0} = 0. \end{cases} \tag{7.57}$$

The function $T(t)$ can be solved that $T(t) \propto e^{-D_{\text{eff}}\lambda^2 t/R_0^2}$, and $\rho(r)$ satisfies the first order spherical Bessel equation. The general solution of $\rho(r)$ is

$$\rho(r) = C_1\left[\frac{\sin(\lambda r/R_0)}{\lambda^2 r^2/R_0^2} - \frac{\cos(\lambda r/R_0)}{\lambda r/R_0}\right] + C_2\left[-\frac{\sin(\lambda r/R_0)}{\lambda r/R_0} - \frac{\cos(\lambda r/R_0)}{\lambda^2 r^2/R_0^2}\right], \tag{7.58}$$

where C_2 must be zero because $\rho(0) = 0$. And the other boundary condition of $\rho(r)$ gives the value of λ. We can see that $\rho(r) = C_1 j_1(\lambda r/R_0)$, where $j_1(x)$ is the first-order spherical Bessel function.

Therefore, the solution of $g(r,t)$ is

$$g(r,t) = \sum_{n=1}^{+\infty} C_n e^{-D_{\text{eff}}\frac{\lambda_n^2}{R_0^2}t}\left[\frac{\sin(\lambda_n r/R_0)}{\lambda_n{}^2 r^2/R_0^2} - \frac{\cos(\lambda_n r/R_0)}{\lambda_n r/R_0}\right], \tag{7.59}$$

where C_n is determined by the initial condition Eq. (7.55),

$$C_n = \frac{\displaystyle\int_0^{R_0} g(r,0)r^2 j_1(\lambda_n r/R_0)\mathrm{d}r}{\displaystyle\int_0^{R_0} r^2 j_1{}^2(\lambda_n r/R_0)\mathrm{d}r}, \tag{7.60}$$

and λ_n is the nth positive solution of the equation

$$G\left[\left(4\lambda^2 - 12\right)\sin\lambda + 12\lambda\cos\lambda\right] + 3K\lambda^2\sin\lambda = 0. \tag{7.61}$$

Equation (7.61) is a transcendental equation, and the eigenvalues need to be solved numerically. If the material parameter K is much larger than G, i.e. $K/G \gg 1$, Eq. (7.61) can be greatly reduced to $\sin\lambda = 0$, giving the eigenvalues

$$\lambda_n = n\pi \quad (n = 1, 2, 3, \ldots). \tag{7.62}$$

Inserting Eq. (7.62) into Eq. (7.60), C_n can be calculated

$$C_n = \frac{6\varepsilon_{\text{eq}}}{1+\varepsilon_{\text{eq}}}\frac{(-1)^n R_0}{\lambda_n}. \tag{7.63}$$

Then Eq. (7.59) becomes

$$g(r,t) = \sum_{n=1}^{+\infty} \frac{6\varepsilon_{eq}}{1+\varepsilon_{eq}} \frac{(-1)^n R_0}{\lambda_n} \left[\frac{\sin(\lambda_n r/R_0)}{\lambda_n{}^2 r^2/R_0^2} - \frac{\cos(\lambda_n r/R_0)}{\lambda_n r/R_0} \right] e^{-D_{eff}\frac{\lambda_n^2}{R_0^2}t}. \quad (7.64)$$

Finally, we have the analytical expression for the swelling of a spherical gel.

7.2.2.2 Cylindrical Gel

We consider a cylindrical gel, which was initially at equilibrium reference state with length L_0 and radius R_0, as shown in Figure 7.2b. We assume that $R_0 \ll L_0$. We choose the cylindrical coordinate system and assume that there is no deformation in the θ-direction. Then, the deformation-free energy becomes

$$A = \int d\mathbf{r} \left\{ \frac{K}{2}\left(\frac{u_r}{r} + \frac{\partial u_r}{\partial r} + \frac{\partial u_z}{\partial z} - 3\varepsilon_{eq} \right)^2 + \frac{G}{3}\left[\left(\frac{\partial u_r}{\partial r} - \frac{u_r}{r} \right)^2 + \left(\frac{u_r}{r} - \frac{\partial u_z}{\partial z} \right)^2 + \left(\frac{\partial u_z}{\partial z} - \frac{\partial u_r}{\partial r} \right)^2 \right] \right\}. \quad (7.65)$$

Inserting Eq. (7.7) into Eq. (7.65), we have

$$\frac{A}{L_0} = \int_0^{R_0} 2\pi r dr \left\{ \frac{K}{2}\left(\frac{u}{r} + \frac{\partial u}{\partial r} + \varepsilon - 3\varepsilon_{eq} \right)^2 + \frac{G}{3}\left[\left(\frac{\partial u}{\partial r} - \frac{u}{r} \right)^2 + \left(\frac{u}{r} - \varepsilon \right)^2 + \left(\varepsilon - \frac{\partial u}{\partial r} \right)^2 \right] \right\}. \quad (7.66)$$

For a thin cylindrical gel of $R_0 \ll L_0$, the relative motion between the polymer and solvent in the z-direction does not take place since such motion causes very large energy dissipation. Therefore, the dissipation function is given by

$$\frac{\Phi}{L_0} = \frac{1}{2}\int_0^{R_0} 2\pi r dr \xi \left(\dot{u}_r - V_{sr} \right)^2. \quad (7.67)$$

As we restrict within small deformation, $\tilde{r} \approx r$ and $\varepsilon \ll 1$, the expression of the velocity of solvent, Eq. (7.22), can be simplified as

$$V_{sr} = \frac{-\frac{1}{2}\dot{\varepsilon} r - \phi \dot{u}}{(1-\phi)}. \quad (7.68)$$

Inserting Eq. (7.68) into Eq. (7.67), the dissipation function of the cylindrical gel becomes

$$\frac{\Phi}{L_0} = \int_0^{R_0} \pi r dr \tilde{\xi} \left(\frac{1}{2}\dot{\varepsilon} r + \dot{u} \right)^2, \quad (7.69)$$

where $\tilde{\xi} = \xi/(1-\phi)^2$ is the same effective friction constant as defined in the spherical gel case.

The dissipation function (Eq. (7.69)) and the time derivative of the free energy (Eq. (7.66)) construct the Rayleighian $\mathcal{R} = \dot{A} + \Phi$ of the swelling of a cylindrical gel. The condition $\delta\mathcal{R}/\delta\dot{u} = 0$ gives the following evolution equation

$$\frac{\partial u}{\partial t} + \frac{1}{2}\frac{d\varepsilon}{dt} r = D_{eff} \frac{\partial}{\partial r}\left[\frac{1}{r}\frac{\partial}{\partial r}(ur) \right], \quad (7.70)$$

and boundary condition at $r = R_0$

$$\left(K + \frac{4G}{3}\right) \left.\frac{\partial u}{\partial r}\right|_{r=R_0} + \left(K - \frac{2G}{3}\right)\left(\frac{u(R_0,t)}{R_0} + \varepsilon\right) = 3K\varepsilon_{\text{eq}}, \tag{7.71}$$

where $D_{\text{eff}} = \left(K + \frac{4G}{3}\right)/\tilde{\xi}$.

Moreover, the condition $\delta\mathcal{R}/\delta\dot{\varepsilon} = 0$ gives

$$\int_0^{R_0} \left\{2K\left(\frac{u}{r} + \frac{\partial u}{\partial r} + \varepsilon - 3\varepsilon_{\text{eq}}\right) + \frac{4G}{3}\left(-\frac{u}{r} + 2\varepsilon - \frac{\partial u}{\partial r}\right) + \tilde{\xi} r\left(\frac{1}{2}\dot{\varepsilon} r + \dot{u}\right)\right\} r\mathrm{d}r = 0. \tag{7.72}$$

The boundary condition at $r = 0$ is given by the cylindrical symmetry

$$u(0,t) = 0. \tag{7.73}$$

Then, Eqs. (7.70), (7.72), and (7.73) give

$$\varepsilon = \frac{u(R_0,t)}{R_0}. \tag{7.74}$$

This indicates that the aspect ratio of the gel radius and the length remains unchanged during swelling: the gel looks to swell isotropically. Such isotropic swelling is only apparent since the individual element of the gel does not swell isotropically. They are deformed anisotropically during the swelling process, as we shall show later. There is another meaning of Eq. (7.74). By comparison of the variational calculation with convention mechanical analysis, actually, Eq. (7.74) is equal to the fact that the total force acting on the unit area in the cross section normal to the z axis, written as $F(t) = 2G\left[\varepsilon(t) - u(R_0,t)/R_0\right]$, is zero [27]. The derivation of such force condition based on the Onsager principle is straightforward and easy.

Finally, Eqs. (7.70), (7.71), (7.73), and (7.74) determine the evolutions of $u(r,t)$ and $\varepsilon(t)$, which are the set of diffusio-mechanical equations of the swelling of a cylindrical gel. The set of equations is also linear and can be solved exactly.

Exact solution for a cylindrical gel. To solve the set of evolution equations, define $g(r,t) = u(r,t) + \frac{1}{2}\varepsilon(t)r - \frac{3}{2}r\varepsilon_{\text{eq}}$. Then, Eq. (7.70) and the boundary conditions become

$$\frac{\partial g}{\partial t} = D_{\text{eff}}\left(\frac{\partial^2 g}{\partial r^2} + \frac{1}{r}\frac{\partial g}{\partial r} - \frac{g}{r^2}\right), \tag{7.75}$$

$$g(0,t) = 0, \tag{7.76}$$

$$\left(K + \frac{4G}{3}\right)\left.\frac{\partial g}{\partial r}\right|_{r=R_0} = -\left(K - \frac{4G}{3}\right)\frac{g(R_0,t)}{R_0}. \tag{7.77}$$

Again, the above set of equations can be solved by the stand method of eigenfunction expansion. The solution of $g(r,t)$ is given by

$$g(r,t) = \sum_{n=1}^{+\infty} C_n J_1(\lambda_n r/R_0) e^{-D_{\text{eff}}\lambda_n^2 t/R_0^2}, \tag{7.78}$$

and the expansion coefficients C_n are determined by the initial condition $g(r,0) = -3r\varepsilon_{\rm eq}/2$, which are given by

$$C_n = \frac{\int_0^{R_0} g(r,0) J_1(\lambda_n r/R_0) r\mathrm{d}r}{\int_0^{R_0} J_1{}^2(\lambda_n r/R_0) r\mathrm{d}r}, \tag{7.79}$$

where the eigenvalues λ_n is the nth positive root of the equation

$$\lambda\left(K + \frac{4G}{3}\right)\left[J_0(\lambda) - J_2(\lambda)\right] + 2\left(K - \frac{4G}{3}\right) J_1(\lambda) = 0, \tag{7.80}$$

where $J_\nu(x)$ is the ν order Bessel function ($\nu = 0, 1, 2$). Here, we only consider positive λ_n.

7.2.2.3 Disk-Shaped Gel

We consider a disk-shaped gel, which was initially at an equilibrium reference state with radius R_0 and half thickness L_0, as shown in Figure 7.2c. We choose the cylindrical coordinate system such that the top and bottom surfaces of the disk are at $\pm L_0$, respectively. The origin $z = 0$ corresponds to the plane of symmetry of the disk. When such a gel is placed in a solution, the gel absorbs the solvent and starts to swell toward a new equilibrium state.

With Eqs. (7.8) and (7.65), the deformation-free energy of a disk-shaped gel is given by

$$\frac{F}{\pi R_0^2} = \int_{-L_0}^{L_0} \mathrm{d}z \left\{ \frac{K}{2}\left(2\varepsilon + \frac{\partial u}{\partial z} - 3\varepsilon_{\rm eq}\right)^2 + \frac{2G}{3}\left[\left(\varepsilon - \frac{\partial u}{\partial z}\right)^2\right]\right\}. \tag{7.81}$$

Let $\boldsymbol{V}_{\rm p}$ and $\boldsymbol{V}_{\rm s}$ be the velocity vectors of a polymer and that of a solvent at point (r, z). The radial and axial components of $\boldsymbol{V}_{\rm p}$ are given by $V_{\rm pr} = \dot{\varepsilon} r$ and $V_{\rm pz} = \dot{u}$. It is worth noting that $V_{\rm pr}$ is much larger that $V_{\rm pz}$ for a thin disk-shaped gel of $L_0 \ll R_0$. This means that the relative motion between a polymer and a solvent in the radial direction does not take place since such motion causes a very large energy dissipation. Therefore, we may assume $V_{\rm pr} = V_{\rm sr}$, and the relative motion takes place only in the z-direction. The energy dissipation function becomes

$$\frac{\Phi}{\pi R_0^2} = \frac{\xi}{2}\int_{-L_0}^{L_0} \mathrm{d}z (V_{\rm pz} - V_{\rm sz})^2. \tag{7.82}$$

For small deformation, the axial component of $\boldsymbol{V}_{\rm s}$ can be simplified from Eq. (7.24) as

$$V_{\rm sz} = \frac{-2z\dot{\varepsilon} - \phi\dot{u}}{1 - \phi}. \tag{7.83}$$

Inserting the expression of $V_{\rm pz}$ and $V_{\rm sz}$ into Φ, the dissipation function becomes

$$\frac{\Phi}{\pi R_0^2} = \frac{\tilde{\xi}}{2}\int_{-L_0}^{L_0} \mathrm{d}z (\dot{u} + 2z\dot{\varepsilon})^2, \tag{7.84}$$

where $\tilde{\xi} = \xi/(1 - \phi)^2$ is the effective friction constant that is the same as the one defined in the spherical and cylindrical gel cases.

The dissipation function (Eq. (7.84)) and the time derivative of the free energy (Eq. (7.81)) construct the Rayleighian $\mathcal{R} = \dot{A} + \Phi$ of the swelling of a disk-shaped gel.

Since the present system has a reflection symmetry with respect to the midplane at $z = 0$, we consider the half part of $z \in [0,\ L_0]$. Then, the conditions $\delta\mathcal{R}/\delta\dot{u} = 0$ and $\delta\mathcal{R}/\delta\dot{\varepsilon} = 0$ give the following evolution equations

$$\tilde{\xi}(\dot{u} + 2z\dot{\varepsilon}) = \left(K + \frac{4G}{3}\right)\frac{\partial^2 u}{\partial z^2}, \tag{7.85}$$

$$\int_0^{L_0} \mathrm{d}z \left[\tilde{\xi}(\dot{u} + 2z\dot{\varepsilon})2z + \left(4K + \frac{4G}{3}\right)\varepsilon + \left(2K - \frac{4G}{3}\right)\frac{\partial u}{\partial z} - 6K\varepsilon_{\mathrm{eq}}\right] = 0, \tag{7.86}$$

and the boundary conditions

$$u(0, t) = 0, \tag{7.87}$$

$$\left(K + \frac{4G}{3}\right)\left.\frac{\partial u}{\partial z}\right|_{z=L_0} = \left(\frac{4G}{3} - 2K\right)\varepsilon + 3K\varepsilon_{\mathrm{eq}}. \tag{7.88}$$

Inserting Eq. (7.85) and the boundary conditions of Eqs. (7.87) and (7.88) into Eq. (7.86), we have

$$\varepsilon(t) = \frac{u(L_0, t)}{L_0}. \tag{7.89}$$

Equation (7.89) is the same as (7.74), which indicates that the aspect ratio of the gel radius and the thickness remains unchanged during swelling. In other words, the gel looks to swell isotropically. It is also the same as the cylindrical gel case that such isotropic swelling is only apparent since the individual element of the gel does not swell isotropically [27, 28].

With the definition of the collective diffusion constant $D_{\mathrm{eff}} = \left(K + \frac{4G}{3}\right)/\tilde{\xi}$, Eq. (7.85) can be written as

$$\dot{u} = D_{\mathrm{eff}}\frac{\partial^2 u}{\partial z^2} - 2z\dot{\varepsilon}. \tag{7.90}$$

Finally, Eqs. (7.89) and (7.90) together with the boundary conditions Eqs. (7.87) and (7.88), and the initial conditions $\varepsilon(0) = 0$ and $u(z, 0) = 0$ are the set of the diffusio-mechanical equations of the swelling of the disk-shaped gel.

Exact solution. To solve the set of evolution equations, we define $g(z, t) = u(z, t) + 2z\varepsilon(t) - 3z\varepsilon_{\mathrm{eq}}$. Then Eq. (7.90) and the boundary conditions become

$$\frac{\partial g}{\partial t} = D_{\mathrm{eff}}\frac{\partial^2 g}{\partial z^2}, \tag{7.91}$$

$$g(0, t) = 0, \tag{7.92}$$

$$\left.\frac{\partial g}{\partial z}\right|_{z=L_0} = \beta\frac{g(L_0, t)}{L_0}, \tag{7.93}$$

where $\beta = 4G/(4G + 3K)$.

Taking L_0 as the unit of length, the solution can be written as

$$g(z, t) = \sum_{n=1}^{+\infty} C_n e^{-D_{\mathrm{eff}}\lambda_n^2 t}\sin(\lambda_n z), \tag{7.94}$$

and the expansion coefficients are

$$C_n = \frac{\int_0^1 g_0(z)\sin(\lambda_n z)\mathrm{d}z}{\int_0^1 \sin^2(\lambda_n z)\mathrm{d}z}, \tag{7.95}$$

where $g_0(z) = -3z\varepsilon_{\mathrm{eq}}$ and λ_n is the nth positive solution of the equation

$$\tan\lambda = \frac{\lambda}{\beta}. \tag{7.96}$$

7.3 Results

Equations (7.64), (7.78), and (7.94) give all swelling features of the gels, including both the overall shape evolution, $R(t)$, and the time evolution of individual elements within the gel, $u(r,t)$ and $u(z,t)$. Moreover, from the eigenvalues, the models give the characteristic swelling time that is a function of G/K. Here, we only show the graphical representation for the apparent shape evolution ($R(t)$) and the swelling time for the three types of gels. More graphical results about the individual element of the gel and other features can be found in recent works [28].

In the gel swelling study, a typical time scale τ_{TF} is defined by

$$\tau_{\mathrm{TF}} = \frac{h_0^2}{D_{\mathrm{eff}}} = \frac{\tilde{\xi}h_0^2}{K + \frac{4G}{3}}, \tag{7.97}$$

where h_0 is the characteristic length. We have $h_0 = R_0$ for spherical gel and cylindrical gel, and $h_0 = L_0$ for disk-shaped gel. We therefore take h_0 as the unit of length and τ_{TF} as the unit of time.

Figure 7.3 shows the $R(t)/R_0$ for a spherical gel (blue-solid line), a cylindrical gel (red-dashed line), and a disk-shaped gel (black-dotted line) for $G/K = 0.5$ and $\varepsilon_{\mathrm{eq}} = 0.3$. Results show that the radius of the gels increases monotonically in time, and finally stops increasing at the equilibrium value $R(t)/R_0 = 1.3$. Figure 7.3 also shows that a spherical gel arrives at its equilibrium shape faster than the other two cases, while a disk-shaped gel arrives at its equilibrium shape is the slowest among the tree gels. Such a feature can also be indicated by the swelling time.

The swelling time with which the system approaches to the final equilibrium state is defined as the longest swelling time in the time evolution equation. According to the exact solutions of Eqs. (7.64), (7.78), and (7.94), this is given by

$$\tau = \frac{h_0^2}{D_{\mathrm{eff}}\lambda_1^2}, \tag{7.98}$$

where λ_1 is the smallest positive eigenvalue. Figure 7.4 shows the rescaled swelling time τ/τ_{TF} as a function of G/K for a spherical gel (blue-solid line), a cylindrical gel (red-dashed line) and a disk-shaped gel (black-dotted line). Results indicate that for all three types of gels, τ/τ_{TF} increases infinitely as G/K increases. Interestingly, Figure 7.4 shows that τ/τ_{TF} of the spherical gel is smaller than that of the other two

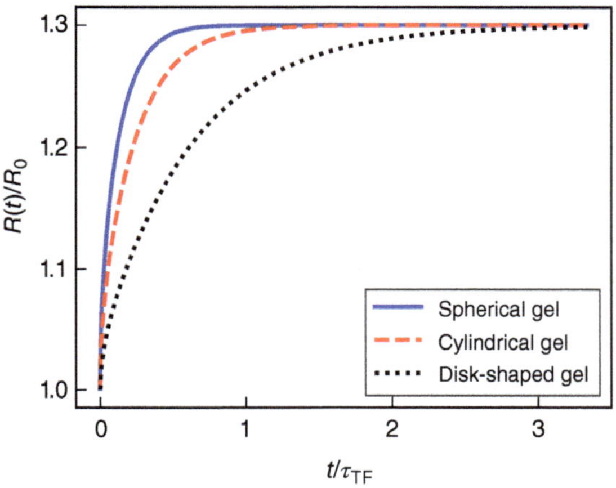

Figure 7.3 The time evolution of the radius of a spherical gel (blue-solid line), a cylindrical gel (red-dashed line), and a disk-shaped gel (black-dotted line). The length scale is the initial radius of the gels, respectively. The time scale τ_{TF} is the conventional characteristic time. For a spherical gel, $\tau_{TF} = \tilde{\xi}R_0^2/(K + 4G/3)$; for cylindrical gel, $\tau_{TF} = \tilde{\xi}R_0^2/(K + 4G/3)$; for disk-shaped gel, $\tau_{TF} = \tilde{\xi}L_0^2/(K + 4G/3)$. For all calculations, $\varepsilon_{eq} = 0.3$ and $G/K = 0.5$.

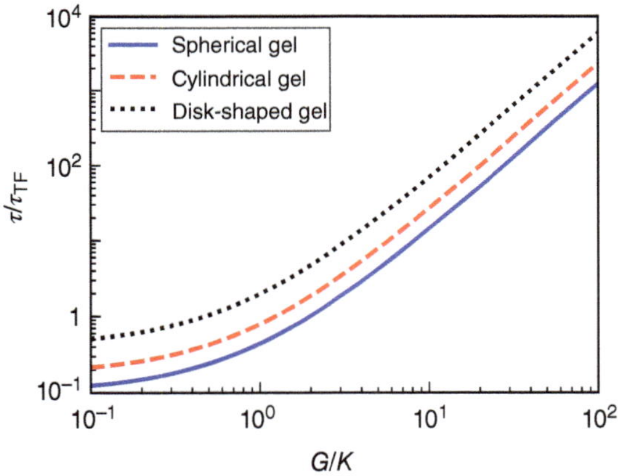

Figure 7.4 The dependence of the longest swelling time τ/τ_{TF} on G/K for a spherical gel (blue-solid line), a cylindrical gel (red-dashed line), and a disk-shaped gel (black-dotted line). For all cases, the time unit is τ_{TF} which is defined as for spherical gel, $\tau_{TF} = \tilde{\xi}R_0^2/(K + 4G/3)$; for cylindrical gel, $\tau_{TF} = \tilde{\xi}R_0^2/(K + 4G/3)$, and for disk-shaped gel, $\tau_{TF} = \tilde{\xi}L_0^2/(K + 4G/3)$. The parameter $\varepsilon_{eq} = 0.3$ for all calculations.

gels, and τ/τ_{TF} of the disk-shaped gel is the largest swelling time for all values of G/K. Such results indicate that a spherical gel swells faster than both cylindrical and disk-shaped gel, and a disk-shaped gel swells slower than both spherical and cylindrical gel.

In the diffusio-mechanical model, the analytical expression for the swelling time can be obtained for the two limits, $G/K \ll 1$ and $G/K \gg 1$. According to Eqs. (7.97) and (7.98), we have

$$\frac{\tau}{\tau_{TF}} = \frac{1}{\lambda_1^2}. \tag{7.99}$$

For $G/K \gg 1$, the divergence of τ/τ_{TF} means that $\lambda_1 \to 0$. Therefore, we can expand the equations that eigenvalues satisfy, i.e. Eqs. (7.61), (7.80), and (7.96), at $\lambda = 0$. Then, we can give analytical expression of the dependence of τ/τ_{TF} on G/K. But for $G/K \ll 1$, this expansion is not valid because λ_1 is not close to 0. In this case, we discuss the swelling time in the limit of $G/K \ll 1$ based on the eigenvalue equations directly.

For the exact solution of a spherical gel, the eigenvalues satisfy Eq. (7.61) that gives

$$\frac{G}{K} = \frac{-3\lambda^2 \sin\lambda}{\left(4\lambda^2 - 12\right)\sin\lambda + 12\lambda\cos\lambda}. \tag{7.100}$$

The smallest positive eigenvalue λ_1 can be solved by expanding the right-hand side of Eq. (7.100) at $\lambda = 0$, which gives

$$\frac{1}{\lambda_1^2} = \frac{2}{21} + \frac{4}{45}\frac{G}{K}. \tag{7.101}$$

Inserting the expression of λ_1 (Eq. (7.101)) for $G/K \gg 1$ and Eq. (7.62) for $G/K \ll 1$ into Eq. (7.99), we have

$$\text{for} \quad G/K \ll 1 \qquad \tau/\tau_{TF} = 1/\pi^2 \approx 0.101, \tag{7.102}$$

$$\text{for} \quad G/K \gg 1 \qquad \tau/\tau_{TF} \approx 4G/45K. \tag{7.103}$$

These results indicate clearly that τ/τ_{TF} increases indefinitely as the shear modulus G increases, and has a fixed value as G becomes small enough to be ignored.

For a cylindrical gel, if $G/K \gg 1$, the eigenvalue equation (Eq. (7.80)) can be reduced to

$$\lambda\frac{J_0(\lambda) - J_2(\lambda)}{J_1(\lambda)} = -2\frac{(3K - 4G)}{(3K + 4G)}. \tag{7.104}$$

Making Taylor expansion of the three Bessel functions in Eq. (7.104), we have

$$2 - \frac{\lambda^2}{2} = -2\frac{(3K - 4G)}{(3K + 4G)}, \tag{7.105}$$

which gives an analytical solution of λ for $G/K \gg 1$ cases

$$\frac{1}{\lambda^2} = \frac{1}{8} + \frac{1}{6}\frac{G}{K}. \tag{7.106}$$

Inserting Eq. (7.106) into Eq. (7.99), we have

$$\text{for} \quad G/K \gg 1 \qquad \tau/\tau_{TF} \approx \frac{1}{6}\frac{G}{K}. \tag{7.107}$$

It is clear that the rescaled swelling time τ/τ_{TF} increases indefinitely as the shear modulus G increases for a cylindrical gel. On the other hand, for $G/K \ll 1$, the numerical result gives that $\tau/\tau_{TF} \approx 2.405$.

For a disk-shaped gel, for $G/K \gg 1$, the eigenvalue equation (Eq. (7.96)) has an approximated form

$$\lambda + \frac{1}{3}\lambda^3 = \frac{\lambda}{\beta}, \tag{7.108}$$

which gives an analytical solution of λ_1 as

$$\frac{1}{\lambda_1^2} = \frac{1}{3\left(\frac{4G+3K}{4G} - 1\right)} \approx \frac{4G}{9K}. \tag{7.109}$$

For $G/K \ll 1$, we have $\beta \to 0$. Therefore, the eigenvalue equation (Eq. (7.96)) simply gives $\lambda_1 \to \pi/2$. Then, the swelling time is given by

$$\text{for} \quad G/K \ll 1 \qquad \tau/\tau_{\mathrm{TF}} = 4/\pi^2 \approx 0.402, \tag{7.110}$$

$$\text{for} \quad G/K \gg 1 \qquad \tau/\tau_{\mathrm{TF}} \approx 4G/9K. \tag{7.111}$$

These results indicate that for a disk-shaped gel, τ/τ_{TF} approaches to a constant as the shear modulus G decreases, but it increases indefinitely as G increases.

It is worth to notice that the reference time τ_{TF} is not constant when G is changed. In fact, a recent work shows that if we choose a reference time, which is independent of G, the swelling time does not vary so much with G [28]. It is, therefore, misleading to say that the swelling time is determined by the collective diffusion constant D_{eff}. Swelling time is essentially estimated by the osmotic bulk modulus (i.e. $\tau \approx h_0^2\tilde{\xi}/K$), and the effect of the shear modulus is small on the swelling time. In fact, it has been proven that for any shape of the gel, the swelling time (i.e. the longest relaxation time) is proportional to $1/K$ in the two limits of $G/K \to 0$ and $G/K \to \infty$ [30].

7.4 Perspective

We have proposed a general model to describe the swelling process of gel by using OVP. Detailed derivation of evolution equations for swelling dynamics are given for a spherical gel, a cylindrical gel, and a disk-shaped gel, separately. Some basic results, such as the time evolution of radius and the characteristic swelling time, are presented here, which are consistent with previous studies, showing the validity of the diffusio-mechanical model. Such analytical studies of gel swelling can provide fundamental understanding of various applications related to gels. However, there are still many cases that have not yet been described by the diffusio-mechanical model, which are more practical, complicated and challenging. Here, we give some insight to broaden the diffusio-mechanical model based on the OVP.

The three swelling models shown in this chapter are limited to small deformation. The current continuum theory of large deformation is mainly based on mechanical analysis, including the kinematics of solid deformation and fluid flow, constitutive laws for polymer and fluid flow and the mechanical equilibrium [19, 31, 32]. Wei Hong et al. [32] proposed a theory of large deformation for polymeric gels in 2008. In their theory, the equations corresponding to force balance and conservation of the number of molecules are linear, while material-specific functions such as the free energy function and the mobility tensor, describing the long-range migration properties of the small molecules, are nonlinear. Huang and coworker [33] presented a comparison between the nonlinear theory by Hong et al. and the linear poroelasticity theory. They showed the consistency of the two theories within the linear regime under the condition of small deformation in the study of an isotropically swollen state of the gel. Chester and Anand [19] proposed a theory using non-Gaussian chain model and combined viscoelastic effects with the standard linear solid model [18]. MacMinn et al. proposed a kinetic theory of large deformation of a soft porous material, and then studied the dynamics of the swelling and drying of a spherical gel with large deformation [34, 35]. Such theories for large deformation provide the basic theoretical framework for the finite element method simulation that has been developed and widely used in the study of gel dynamics [36–38]. The diffusio-mechanical model is exactly consistent with these previous theories, but the calculation based on the OVP is straightforward and much simple. Moreover, the diffusio-mechanical model shown in this chapter can be used to study large deformation of gels by using a more general free energy such as the Flory–Rehner's model instead of the quadratic model used here. A classical challenging of gel swelling is that how to take into account the mechanical instability in all existing models since inhomogeneous instead of homogeneous swelling of gel is more observed in nature.

The OVP can not only provide a systematic way of deriving models for soft matter dynamic problems but also can be used as an approximate method to solve the time evolution equations. Though the diffusio-mechanical equations can be solved analytically, the solution is not simple and requires some numerical calculation in getting the final result. Moreover, if the gel has an irregular shape, the corresponding time evolution equations have no exact solution but need to be solved numerically. OVP as an approximate method provides a different strategy to solve the problem. The main idea is that one can assume an approximate form of the displacement vector $u(r,t)$. For example, the z-direction displacement vector of a disk-shaped gel can be assumed by $u(z,t) = \beta_1(t)z + \beta_2(t)z^2$. Then, the deformation of the gel is now described by ε, β_1, and β_2 in the variational method instead of ε and $u(z)$ in the exact solution. Then, both the time derivative of the free energy and the dissipation energy become functions of $\dot{\varepsilon}$, $\dot{\beta}_1$, and $\dot{\beta}_2$, resulting in a new Rayleighian. Following the same minimization process with respect to $\dot{\varepsilon}$, $\dot{\beta}_1$, and $\dot{\beta}_2$, we have a new set of evolution equations. The advantage of this strategy is that one can reduce the partial differential equations to ordinary differential equations, which simplifies a lot in solving the time evolution equations. While this method is not exact, it gives an explicit analytical solution that is in good agreement with the exact solution, as shown by a recent work on the swelling of a disk-shaped gel [28]. This method has been successfully used in the evaporation of liquid droplets [39–42] but is in the beginning stage in the study of polymer network. It is interesting to develop this method to solve dynamical models for various gel dynamic problems.

Another important feature of swelling dynamics is that the current theory a macroscopic theory without considering any inner structures of the polymer network. However, the defects in the gel network have an important effect on the mechanical properties of the gel. For example, cyclic defects can reduce gel stiffness, extensibility, and toughness. However, the understanding of the defect structure inside the gel is still lacking, and quantifying cyclic defects is difficult due to the limitations of experimental technology. There have been preliminary studies of defects in gels. Zhong et al. [43] measured the shear elastic modulus and counted the number of topological "loop" defects in the gels. They gave a real elastic network theory that describes how loop defects affect bulk elasticity. In order to predict the mechanical response of a polymer network based on its structure, a more precise and refined theory of the network's defects is still needed. It is challenging to take into account such inner structures by using the diffusio-mechanical model.

Viscoelasticity plays an important role in diversifying swelling phenomena. Various experiments observed rheological behaviors of polymer networks [44, 45] and anomalous diffusion process of solvent in polymer gel [46, 47]. It is believed that such phenomena are contributed by the stress relaxation in polymer gel. One way to analyze linear viscoelasticity is to study the combination of springs and dashpots in series and/or parallel (such as Maxwell element and Kelvin element) and regard their stress–strain equation as the constitutive relation of the gel [48]. Since there is infinite number of possible combinations of springs and dashpots, such a method is quite powerful and can describe almost all anomalous diffusion in gels [18, 49]. Although it is intuitive to consider the springs and dashpots rather than viscoelastic gels, physical nature of the method is still lacking. On the other hand, classical continuum mechanics or thermodynamic methods [50–52] often consist of coupled high-order PDEs or even complex equations of tensors, which can only be numerically solved and cannot explain the physical mechanisms directly. Therefore, we expect that the OVP calculation of gel swelling will simplify the conventional theoretical models and improve upon the understanding of the physical mechanism on how viscoelastic properties controls the swelling behaviors of polymer gels.

7.5 Conclusion

In this chapter, we review the history of the diffusio-mechanical coupling theory of gel swelling. The general dynamical models for a spherical gel, a cylindrical gel and a disk-shaped gel are given based on OVP calculations, and then the behaviors of the three types of gels for small deformation are discussed, separately. We show detailed derivations of the evolution equations of displacement vector of gels undergoing small deformation and the corresponding analytical solutions, and present the basic results such as the time evolution of radius and the characteristic swelling time. We also discuss the possible extension that the diffusio-mechanical theory can be applied to the studies of large deformation, polymer network defects, and viscoelasticity for gel swelling.

Acknowledgments

This work was supported in part by the NSFC-ISF Research Program, jointly funded by the National Natural Science Foundation of China (NSFC) under Grant No. 21961142020 and the Israel Science Foundation (ISF) under Grant No. 3396/19, NSFC Grants No. 21822302 and 22073004 and ISF Grant Nos. 213/19, the Fundamental Research Funds for the Central University under Grant No. YWF-22-K-101. We also acknowledge the support of the High-Performance Computing Center of Beihang University.

References

1 Zwieniecki, M.A., Melcher, P.J., and Michele Holbrook, N. (2001). Hydrogel control of xylem hydraulic resistance in plants. *Science* 291: 1059–1062.
2 Fratzl, P. and Barth, F.G. (2009). Biomaterial systems for mechanosensing and actuation. *Nature* 462: 442–448.
3 Smiles, D.E. (2000). Hydrology of swelling soils: a review. *Australian Journal of Soil Research* 38: 501–521.
4 Goriely, A., Weickenmeier, J., and Kuhl, E. (2016). Stress singularities in swelling soft solids. *Physical Review Letters* 117: 138001.
5 Liu, Z. and Calvert, P. (2000). Multilayer hydrogels as muscle-like actuators. *Advanced Materials* 12: 288–291.
6 Gonuguntla, S., Lim, W.C., Leong, F.Y. et al. (2021). Performing calculus: asymmetric adaptive stimuli-responsive material for derivative control. *Science Advances* 7: eabe5698.
7 Liu, Y., Sun, A., Sridhar, S. et al. (2021). Spatially and reversibly actuating soft gel structure by harnessing multimode elastic instabilities. *ACS Applied Materials & Interfaces* 13: 36361–36369, PMID: 34291634.
8 Eddington, D.T. and Beebe, D.J. (2004). Flow control with hydrogels. *Advanced Drug Delivery Reviews* 56: 199–210.
9 Beebe, D.J., Moore, J.S., Bauer, J.M. et al. (2000). Functional hydrogel structures for autonomous flow control inside microfluidic channels. *Nature* 404: 588–590.
10 Doi, M. (2009). Gel dynamics. *Journal of the Physical Society of Japan* 78: 052001.
11 Doi, M. (2013). *Soft Matter Physics*. Oxford University Press.
12 Hong, W., Zhao, X., and Suo, Z. (2010). Large deformation and electrochemistry of polyelectrolyte gels. *Journal of the Mechanics and Physics of Solids* 58: 558–577.
13 Doi, M., Takahashi, K., Yonemoto, T., and Yamaue, T. (2013). Electro-mechanical coupling in ionic gels. *Reactive and Functional Polymers* 73: 891–893.
14 Parson, W.W. (1978). Thermodynamics of the primary reactions of photosynthesis. *Photochemistry and photobiology* 28: 389–393.
15 Han, Y., Hong, W., and Faidley, L. (2011). Coupled magnetic field and viscoelasticity of ferrogel. *International Journal of Applied Mechanics* 3: 259–278.

16 Drozdov, A. (2015). Volume phase transition in thermo-responsive hydrogels: constitutive modeling and structure–property relations. *Acta Mechanica* 226: 1283–1303.

17 Li, J., Suo, Z., and Vlassak, J.J. (2014). A model of ideal elastomeric gels for polyelectrolyte gels. *Soft Matter* 10: 2582–2590.

18 Chester, S.A. (2012). A constitutive model for coupled fluid permeation and large viscoelastic deformation in polymeric gels. *Soft Matter* 8: 8223–8233.

19 Chester, S.A. and Anand, L. (2010). A coupled theory of fluid permeation and large deformations for elastomeric materials. *Journal of the Mechanics and Physics of Solids* 58: 1879–1906.

20 Barrière, B. and Leibler, L. (2003). Kinetics of solvent absorption and permeation through a highly swellable elastomeric network. *Journal of Polymer Science Part B: Polymer Physics* 41: 166–182.

21 Tanaka, T. and Fillmore, D.J. (1979). Kinetics of swelling of gels. *The Journal of Chemical Physics* 70: 1214–1218.

22 Peters, A. and Candau, S. (1986). Kinetics of swelling of polyacrylamide gels. *Macromolecules* 19: 1952–1955.

23 Peters, A. and Candau, S. (1988). Kinetics of swelling of spherical and cylindrical gels. *Macromolecules* 21: 2278–2282.

24 Li, Y. and Tanaka, T. (1990). Kinetics of swelling and shrinking of gels. *The Journal of Chemical Physics* 92: 1365–1371.

25 Wang, C., Li, Y., and Hu, Z. (1997). Swelling kinetics of polymer gels. *Macromolecules* 30: 4727–4732.

26 Doi, M. and Onuki, A. (1992). Dynamic coupling between stress and composition in polymer solutions and blends. *Journal de Physique* II (2): 1631–1656.

27 Yamaue, T. and Doi, M. (2005). The stress diffusion coupling in the swelling dynamics of cylindrical gels. *The Journal of Chemical Physics* 122: 084703.

28 Man, X. and Doi, M. (2021). Swelling dynamics of a disk-shaped gel. *Macromolecules* 54: 4626–4632.

29 Rubinstein, M. and Colby, R.H. (2003). *Polymer Physics*. New York: Oxford University Press.

30 Doi, M. and Yamaue, T. (2005). Variational bounds for the relaxation times of swelling gels. *Physical Review E* 71: 041404.

31 Duda, F.P., Souza, A.C., and Fried, E. (2010). A theory for species migration in a finitely strained solid with application to polymer network swelling. *Journal of the Mechanics and Physics of Solids* 58: 515–529.

32 Hong, W., Zhao, X., Zhou, J., and Suo, Z. (2008). A theory of coupled diffusion and large deformation in polymeric gels. *Journal of the Mechanics and Physics of Solids* 56: 1779–1793.

33 Bouklas, N. and Huang, R. (2012). Swelling kinetics of polymer gels: comparison of linear and nonlinear theories. *Soft Matter* 8: 8194–8203.

34 MacMinn, C.W., Dufresne, E.R., and Wettlaufer, J.S. (2016). Large deformations of a soft porous material. *Physical Review Applied* 5: 044020.

35 Bertrand, T., Peixinho, J., Mukhopadhyay, S., and MacMinn, C.W. (2016). Dynamics of swelling and drying in a spherical gel. *Physical Review Applied* 6: 064010.

36 Lucantonio, A., Nardinocchi, P., and Teresi, L. (2013). Transient analysis of swelling-induced large deformations in polymer gels. *Journal of the Mechanics and Physics of Solids* 61: 205–218.

37 Bouklas, N., Landis, C.M., and Huang, R. (2015). A nonlinear, transient finite element method for coupled solvent diffusion and large deformation of hydrogels. *Journal of the Mechanics and Physics of Solids* 79: 21–43.

38 Chester, S.A., Di Leo, C.V., and Anand, L. (2015). A finite element implementation of a coupled diffusion-deformation theory for elastomeric gels. *International Journal of Solids and Structures* 52: 1–18.

39 Man, X. and Doi, M. (2016). Ring to mountain transition in deposition pattern of drying droplets. *Physical Review Letters* 116: 066101.

40 Man, X. and Doi, M. (2017). Vapor-induced motion of liquid droplets on an inert substrate. *Physical Review Letters* 119: 044502.

41 Jiang, Z., Yang, X., Wu, M., and Man, X. (2020). The drying of liquid droplets. *Chinese Physics B* 29: 096803.

42 Yang, X., Jiang, Z., Lyu, P. et al. (2021). Deposition pattern of drying droplets. *Communications in Theoretical Physics* 73: 047601.

43 Zhong, M., Wang, R., Kawamoto, K. et al. (2016). Quantifying the impact of molecular defects on polymer network elasticity. *Science* 353: 1264–1268.

44 Celli, J.P., Turner, B.S., Afdhal, N.H. et al. (2007). Rheology of gastric mucin exhibits a pH-dependent sol-gel transition. *Biomacromolecules* 8: 1580–1586.

45 García, M.C., Alfaro, M.C., and Muñoz, J. (2016). Rheology of sheared gels based on low acyl-gellan gum. *Food Science and Technology International* 22: 325–332.

46 Thomas, N. and Windle, A. (1978). Transport of methanol in poly(methyl methacrylate). *Polymer* 19: 255–265.

47 Krüger, K.-M. and Sadowski, G. (2005). Fickian and non-fickian sorption kinetics of toluene in glassy polystyrene. *Macromolecules* 38: 8408–8417.

48 Brinson, H.F. and Brinson, L.C. (2008). *Polymer Engineering Science and Viscoelasticity: An Introduction*. Springer.

49 Borrmann, D., Danzer, A., and Sadowski, G. (2021). Generalized diffusion–relaxation model for solvent sorption in polymers. *Industrial and Engineering Chemistry Research* 60: 15766–15781.

50 Aifantis, E. (1980). On the problem of diffusion in solids. *Acta Mechanica* 37: 265–296.

51 Lustig, S., Caruthers, J., and Peppas, N. (1992). Continuum thermodynamics and transport theory for polymer-fluid mixtures. *Chemical Engineering Science* 47: 3037–3057.

52 Govindjee, S. and Simo, J.C. (1993). Coupled stress-diffusion: Case II. *Journal of the Mechanics and Physics of Solids* 41: 863–887.

8

Theoretical and Computational Perspective on Hopping Diffusion of Nanoparticles in Cross-linked Polymer Networks

Ting Ge

University of South Carolina, Department of Chemistry and Biochemistry, 631 Sumter Street, Columbia, SC 29208, USA

8.1 Introduction

The dynamics of nanoparticles in a polymer network is a fundamental process present in the fabrication and processing of functional nanoparticle polymer composites [1, 2], the particle-based micro-rheology [3–5] of both synthetic polymeric materials and naturally occurring biopolymers, as well as the delivery of drug carriers and diagnostic imaging agents at the nanoscale in biopolymeric environments [6–8]. A precise manipulation of the distribution of nanoparticles in both spatial and time domains, which is critical for many technological applications, requires an in-depth understanding of the length- and time-dependent features of nanoparticle dynamics in the polymer matrix. This chapter focuses on a specific topic – *the hopping diffusion of nanoparticles in cross-linked polymer networks*. The structure of a cross-linked polymer network is often characterized simply by the average mesh size between cross-links, although there are various reaction schemes that lead to the formation of a cross-linked network, such as the random cross-linking of precursor polymers and polymerization of multifunctional monomers [9, 10]. A cross-linked polymer network exhibits the viscoelasticity of a polymeric liquid at length scales smaller than the network mesh size and at the accompanying short time scales. However, it acts as an elastic solid at larger length scales and longer time scales. The coupling of nanoparticles to the network elasticity gives rise to hopping diffusion [11, 12], a distinctive mode of nanoparticle dynamics in cross-linked polymers. As illustrated in Figure 8.1, hopping occurs when the nanoparticle size is moderately larger than the average spacing between network constraints and relies on the fluctuation of the network strands that enables the spacing to exceed the particle size. For a successful hopping event, there is a free energy barrier that needs to be overcome. This hopping diffusion mechanism differs qualitatively from the nanoparticle–polymer coupling that dictates the diffusion of nanoparticles in uncrosslinked polymer melts [13–15].

This chapter reviews the hopping diffusion of nanoparticles in cross-linked polymer networks from the perspective of theory and molecular simulations.

Dynamics and Transport in Macromolecular Networks: Theory, Modeling, and Experiments, First Edition. Edited by Li-Tang Yan.

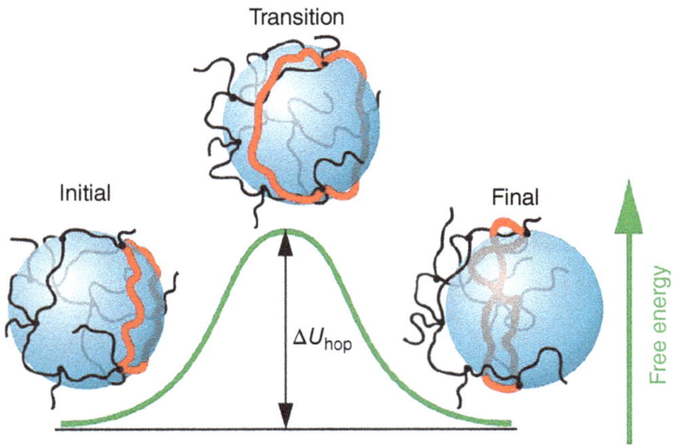

Figure 8.1 Illustration of one individual hopping event for a large nanoparticle in a cross-linked polymer network. The hopping involves a free energy barrier due to the elastic deformation of network strands around the nanoparticle. Source: Adapted from Ref. [12].

Section 8.2 revisits the pioneering scaling theory of hopping diffusion [12], which was developed by Cai, Paniukov, and Rubinstein in the 2010s, with the backbone of the theory delineated. In the same section, another pioneering theoretical work, the nonscaling microscopic theory developed by Dell and Schweizer [11], is briefly reviewed and compared with the scaling theory. Section 8.3 examines computer simulations on hopping diffusion in cross-linked networks and recent theoretical advances addressing network strand stiffness and geometry effects on nanoparticle hopping. Section 8.4 discusses open questions and possible directions for future research. This chapter ends with concluding remarks in Section 8.5.

8.2 2010s' Theories of Nanoparticle Hopping Diffusion

8.2.1 Scaling Theory by Cai, Paniukov, and Rubinstein

Before revisiting the scaling theory by Cai et al., one essential scaling relation for the entropic elasticity of a single polymer chain is reviewed first. The scaling relation specifies the dependence of the elastic deformation energy $U_{\text{elastic}}^{\text{chain}}$ of a polymer chain on the stretch factor R/R_0, where R and R_0 are the average end-to-end size of a stretched polymer chain and that of the undeformed chain in the Gaussian random walk conformation, respectively. According to the classical theory of entropic rubber elasticity in the linear response regime [16], $U_{\text{elastic}}^{\text{chain}} \approx k_B T(R/R_0)^2 \approx \left(k_B T/R_0^2\right)R^2$, which corresponds to a spring with spring constant $K^{\text{chain}} \approx k_B T/R_0^2$. Here, and throughout this chapter, $\approx$ indicates that the equation is on the scaling level with the numerical pre-factor of order unity dropped. Based on the random walk statistics of a Gaussian chain, the equilibrium end-to-end size of a polymer chain consisting of N Kuhn monomers of size b is $R_0 \approx Nb^2$. Therefore, the spring constant may be further expressed as $K^{\text{chain}} \approx k_B T/Nb^2$. The theoretical result allows one to regard an individual polymer chain as a spring with spring constant K^{chain}.

8.2.1.1 Confinement by Network as Attachment to Virtual Chains

Cai et al. first derived the elastic deformation energy $U_{\text{elastic}}^{\text{elementary}}$ induced by embedding a large nanoparticle in an "elementary" network, which consists of *nonoverlapping* network strands between cross-links. The mesh size of a cross-linked network or, equivalently, the average end-to-end size of a network strand is $a_x \approx bN_x^{1/2}$, where b is the size of a statistical Kuhn monomer and N_x is the number of Kuhn monomers per elementary network strand. The elastic energy of displacing d^3/a_x^3 network strands from the pervaded volume of the embedded nanoparticle of volume d^3 is estimated to be $U_{\text{in}} \approx k_BT\left(d^3/a_x^3\right)$, which corresponds to k_BT per network strand displaced. The elastic energy of deforming the network strands outside the displaced ones is estimated to be $U_{\text{out}} \approx k_BT\left(d^3/a_x^3\right)$, which is on the same order as U_{in}. As a result, the total elastic deformation energy $U_{\text{elastic}}^{\text{elementary}} \approx U_{\text{in}} + U_{\text{out}} \approx k_BT\left(d^3/a_x^3\right)$. Given the scaling relation for the dependence of $U_{\text{elastic}}^{\text{elementary}}$ on the confinement ratio d/a_x, Cai et al. then derived the increase in $U_{\text{elastic}}^{\text{elementary}}$ as the position of a nanoparticle surface is displaced from $d/2$ to $d/2 \pm \delta r$ where $\pm$ indicates outward and inward radial displacements. The increase is $\Delta U_{\text{elastic}}^{\text{elementary}} \approx \left(\partial^2 U_{\text{elastic}}^{\text{elementary}}/\partial d^2\right)(\delta r)^2 \approx k_BT\left(d/a_x^3\right)(\delta r)^2$. Note that the first-order derivative $\partial U_{\text{elastic}}^{\text{elementary}}/\partial d$ vanishes for the nanoparticle at equilibrium. The strength of the elastic restoring force opposing the nanoparticle displacement is $f_s \approx \partial \Delta U_{\text{elastic}}^{\text{elementary}}/\partial \delta r \approx k_BT\left(d/a_x^3\right)\delta r$, which depends linearly on δr. Essentially, the detailed calculations by Cai et al. showed that the motion of the embedded nanoparticle is subject to a harmonic potential.

Cai et al. further approximated $\Delta U_{\text{elastic}}^{\text{elementary}}$ as the elastic deformation energy arising from a virtual spring-like chain that fixes the nanoparticle to the center of a confinement cage and consists of n_{cage} monomers. For the deformation of the virtual chain, $U_{\text{elastic}}^{\text{virtual-chain}} \approx K^{\text{virtual-chain}}(\delta r)^2 \approx (k_BT/n_{\text{cage}}b^2)(\delta r)^2$. From equating $U_{\text{elastic}}^{\text{virtual-chain}}$ and $\Delta U_{\text{elastic}}^{\text{elementary}}$, Cai et al. obtained $n_{\text{cage}} \approx N_x(a_x/d)$ for modeling the confinement due to one cage as the attachment to a virtual chain consisting of n_{cage} monomers.

Following the description of a large nanoparticle in an elementary network, Cai et al. modeled a large nanoparticle in a solvent-free dry network. Unlike an elementary network with no overlapping of network strands, each network strand in a dry network overlaps with multiple other strands in its pervaded volume. The number of network strands overlapping with one another in the pervaded volume of one strand is $P \approx a_x^3/(N_xb^3) \approx N_x^{1/2}$. Accordingly, Cai et al. modeled a dry network as P overlapping elementary networks, as illustrated in Figure 8.2b. A large nanoparticle is trapped simultaneously by P cages from P overlapping elementary networks. Equivalently, the nanoparticle is connected by P virtual chains to the centers of the corresponding P cages, as shown in Figure 8.2c. Suppose the nanoparticle displacement vector with respect to the j-th cage center is $\delta\vec{r}_j$, the balance of P elastic restoring force at equilibrium results in $\sum_{j=1}^{P}\vec{f}_{s,j} = (k_BT/n_{\text{cage}}b^2)\sum_{j=1}^{P}\delta\vec{r}_j = 0$ and hence $\sum_{j=1}^{P}\delta\vec{r}_j = 0$. In short, the network confinement on a nanoparticle is modeled as the attachment of the nanoparticle to P virtual chains consisting of n_{cage} monomers each and balancing one another.

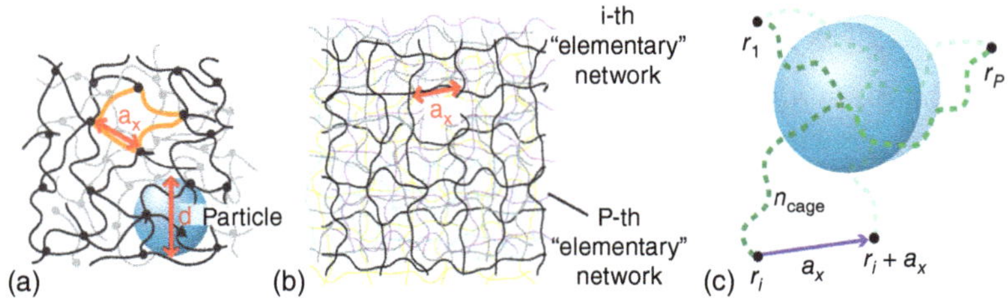

Figure 8.2 Illustrations of (a) a large nanoparticle of diameter d embedded in a cross-linked polymer network with average mesh size a_x, (b) the model of the network as P overlapping elementary networks, and (c) mapping the confinement by the network on the nanoparticle to the attachment of the nanoparticle to P virtual chains (connected to the cage centers at $\vec{r}_1, \ldots, \vec{r}_p$) and relating a hopping event to the replacement of one virtual chain (connected to the cage center at $\vec{r}_i$) by a neighboring virtual chain (connected to the cage center at $\vec{r}_i + \vec{a}_x$). Source: Adapted from Ref. [12].

8.2.1.2 Hopping Diffusion as Successive Individual Hopping Events

The hopping diffusion of a large nanoparticle within a cross-linked polymer network is a random walk where each step corresponds to a hopping event. Cai et al. showed that while a nanoparticle jumps from one cage to an adjacent cage over a distance a_x within one elementary network, the hopping distance in the real space Δr is on the order of the monomer size b, which is smaller than a_x. This result was derived from the balance of P elastic restoring force after the hopping $\sum_{j=1}^{P} \widetilde{\delta \vec{r}_j} = 0$, where $\widetilde{\delta \vec{r}_j} = \delta \vec{r}_j + \Delta \vec{r}$ for all cages except the one in which hopping takes place. Suppose the i-th elementary network is the one for the jump from one cage to a neighboring one (see Figure 8.2c), then the displacement of the nanoparticle in the i-th elementary network is $\widetilde{\delta \vec{r}_i} = \delta \vec{r}_i + \Delta \vec{r} + \vec{a}_x$. Note the extra contribution $\vec{a}_x$ from the cross-cage jumping. The new restoring force balance is then $\sum_{j=1, j\neq i}^{P} \widetilde{\delta \vec{r}_j} + \widetilde{\delta \vec{r}_i} = \sum_{j=1, j\neq i}^{P} (\delta \vec{r}_j + \Delta \vec{r}) + (\delta \vec{r}_i + \Delta \vec{r} + \vec{a}_x) = P\Delta \vec{r} + \vec{a}_x = 0$. As a result, the hopping step size $\Delta r \approx a_x/P \approx b$.

An embedded nanoparticle initiates an attempt to hop at the time scale set by the Rouse time of a network strand $\tau_x \approx \tau_0 N_x^2$, where τ_0 is the monomeric relaxation time. However, successful hopping needs to overcome an elastic energy barrier of height at least $\Delta U \approx k_B T(d/a_x)^2$, which corresponds to $U_{\text{elastic}}^{\text{chain}}$ of a network strand stretched from the undeformed size a_x to the nanoparticle size d. Note that the classical entropic rubber elasticity theory is used here. The probability for the network strand size x to exceed d is $P(x > d) \approx \int_d^{\infty} dx \left[\exp\left(-x^2/a_x^2\right)/a_x\right]$, where $\exp\left(-x^2/a_x^2\right)/a_x$ corresponds to the normalized probability density function. As a result, the waiting time for a successful hopping event is $\tau_w \approx \tau_x/P(x > d)$, which is longer than the average time τ_x for an attempt. Cai et al. estimated that $\tau_w \approx \tau_x (d/a_x) \exp\left(d^2/a_x^2\right)$.

Knowing the hopping step size b and the waiting time τ_w, the mean squared displacement of a large nanoparticle over a time $t > \tau_w$ is $\langle r^2(t)\rangle_{\text{hop}} \approx b^2(t/\tau_w)$, which corresponds to a random walk consisting of t/τ_w successive hopping events. The diffusion coefficient due to hopping diffusion is $D_{\text{hop}} \approx b^2/\tau_w \approx (b^2/\tau_x)(a_x/d)\exp\left(-d^2/a_x^2\right)$.

While hopping diffusion begins at the time scale τ_w, its contribution to nanoparticle mean squared displacement $\langle r^2(t)\rangle_{\text{hop}}$ is not manifested until the hopping diffusion time $\tau_{\text{hop}} > \tau_w$. This is because at time scales $\tau_x < t < \tau_{\text{hop}}$ the mean squared displacement due to the fluctuation of the nanoparticle $\langle r^2(t)\rangle_{\text{fluctuation}}$ is larger than $\langle r^2(t)\rangle_{\text{hop}}$ and therefore dominates the overall mean squared displacement $\langle r^2(t)\rangle$. Cai et al. modeled the P virtual chains connected to the nanoparticle as parallel springs with a composite spring constant $PK^{\text{virtual - chain}} \approx k_BT/(n_{\text{cage}}/P)b^2$. The number of monomers in the composite virtual chain is n_{cage}/P, and the mean squared end-to-end size of the composite virtual chain determines the fluctuation of the trapped nanoparticle, i.e. $\langle r^2(t)\rangle_{\text{fluctuation}} \approx (n_{\text{cage}}/P)b^2 \approx a_x^2/(dP)$ for $\tau_x < t < \tau_{\text{hop}}$. The hopping diffusion time τ_{hop} is determined by equating $\langle r^2(\tau_{\text{hop}})\rangle_{\text{fluctuation}}$ and $\langle r^2(\tau_{\text{hop}})\rangle_{\text{hop}}$, and the result is $\tau_{\text{hop}} \approx \tau_x\left(a_x^2/Pb^2\right)\exp\left(d^2/a_x^2\right)$.

Figure 8.3 illustrates $\langle r^2(t)\rangle$ of a large nanoparticle with $d > a_x$ in a dry cross-linked network. The plateau at time scales $\tau_x < t < \tau_{\text{hop}}$ indicates $\langle r^2(t)\rangle \approx \langle r^2(t)\rangle_{\text{fluctuation}} \approx a_x^3/(dP)$. The solid line at time scales $t > \tau_{\text{hop}}$ with log–log slope 1 indicates the manifested hopping diffusive regime, whereas the dashed line at time scales $\tau_w < t < \tau_{\text{hop}}$ indicates $\langle r^2(t)\rangle_{\text{hop}}$ that is masked by $\langle r^2(t)\rangle_{\text{fluctuation}}$. Prior to the Rouse time of a network strand τ_x, there is a crossover from the ballistic motion of the nanoparticle due to the initial momentum to the sub-diffusive motion due to the progressive coupling of nanoparticle motion and the relaxation dynamics of network strands. This is indicated by the crossover from the regime at time scales $t < \tau_{\text{bal}}$ with log–log slope 2 to the subsequent regime at time scales $\tau_{\text{bal}} < t < \tau_x$ with log–log slope 1/2. Readers are advised to read the original scaling theory papers by Cai et al. [14] as well as a recent

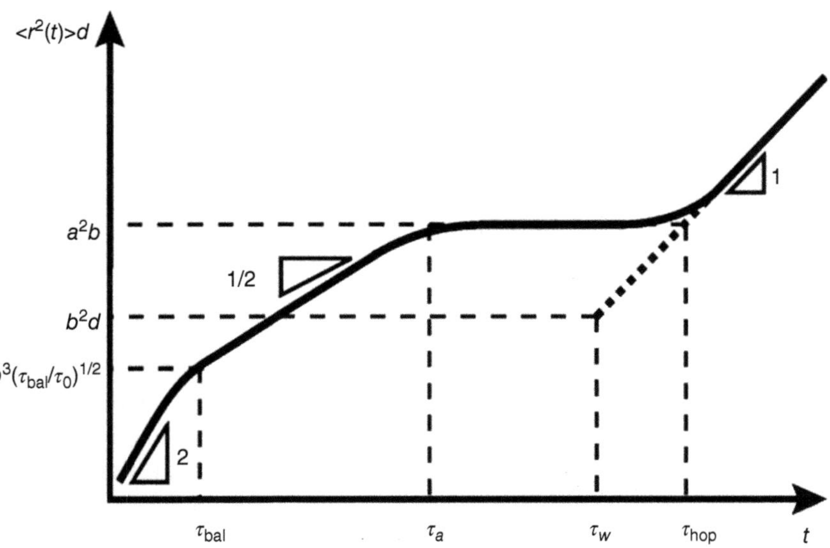

Figure 8.3 The product of the nanoparticle mean squared displacement $\langle r^2(t)\rangle$ and the nanoparticle diameter d as a function of time as predicted theoretically by Cai et al. τ_a and a are interpreted as τ_x and a_x, respectively, for a network dominated by cross-links. Source: Adapted from Ref. [12].

perspective article [15] for details regarding the ballistic regime and sub-diffusive regime at short time scales.

8.2.1.3 Beyond Homogeneous, Entanglement-Free, and Dry Cross-linked Networks

The scaling theory developed by Cai et al. for a large nanoparticle in a cross-linked polymer network is applicable not only to a perfect network with a monodisperse network strand length N_x but also to a network with polydispersity in N_x provided that the nanoparticle is smaller than a crossover size d_m. Assuming a Poisson distribution of N_x, Cai et al. determined that the variation of the hopping diffusion energy barrier due to the polydispersity is $\delta(\Delta U) \approx \pm k_B T d^2/a_x^2 \bar{l}$, where $\bar{l} \approx \ln P$ is the average number of network strands that constitute a loop wrapping the particle. The crossover size above which the magnitude of $\delta(\Delta U)$ is larger than the thermal fluctuation $k_B T$ and thus significantly affecting nanoparticle hopping is $d_m \approx a_x \bar{l}^{1/2}$. Note that d_m is the average size of the loop consisting of $\bar{l}$ network strands of size a_x. For $d < d_m$, the hopping diffusion in the polydisperse network is almost the same as that in the monodisperse network. For $d > d_m$, the particle is large enough to be significantly affected by the polydispersity and hops through cages with relatively larger N_x or equivalently hops through the regions with relatively lower densities of cross-links.

Sufficiently long network strands get entangled with each other. The dynamics of entangled network strands are confined in a tube-like region as envisioned first by Edwards [17]. The diameter of the tube is $a_e \approx bN_e^{1/2}$, where N_e is the number of monomers between neighboring entanglements. Note that $bN_e^{1/2}$ is the average end-to-end size of an entanglement strand consisting of N_e monomers. If the density of entanglements that are trapped between the cross-links is higher than the density of cross-links, the polymer network is dominated by the entanglements and the effective network mesh size becomes a_e, which is smaller than a_x. Intuitively, the confinement of a large nanoparticle by the entanglement network is softer compared to that by the cross-linked network, as entanglements are not permanent but can slide between polymer chains. In other words, the deformation of an entanglement network may be non-affine to accommodate the embedded large nanoparticle. Indeed, the softer confinement by entanglements was captured by Cai et al. in their theory.

Cai et al. developed the theory for a large nanoparticle in an entanglement network based on the non-affine deformation of an entanglement network. The first step was to derive the number of monomers n_{loop} in a loop that wraps around the large nanoparticle. Instead of balancing $U_{\mathrm{elastic}}^{\mathrm{virtual-chain}}$ and $\Delta U_{\mathrm{elastic}}^{\mathrm{elementary}}$, which was used in the mapping of a cross-linked elementary network to a virtual chain, the tension of the loop $K_e^{\mathrm{loop}} d \approx (k_B T/n_{\mathrm{loop}} b^2)d$ was balanced by the tension in the confining tube $k_B T/a_e$. As a result, $n_{\mathrm{loop}} \approx a_e d/b^2$, which differs from N_e due to the non-affine deformation. This result describes the mapping of the confinement by an entanglement network to the confinement by a loop consisting of n_{loop} monomers.

According to Cai et al., the hopping step size of a large nanoparticle in an entanglement network is still on the order of b, but the entropic energy barrier for the hopping is $\Delta U_e \approx k_B T d^2/(n_{\text{loop}} b^2) \approx k_B T d/a_e$. The dependence of ΔU_e on d/a_e is linear, in contrast to the quadratic dependence $(d/a_e)^2$ for ΔU in a cross-linked network, demonstrating the softer confinement of a large nanoparticle in an entanglement network.

The theoretical description of the hopping diffusion was revised based on the new entropic energy barrier ΔU_e. The probability for the loop size to be larger than d is $P(x > d) \approx \int_d^\infty dx \exp(-x/a_e) \approx \exp(-d/a_e)$. The hopping diffusion waiting time $\tau_w^e \approx \tau_e/P(x > d) \approx \tau_e \exp(d/a_e)$, where $\tau_e \approx \tau_0 N_e^2$ is the Rouse time of an entanglement strand and the time scale at which the nanoparticle initiates the attempt to hop. The diffusion coefficient due to hopping diffusion is $D_{\text{hop}}^e \approx b^2/\tau_w^e \approx (b^2/\tau_e)\exp(-d/a_e)$. The particle mean squared displacement $\langle r^2(t)\rangle \approx \langle r^2(t)\rangle_{\text{fluctuation}}^e \approx a_e^2 b/d$ at time scales $\tau_e < t < \tau_{\text{hop}}^e$, and $\langle r^2(t)\rangle \approx \langle r^2(t)\rangle_{\text{hop}}^e \approx D_{\text{hop}}^e t$ at time scales longer than the hopping diffusion time $\tau_{\text{hop}}^e \approx \tau_e \left(a_e^2 b/d\right)\exp(d/a_e)$. Despite the different expressions of relevant physical quantities, the time dependence of $\langle r^2(t)\rangle$ can still be schematically shown using Figure 8.3 with a and τ_a interpreted as a_e and τ_e, respectively.

Figure 8.4 summarizes the dependence of the nanoparticle diffusivity on the particle size in a cross-linked polymer network, as predicted by Cai et al. For $d < a_e$, the particle diffusion is coupled to the local viscosity of unentangled polymer sections whose sizes are comparable to d, and the diffusion coefficient decreases with increasing d as $D \sim d^{-3}$. As d exceeds a_e, the particle experiences the trapped entanglements, and there is a hopping diffusion through an entanglement-dominated network with $D \sim \exp(-d/a_e)$. As d further exceeds a_x^2/a_e, the hopping diffusion occurs in a cross-link-dominated network with $D \sim \exp\left(-d^2/a_x^2\right)$.

Cai et al. also extended their scaling theory for the hopping diffusion in a solvent-free dry network to describe the hopping diffusion in a gel consisting

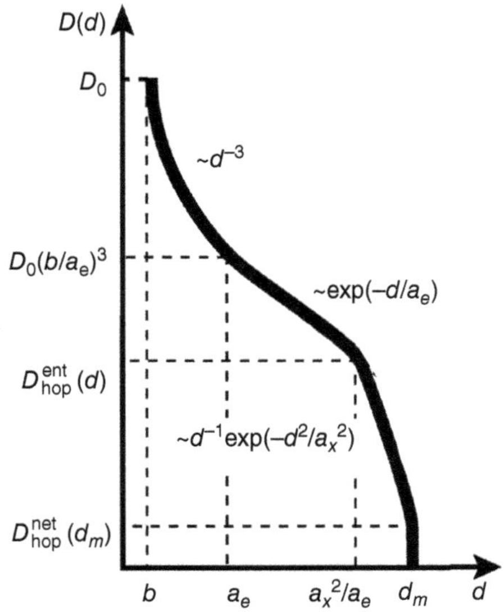

Figure 8.4 The diffusion coefficient D of the nanoparticle as a function of the nanoparticle diameter d as predicted theoretically by Cai et al. Source: Adapted from Ref. [12].

of a cross-linked network and solvent molecules. The extension was based on replacing important physical properties of the dry system with their counterparts in a concentrated polymer solution [16]. The hopping step size b in the dry network was replaced with the correlation length ξ in a polymer solution, which is the average spacing between a pair of interchain monomers. The dependence of ξ on the volume fraction of polymers in the solution ϕ is $\xi(\phi) \approx b\phi^{-\nu/(3\nu-1)}$, where the scaling exponent $\upsilon = 1/2$ and $3/5$ for θ solvent and athermal or good solvent, respectively. Likewise, the network strand size a_x, the entanglement length N_e, and the entanglement spacing a_e, as well as their respective relaxation times τ_x and τ_e, were all replaced by the corresponding ϕ- and ν-dependent expressions. Readers are advised to read the original paper [12] for the comprehensive list of the results derived from the extended theory for hopping diffusion in polymer gels.

8.2.2 Microscopic Theory by Dell and Schweizer

Nonscaling theoretical approach to nanoparticle dynamics in polymer networks has also been developed in the 2010s. Dell and Schweizer combined statistical mechanics and polymer physics, specifically a nonlinear Langevin equation and PRISM theory, to construct a microscopic theory for the localization and activated hopping of large nanoparticles in cross-linked polymer networks [11]. Both the scaling theory by Cai, Paniukov, and Rubinstein and the nonscaling microscopic theory by Dell and Schweizer can be applied to model the hopping of nonsticky hard nanoparticles in solvent-free dry networks of cross-linked polymers. However, the two theories are qualitatively different. While the Cai–Paniukov–Rubinstein theory predicted that the hopping step size is on the order of Kuhn monomer size b and therefore independent of particle size d, the Dell–Schweizer theory predicted that the hopping step size grows as the ratio of particle size d and network mesh size a increases, i.e. as the confinement increases, and it also exhibits a dependence on d. Another difference lies in the treatment of polymer entanglements and cross-links. Cai et al. demonstrated that the confinement by entanglements is weaker than the confinement of cross-links, because of the non-affine deformation of the entangled network upon the embedding of a large nanoparticle. By contrast, Dell et al. approximated entanglements as effective cross-links. Quantitatively, the Cai–Paniukov–Rubinstein theory predicted the hopping energy barrier is $\Delta U \approx k_B T(d/a_x)^2 {\sim} d^2$ when cross-links dominate the confinement whereas $\Delta U_e \approx k_B T d/a_e {\sim} d$ when entanglements dominate. The Dell–Schweizer theory predicted that the hopping energy barrier asymptotically depends on the particle volume and is proportional to d^3 regardless of whether cross-links or entanglement dominate. Despite the differences, both the Cai–Paniukov–Rubinstein theory and the Dell–Schweizer theory describe the origin of the hopping energy barrier as purely entropic and thus proportional to the thermal energy unit $k_B T$. As a result, the Boltzmann factor $\exp(-\Delta U/k_B T)$ for the reduction in the rate of successful hopping is independent of temperature and only a function of the geometric confinement ratio d/a.

8.3 Recent Computational and Theoretical Work

Molecular simulation has been established as a research methodology complementing theory and experiments. Coarse-grained molecular simulations based on the bead-spring representation of a polymer [18] have been well developed and employed to investigate various physical properties of polymers. The bead-spring polymer model captures the strong covalent backbone of a polymer chain as well as the weaker interchain interactions of the van der Waals type, allowing an efficient simulation of the universal physics aspect of various polymers with differing chemistry details.

For the coarse-grained simulations of nanoparticles in a cross-linked polymer network, the bead-spring model may be used to simulate the polymers. To generate a cross-linked network, one can either manually construct a perfect network with prescribed cross-link density and functionality or obtain it from cross-linking a melt of polymer chains as in experiments [10]. Two methods of cross-linking may be used: one is random cross-linking pairs of monomers in contact and the other is end-linking of linear precursor chains. Compared to the manually constructed perfect network, the networks obtained from cross-linking are not perfect and contain a small fraction of defects such as dangling ends and intrachain loops.

Embedded nanoparticles are modeled either as rigid clusters of atoms [19–23] with sizes identical to the monomers or as smooth particles [24] whose interactions with each other and with the monomers are described based on integrating the interatomic potentials. Alternatively, the interactions between a nanoparticle and a monomer as well as between two nanoparticles may be modeled using an expanded Lennard–Jones potential [25]. The state-of-the-art large-scale molecular simulations of nanoparticles in a polymer matrix are capable of simulating $O(100)$ nanoparticles with diameters up to 15 times the Kuhn monomer size. The unbalanced workload of simulating large nanoparticles and polymer chains in a hybrid system has been alleviated by recently developed algorithms [26]. Additionally, Monte Carlo (MC)-type algorithms have been utilized to accelerate the pure molecular dynamics (MD) or dissipative particle dynamics (DPD) simulations.

8.3.1 Evaluating Cai–Paniukov–Rubinstein and Dell–Schweizer Theories by Simulations

Sorichetti et al. have performed systematic coarse-grained MD simulations to study the static and dynamic properties of spherical nanoparticles in a polydisperse network of cross-linked polymers [25]. The networks were generated using a specific cross-linking procedure and possessed heterogeneity in the network structure as demonstrated by the broad distribution of pore diameters. The network strands were chosen to be short and therefore unentangled. Both purely repulsive and weakly attractive nanoparticle–polymer interactions were simulated. The particle diameter d was systematically increased with respect

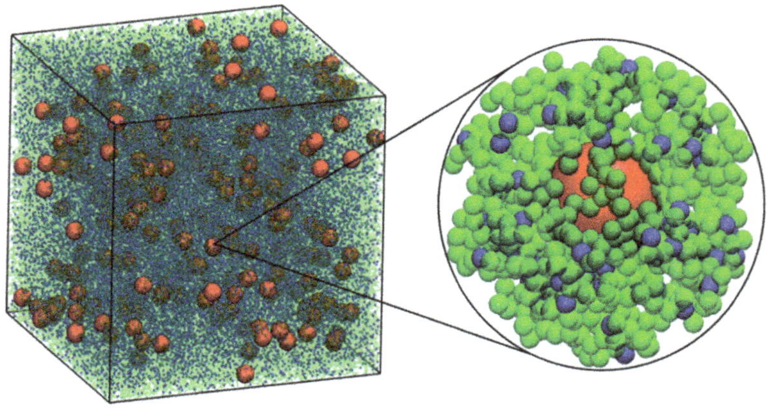

Figure 8.5 Snapshot of nanoparticles in a cross-linked polymer network in the molecular simulations by Sorichetti et al. The nanoparticles, monomers, and cross-links are indicated by the red, green, and blue spheres, respectively. Source: Adapted from Ref. [25].

to the dynamic localization length λ in the network, which was determined as the plateau value of the root-mean squared displacement of the cross-links. A snapshot of the simulated nanoparticles in a cross-linked network is shown in Figure 8.5.

The mean squared displacements of nanoparticles in the simulations by Sorichetti et al. are shown in Figure 8.6. With increasing confinement, which is measured by the ratio d/λ, one observes the suppressed motion of trapped nanoparticles at intermediate time scales and the hopping diffusion in the terminal regime, as predicted by Cai et al. Note that the ballistic motion and sub-diffusive motion prior to the network confinement are also observed in the simulations. Figure 8.7 shows the squared displacements of three preventative nanoparticles in the simulations and the comparison of them with the cross-link mean squared displacement. Within the time window of observation, particles A and B can hop around, while particle C is trapped with no hopping.

Sorichetti et al. computed the terminal hopping diffusion coefficient of the nanoparticles D_N as a function of the confinement parameter $C = d/\lambda$. The decrease of D_N with C is shown in Figure 8.8. They further tested the predictive power of the hopping diffusion theories developed by Cai et al. and Dell et al. Since the network strands are unentangled, the confinement on nanoparticle motion is solely from the cross-links. Motivated by the result $D_{\text{hop}} \approx (b^2/\tau_x)(a_x/d)\exp\left(-d^2/a_x^2\right)$ in the Cai–Paniukov–Rubinstein theory, Sorichetti et al. used $D_N = A\exp[-(BC)^2]/BC$ with adjustable parameters A and B to fit their simulation data. As shown in Figure 8.8a, the fitting works well. They also fit their simulation data to an alternative function $D_N = A' \exp[-(B' C)^\delta]$ with adjustable parameters A', B', and δ, which was motivated by the Dell–Schweizer theory. The corresponding fitting shown in Figure 8.8b is equally good as the one in Figure 8.8a. As a result, Sorichetti et al. concluded that the two theories both give a very good description of the C-dependence of D_N in their simulations.

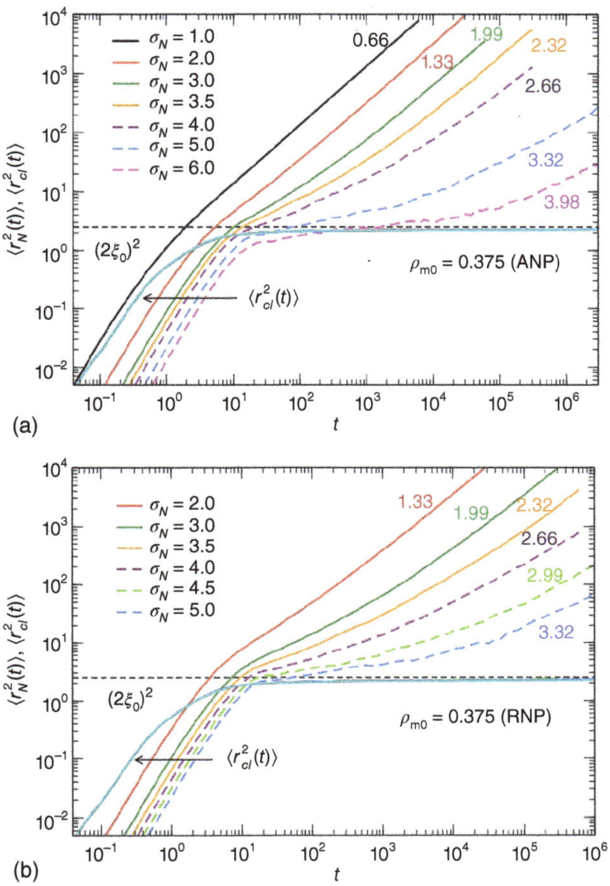

Figure 8.6 The mean squared displacements of the nanoparticles $\langle r_N^2(t) \rangle$ and of the cross-links $\langle r_{cl}^2(t) \rangle$ for indicated nanoparticle diameters σ_N in the simulations by Sorichetti et al. Nanoparticle–polymer interactions are weakly attractive and purely repulsive in (a) and (b), respectively. Source: Adapted from Ref. [25].

8.3.2 Exploring New Aspects of Cross-linked Networks – Stiffness and Geometry

Xu et al. used DPD simulations to study the diffusion of nanoparticles in a network of cross-linked semiflexible polymers [27], where the large bending rigidity of a network strand competes with the random walk conformational entropy, resulting in an almost straight conformation of the strand. A bending potential between two consecutive bonds along the polymer chain was used to introduce the bending rigidity and increase the persistence length l_p. A perfect tetrafunctional network with a diamond-like topology of cross-linked strands was manually constructed, as illustrated in Figure 8.9. A spherical nanoparticle is carved out of beads on a FCC lattice and weakly interacts with the polymers. The contour length L_c of the minimum loop that consists of six network strands around the nanoparticle was

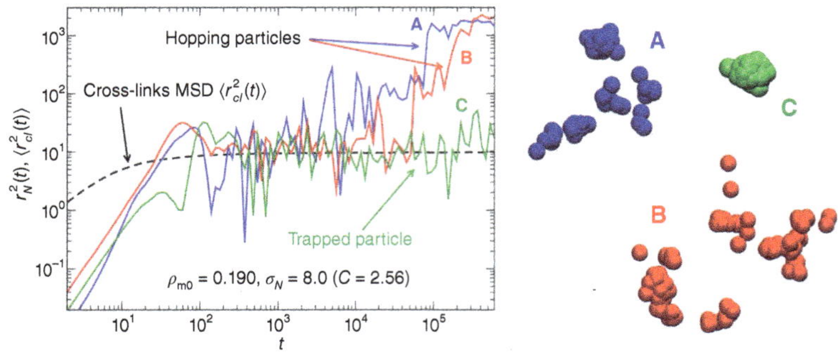

Figure 8.7 The squared displacements of three selective nanoparticles $r_N^2(t)$ and the mean squared displacements of the cross-links $\left\langle r_{cl}^2(t)\right\rangle$ in the simulations by Sorichetti et al. Trajectories of the three nanoparticles (A, B, and C) over time are shown explicitly. Source: Adapted from Ref. [25].

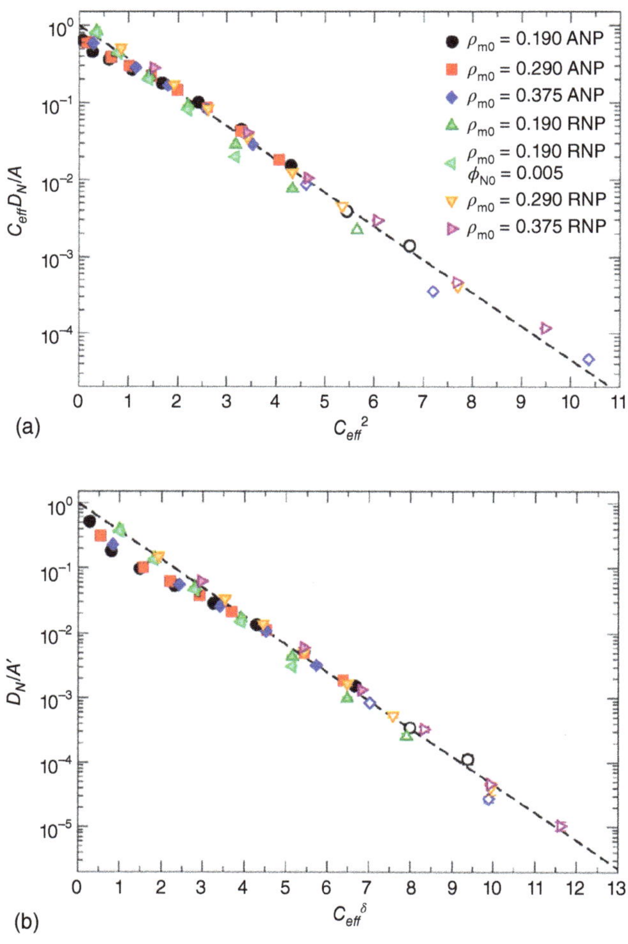

Figure 8.8 Comparisons of the results (symbols) from the simulations by Sorichetti et al. with the predictions (dashed line) of (a) the Cai–Paniukov–Rubinstein theory and (b) the Dell–Schweizer theory. Source: Adapted from Ref. [25].

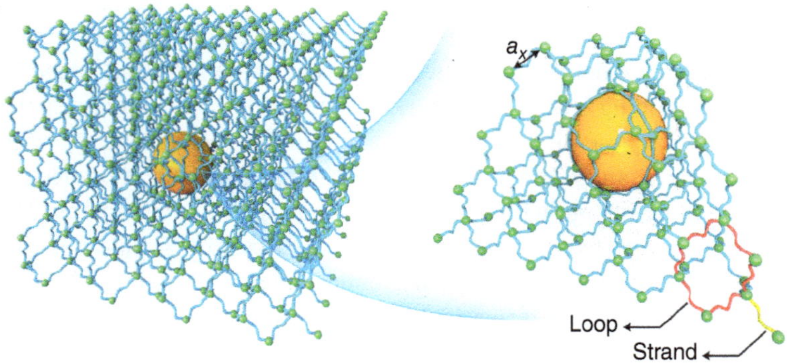

Figure 8.9 Schematic illustration of a nanoparticle in a semiflexible tetrafunctional network, which is a model system used in the simulations by Xu et al. A network strand and a six-strand loop are highlighted in yellow and red, respectively. Source: Adapted from Ref. [27].

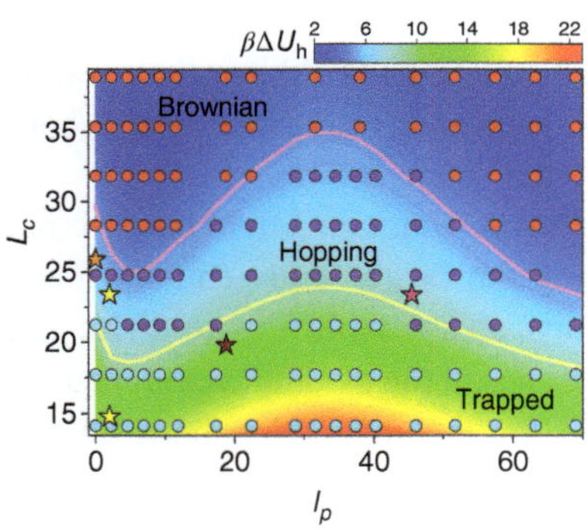

Figure 8.10 Diagram illustrating different regimes for the dynamics of nanoparticles in semiflexible tetrafunctional networks. L_c and l_p indicate the contour length of a six-strand loop and the polymer persistence length, respectively. The contour map indicates the energy barrier for nanoparticle hopping. Source: Adapted from Ref. [27].

varied and combined with the variation of l_p to systematically generate various semiflexible polymer network structures. A diagram indicating features of nanoparticle dynamics in the (l_p, L_c) parameter space has been constructed based on the simulation results. As shown in Figure 8.10, for fixed l_p and particle diameter d, the nanoparticle dynamics transit from Brownian diffusion with no confinement at large L_c to hopping diffusion with weak confinement at intermediate L_c and eventually to fluctuation in trapped states with strong confinement at small L_c. This type of transition with increasing confinement in semiflexible networks is similar to that in the flexible network [25]. A distinctively new feature of nanoparticle dynamics in semiflexible polymer networks is the non-monotonic change with increasing l_p at fixed L_c and d. Specifically, as shown in Figure 8.10, the hopping diffusion is more pronounced at intermediate l_p compared to both softer networks with smaller l_p and stiffer networks with larger l_p.

Xu et al. developed a theory to account for the non-monotonic variation with increasing l_p. They derived the bending energy of semiflexible network strand using the railway track model [28, 29] of worm-like chains (WLCs) and added its contribution to the hopping energy barrier. The expanded hopping energy barrier $\Delta U = k_B T[A\xi^2 + B\xi\kappa^2 \exp(-\kappa^2)]$, where A and B are numerical coefficients, $\xi = d/a_x$ is the confinement ratio, and κ is a dimensionless parameter essentially reflecting the strength of bending energy with respect to the entropic binding energy. The quadratic term is the same as the entropic hopping barrier in flexible polymer networks. The new linear term arises from the bending energy of semiflexible polymers. The competition of the two terms determines whether $\Delta U \sim \xi^2$ or $\Delta U \sim \xi$. The linear term depends non-monotonically on κ, with a crossover from κ^2 to $\exp(-\kappa^2)$ as κ increases. The linear term can outcompete the quadratic term when it rises as κ^2 but eventually yield to the quadratic term when it decays as $\exp(-\kappa^2)$. As a result, $\Delta U \sim \xi$ emerges only in the networks with intermediate bending stiffness, while $\Delta U \sim \xi^2$ for both softer and stiffer networks. The dependence of ΔU on κ explains the non-monotonic variation with increasing l_p in Figure 8.10. Furthermore, Xu et al. determined ΔU based on measuring the residence time t_c of a nanoparticle in a network cell and the relation $t_c \sim \exp(-\Delta U/k_B T)$. Figure 8.11a shows their results of t_c as functions of ξ, and Figure 8.11b shows the scaling of the ΔU with ξ, both corroborating their theory.

Simulations enable access to microscopic details that are often hard if not impossible to obtain in experiments. By varying the degree and genus of a cell within a perfect polymer network in simulations, as shown in Figure 8.12a, Dai et al. delineated the effects of cell topological features on the dynamics of embedded nanoparticles [30]. The free energy landscape for the interaction between the nanoparticle and the surrounding network cell was explicitly constructed (see Figure 8.12b), and its variations with particle diameter d and cell topology were identified. Furthermore, the topology-mediated nanoparticle dynamics, which are quantified by the time dependence of the mean squared displacement and the d-dependence of the terminal diffusion coefficient, were studied. While the asymptotic behaviors of the dynamics in the simulations are captured by the

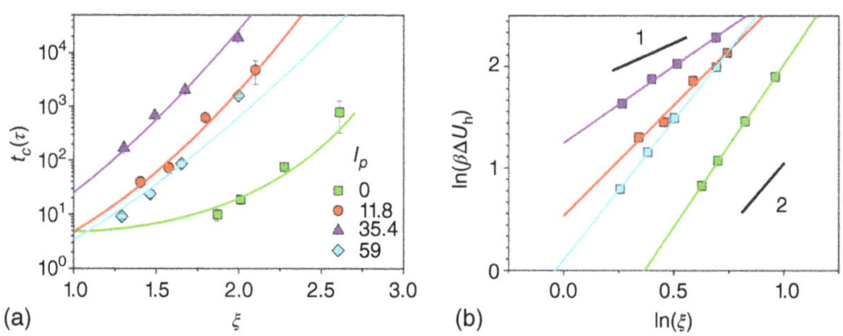

Figure 8.11 (a) The residence time t_c of a nanoparticle in a network cell as a function of the confinement ratio $\xi = d/a_x$ and (b) the hopping diffusion energy barrier as a function of ξ on the log–log scale in the simulations by Xu et al. Source: Adapted from Ref. [27].

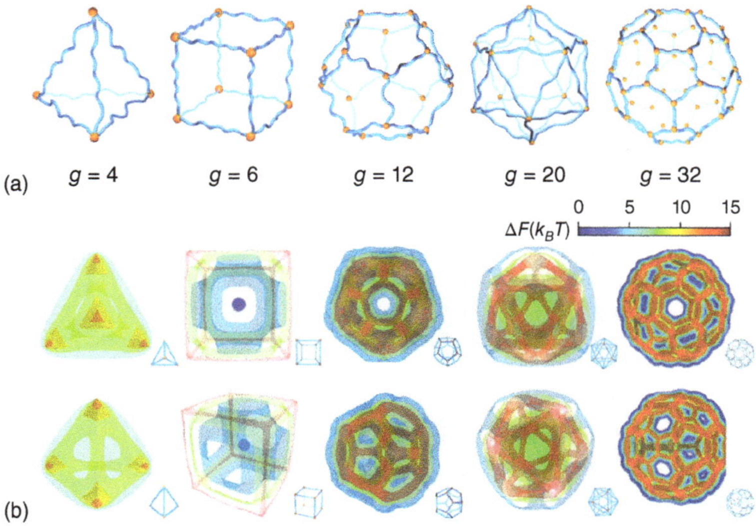

Figure 8.12 (a) Schematic illustration of the network cells with different geometries (g denotes the genus of the cell) and (b) the corresponding isosurfaces of the free energy change ΔF of a particle in the networks. Source: Adapted from Ref. [30].

scaling theory of Cai et al., the simulations and accompanying analytical theory reveal the topology-related details that are often in the crossover regimes between well-defined scaling regimes. One example is the crossover from $D{\sim}d^{-3}$ to exponentially suppressed D as the particle size increases with respect to the network cell size. Another one is the crossover from the inertia-dominated ballistic motion to the thermal motion of nanoparticles with increasing time, where various modes such as overdamped, harmonic, and underdamped modes may appear. Readers are advised to read the original paper by Dai et al. for the rich details about the geometry effects.

8.4 Open Questions and Future Research Directions

8.4.1 Network Strands with Nonlinear Architectures

Polymer networks may be created by cross-linking polymers with nonlinear architectures. As a result, the conventional linear network strands are replaced with nonlinear polymers. The research on the physical properties of polymers beyond linear architectures has been driven by advances in polymer chemistry that enable the precise control of polymer architectures. In the context of large nanoparticle dynamics in a cross-linked network, one would expect the nonlinear architecture changes the hopping energy barrier and thus affect the hopping diffusion. Two systems for future research envisioned by the author are nanoparticle hopping in cross-linked ring polymers and cross-linked bottlebrush polymers, as illustrated in Figure 8.13a,b, respectively.

Non-concatenated ring polymers have been at the frontier of topological polymer chemistry and physics [31–34]. The distinctive static and dynamic properties of ring

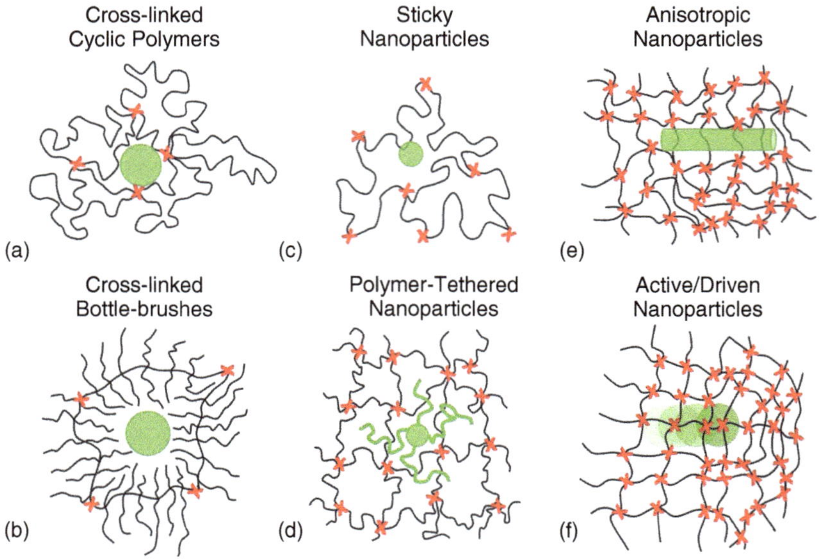

Figure 8.13 Schematic illustration of (a) a large nanoparticle trapped in cross-linked ring polymers, (b) a large nanoparticle trapped in cross-linked bottlebrush polymers, (c) a small sticky nanoparticle in cross-linked linear polymers, (d) a small nanoparticle with grafted polymer chains that get entangled with the surrounding network strands, (e) a thick nanorod trapped in a cross-linked linear polymer network, and (f) a large nanoparticle embedded in a cross-linked linear polymer network and driven by an external force.

polymer in concentrated solutions and the melt state have been studied extensively. Cross-linked ring polymer networks are less explored, but existing experiments of cross-linked ring polymer gels have demonstrated a lower modulus and larger swelling ratio compared to cross-linked linear polymer gels [35]. Wang et al. used large-scale molecular simulations to demonstrate that cross-linked ring polymer melts, or ring elastomers, are significantly softer and more stretchable compared to conventional elastomers made of linear polymers [36]. They identified the microscopic origin of ring elastomers' superior softness and stretchability as the lack of polymer entanglements at the same cross-linking density. Simulations have also shown that there is no entanglement network in a non-concatenated ring polymer melt that can confine the motion of a large nanoparticle [37, 38]. One thing to check for the nanoparticles in cross-linked ring polymers would be the lack of confinement on the particles by polymer entanglements. This new system would also serve as a good model of the diffusion of nanoparticles in a de-swollen polymer gel, in which upon solvent evaporation the network strands collapse to the conformations resembling that of non-concatenated ring polymers.

The bottlebrush polymer consists of a linear polymer backbone with densely grafted side chains, as illustrated in Figure 8.13b. Compared to linear polymers, it is harder for the bulky bottlebrush polymers to significantly overlap with each other and develop mutual entanglements. As a result, cross-linked bottlebrush polymer networks can be entanglement-free and possess a much lower modulus than the limit set by the entanglements in conventional linear polymers [39–41]. For the new system of nanoparticles in cross-linked bottlebrush polymers, one would expect the hopping energy barrier to be free from the entanglement effects and controlled solely by the architectural features of a bottlebrush polymer, such as the backbone contour length, side chain length, and grafting density. Additionally, it has been shown that a bottlebrush polymer may be coarse-grained as a worm-like cylinder with the backbone persistence length controlled by the side chain length and grafting density [42]. As such, the effects of network strand stiffness on the hopping diffusion in the conventional linear polymer network may be extended to study the corresponding hopping diffusion in the bottlebrush polymer network.

8.4.2 Sticky and Polymer-Tethered Nanoparticles

Theoretical and computational research on nanoparticle dynamics in cross-linked polymer networks has focused on purely repulsive or weakly attractive interactions (on the order of $k_B T$ per monomer) between nanoparticles and surrounding polymers. One direction for future research would be the study of sticky nanoparticles ($>10k_B T$ per monomer) with strong attractive nanoparticle polymer interactions. The stickiness is anticipated to move the nanoparticle and surrounding network strands together and thus allow the network confinement to affect particles even if they are smaller than the network mesh size, as illustrated in Figure 8.13c. Previous experiments, simulations, and theory have suggested sticky particles participate in the dynamics of melt chains and move together with them in a mechanism termed "vehicle" transport [19, 43, 44]. Alternatively, a sticky particle may be approximated as a larger particle with a bound layer of adsorbed polymer chains as a shell surrounding the nanoparticle core [5]. Future research is needed to elucidate the microscopic mechanism of sticky nanoparticle motion in cross-linked polymers.

Another related direction for future research would be the study of polymer-tethered nanoparticles in cross-linked networks. As illustrated in Figure 8.13d, the entanglements between the tethered polymer chains and network strands would couple a particle smaller than the network mesh size to the network. For tethered nanoparticles in melts of linear polymer chains, recent molecular simulations and theories have identified the conditions and mechanisms of nanoparticles participating in the dynamics of permanently grafted polymer chains [45, 46]. Molecular simulations of tethered nanoparticles in cross-linked polymers may be readily performed. The theory for the dynamics of tethered nanoparticles in polymer melts provides some hints for the development of a new theory for tethered nanoparticle diffusion in cross-linked polymers.

8.4.3 Nanoparticles with Anisotropic Shape

Another direction for future theoretical and computational research on nanoparticle dynamics in cross-linked polymer networks would be the extension from spherical nanoparticles to highly anisotropic nanoparticles such as nanorods. For the motion of nanorods, there is an easy direction along the rod axis and two orthogonal hard directions normal to the rod axis. Translation motions in the easy and hard directions within the frame of the rod may be coupled in the lab frame by the rotation of the rod. Therefore, the nanorod dynamics are enriched by the different translational and rotational modes. For nanorods in polymer melts, recent molecular simulations on monomerically thin nanorods in both unentangled and entangled polymer melts have been performed by Wang et al. [22]. Different modes exhibit dependencies on the nanorod length l with respect to the structural length scales in the melt, such as the size R of unentangled melt chains and the entanglement mesh size a_e of entangled melt chains. The rod reptation picture [47], which was proposed by de Gennes for thin rods in entangled polymers, was confirmed in the simulations. For thin nanorods with thickness smaller than the network mesh size, one would also expect a similar reptation across a cross-linked network by taking advantage of the easy direction along the rod axis. For thin nanorods with thickness moderately larger than the network mesh size, as illustrated in Figure 8.13e, one would expect a hopping diffusion along the rod axis. These conjectures may be evaluated in future computer simulations and quantitatively described by developing new theories.

8.4.4 Active Nanoparticles – Nonequilibrium Effects

Future research may also explore the motion of active nanoparticles in cross-linked polymer networks. Multiple means [48–56] may be utilized to enable activeness in the motion of nanoparticles within a polymer matrix, including external fields such as a magnetic field, an electric field, near-infrared light, and ultrasound or by chemistry-based mechanisms of self-propulsion. Overcoming the randomness of the passive nanoparticle dynamics that are coupled with the thermally driven dynamics of polymers, activeness leads to definite and directed motion, facilitating the control of nanoparticles in functional nanocomposites and particle-based nanomotors in bio-imaging and drug delivery.

Recent molecular simulations and accompanying scaling theory [23] for force-driven thin nanorods in an unentangled polymer melt have delineated the force-driven ballistic motion of nanorods with reduced frictional coefficient, where the external force renews the local viscous polymeric environment at a rate faster enough that the nanorod is only coupled to a fraction of the bulk melt viscosity. For nanoparticles in a cross-linked network, one conjecture is that the external driving force may effectively lower the hopping energy barrier and the force-driven ballistic motion would facilitate the hop from one cage to another. See Figure 8.13f for an illustration of a force-driven nanoparticle embedded in a cross-linked polymer network. Simulations may be performed to test the conjectured picture, and an accompanying theory may be developed. Furthermore, one may also extend the research to study the nanoparticle motion in cross-linked polymer networks under large deformation. The hopping energy landscape would be altered from the network side (rather than from the particle side as in the case of active nanoparticles) and affect the particle diffusion in a qualitatively different way.

8.5 Concluding Remarks

Theoretical studies of nanoparticle hopping in cross-linked polymer networks began with the prototypical case of nonsticky spherical nanoparticles in cross-linked flexible polymers. Two qualitatively different theories have been formulated: the scaling theory by Cai, Paniukov, and Rubinstein and the nonscaling microscopic theory by Dell and Schweizer. Computer simulations that offer detailed information on the microscopic picture have been used by Sorichetti et al. to evaluate the two theories and substantiate the partial success of both. Computer simulations have also enabled the examination of important effects beyond the prototypical case, including the effects of polymer chain stiffness and network geometric details, as revealed by Xu et al. and Dai et al., respectively. More details of the prototypical theories need careful inspection by computer simulation. For instance, the small hopping diffusion step size independent of the nanoparticle size and the softer confinement by entanglements with respect to cross-links, both of which are the predictions of the Cai–Paniukov–Rubinstein theory, need to be assessed by simulations. The entropic origin of the hopping energy barrier as described in both the Cai–Paniukov–Rubinstein and the Dell–Schweizer theories also needs to be verified. One route to verifying the entropic origin is via the examination of the temperature dependence of hopping diffusion, as a purely entropic hopping energy barrier would give no temperature dependence for the reduction in the rate of successful hopping.

Going beyond the prototypical case, qualitatively distinctive systems may be simulated and theoretically modeled. As speculated in Session 4, future research directions may aim at resolving the roles of nonlinear network strand architectures, sticky nanoparticle–polymer interactions, polymers tethered to nanoparticle surfaces, anisotropic nanoparticle shape, activeness as well as nonequilibrium effects, each adding a new dimension to the hopping diffusion of nanoparticle in polymer networks and mirroring the rich physics embodied in complicated real systems in polymer engineering and biomedical applications. It is hoped that not only theoretical and computational researchers but also experimentalists in diverse fields would be inspired to pursue the speculated research directions, conceive new research ideas, and make impactful interdisciplinary progress.

Acknowledgments

T.G. is grateful to M. Rubinstein and G. S. Grest for introducing the reviewed topic to him. T.G. would like to thank L.-T. Yan for inviting him to write up this chapter. T.G. also would like to thank S. Paniukov, D. Vlassopoulos, T. P. Lodge, S. K. Kumar, B. C. Benicewicz, K. I. Winey, R. Composto, R. Leheny, R. Riggleman, K. S. Schweizer, P. Olmsted, E. Kumacheva, L.-H. Cai, S.-W. Cheng, C. Aponte-Rivera, S.-F. Cheng, T. C. O'Connor, J. Wang, J. Clemmer, and P. A. Taylor for discussions and communications regarding the topic on different occasions. T.G. acknowledges start-up funds from the University of South Carolina. This work was supported in part by the National Science Foundation EPSCoR Program under NSF Award

No.OIA-1655740. Any opinions, findings, and conclusions or recommendations expressed in this material are those of the authors and do not necessarily reflect those of the National Science Foundation. This work is partially supported by an ASPIRE grant from the Office of the Vice President for Research at the University of South Carolina.

References

1 Kumar, S.K., Benicewicz, B.C., Vaia, R.A., and Winey, K.I. (2017). 50th anniversary perspective: are polymer nanocomposites practical for applications? *Macromolecules* 50: 714–731.

2 Bailey, E.J. and Winey, K.I. (2020). Dynamics of polymer segments, polymer chains, and nanoparticles in polymer nanocomposite melts: a review. *Progress in Polymer Science* 105: 101242.

3 Mackintosh, F.C. and Schmidt, C.F. (1999). Microrheology. *Current Opinion in Colloid & Interface Science* 4: 300–307.

4 Molaei, M., Atefi, E., and Crocker, J.C. (2018). Nanoscale rheology and anisotropic diffusion using single gold nanorod probes. *Physical Review Letters* 120: 118002.

5 Park, J., Bailey, E.J., Composto, R.J., and Winey, K.I. (2020). Single-particle tracking of nonsticky and sticky nanoparticles in polymer melts. *Macromolecules* 53: 3933–3939.

6 Suk, J.S., Suh, J., Choy, K. et al. (2006). Gene delivery to differentiated neurotypic cells with RGD and HIV Tat peptide functionalized polymeric nanoparticles. *Biomaterials* 27: 5143–5150.

7 Lai, S.K., Wang, Y.Y., and Hanes, J. (2009). Mucus-penetrating nanoparticles for drug and gene delivery to mucosal tissues. *Advanced Drug Delivery Reviews* 61: 158–171.

8 Blanco, E., Shen, H., and Ferrari, M. (2015). Principles of nanoparticle design for overcoming biological barriers to drug delivery. *Nature Biotechnology* 33: 941–951.

9 Danielsen, S.P.O., Beech, H.K., Wang, S. et al. (2021). Molecular characterization of polymer networks. *Chemical Reviews* 121: 5042–5092.

10 Lodge, T.P. and Hiemenz, P.C. (2021). *Polymer Chemistry*. CRC Press.

11 Dell, Z.E. and Schweizer, K.S. (2014). Theory of localization and activated hopping of nanoparticles in cross-linked networks and entangled polymer melts. *Macromolecules* 47: 405–414.

12 Cai, L.H., Panyukov, S., and Rubinstein, M. (2015). Hopping diffusion of nanoparticles in polymer matrices. *Macromolecules* 48: 847–862.

13 Wyart, F.B. and De Gennes, P.G. (2000). Viscosity at small scales in polymer melts. *European Physical Journal E* 1: 93–97.

14 Cai, L., Panyukov, S., and Rubinstein, M. (2011). Mobility of nonsticky nanoparticles in polymer liquids. *Macromolecules* 44: 7853–7863.

15 Ge, T. Scaling perspective on dynamics of nanoparticles in polymers: length- and time-scale dependent nanoparticle – polymer coupling. *Macromolecules* ASAP (2023). doi:https://doi.org/10.1021/acs.macromol.3c00260
16 Rubinstein, M. and Colby, R.H. (2003). *Polymer Physics*. Oxford University Press.
17 Doi, M. and Edwards, S.F. (1988). *The Theory of Polymer Dynamics*. Clarendon Press.
18 Kremer, K. and Grest, G. (1990). Dynamics of entangled linear polymer melts: a molecular-dynamics simulation. *Journal of Chemical Physics* 92: 5057–5086.
19 Carroll, B., Bocharova, V., Carrillo, J.M.Y. et al. (2018). Diffusion of sticky nanoparticles in a polymer melt: crossover from suppressed to enhanced transport. *Macromolecules* 51: 2268–2275.
20 Karim, M., Kohale, S.C., Indei, T. et al. (2012). Determination of viscoelastic properties by analysis of probe-particle motion in molecular simulations. *Physical Review E – Statistical, Nonlinear, and Soft Matter Physics* 86: 1–6.
21 Karim, M., Indei, T., Schieber, J.D., and Khare, R. (2016). Determination of linear viscoelastic properties of an entangled polymer melt by probe rheology simulations. *Physical Review E* 93: 1–12.
22 Wang, J., O'Connor, T.C., Grest, G.S. et al. (2021). Diffusion of thin nanorods in polymer melts. *Macromolecules* 54: 7051–7059.
23 Zhang, S., Wang, J., and Ge, T. (2022). Force-driven active dynamics of thin nanorods in unentangled polymer melts. *Soft Matter* 18: 6582–6591.
24 Everaers, R. and Ejtehadi, M.R. (2003). Interaction potentials for soft and hard ellipsoids. *Physical Review E* 67: 041710.
25 Sorichetti, V., Hugouvieux, V., and Kob, W. (2021). Dynamics of nanoparticles in polydisperse polymer networks: from free diffusion to hopping. *Macromolecules* 54: 8575–8589.
26 Monti, J.M., Clemmer, J.T., Srivastava, I. et al. (2022). Large-scale frictionless jamming with power-law particle size distributions. *Physical Review E* 106: 1–13.
27 Xu, Z., Dai, X., Bu, X. et al. (2021). Enhanced heterogeneous diffusion of nanoparticles in semiflexible networks. *ACS Nano* 15: 4608–4616.
28 Everaers, R., Bundschuh, R., and Kremer, K. (1995). Fluctuations and stiffness of double-stranded polymers: railway-track model. *Epl* 29: 263–268.
29 Liverpool, T.B., Golestanian, R., and Kremer, K. (1998). Statistical mechanics of double-stranded semiflexible polymers. *Physical Review Letters* 80: 405–408.
30 Dai, X., Zhang, X., Gao, L. et al. (2022). Topology mediates transport of nanoparticles in macromolecular networks. *Nature Communications* 13: 1–8.
31 Haque, F.M. and Grayson, S.M. (2020). The synthesis, properties and potential applications of cyclic polymers. *Nature Chemistry* 12: 433–444.
32 Kapnistos, M., Lang, M., Vlassopoulos, D. et al. (2008). Unexpected power-law stress relaxation of entangled ring polymers. *Nature Materials* 7: 997–1002.
33 Morgese, G., Trachsel, L., Romio, M. et al. (2016). Topological polymer chemistry enters surface science: linear versus cyclic polymer brushes. *Angewandte Chemie* 128: 15812–15817.

34 Ge, T., Panyukov, S., and Rubinstein, M. (2016). Self-similar conformations and dynamics in entangled melts and solutions of nonconcatenated ring polymers. *Macromolecules* 49: 708–722.

35 Zhang, K., Lackey, M.A., Cui, J., and Tew, G.N. (2011). Gels based on cyclic polymers. *Journal of the American Chemical Society* 133: 4140–4148.

36 Wang, J., O'Connor, T.C., Grest, G.S., and Ge, T. (2022). Superstretchable elastomer from cross-linked ring polymers. *Physical Review Letters* 128: 237801.

37 Ge, T., Kalathi, J.T., Halverson, J.D. et al. (2017). Nanoparticle motion in entangled melts of linear and nonconcatenated ring polymers. *Macromolecules* 50: 1749–1754.

38 Ge, T., Grest, G.S., and Rubinstein, M. (2018). Nanorheology of entangled polymer melts. *Physical Review Letters* 120: 57801.

39 Cai, L., Kodger, T.E., Guerra, R.E. et al. (2015). Soft poly (dimethylsiloxane) elastomers from architecture-driven entanglement free design. *Advanced Materials* 27: 5132–5140.

40 Daniel, W.F.M., Burdyńska, J., Vatankhah-Varnoosfaderani, M. et al. (2016). Solvent-free, supersoft and superelastic bottlebrush melts and networks. *Nature Materials* 15: 183–189.

41 Sheiko, S.S. and Dobrynin, A.V. (2019). Architectural code for rubber elasticity: from supersoft to superfirm materials. *Macromolecules* 52: 7531–7546.

42 Dutta, S., Pan, T., and Sing, C.E. (2019). Bridging simulation length scales of bottlebrush polymers using a wormlike cylinder model. *Macromolecules* 52: 4858–4874.

43 Yamamoto, U., Carrillo, J.M.Y., Bocharova, V. et al. (2018). Theory and simulation of attractive nanoparticle transport in polymer melts. *Macromolecules* 51: 2258–2267.

44 Bailey, E.J., Griffin, P.J., Composto, R.J., and Winey, K.I. (2019). Multiscale dynamics of small, attractive nanoparticles and entangled polymers in polymer nanocomposites. *Macromolecules* 52: 2181–2188.

45 Ge, T. and Rubinstein, M. (2019). Mobility of polymer-tethered nanoparticles in unentangled polymer melts. *Macromolecules* 52: 1536–1545.

46 Ge, T., Rubinstein, M., and Grest, G.S. (2020). Effects of tethered polymers on dynamics of nanoparticles in unentangled polymer melts. *Macromolecules* 53: 6898–6906.

47 de Gennes, P.G. (1981). Motions of one stiff molecule in an entangled polymer melt. *Journal de physique Paris* 42: 473–477.

48 Kagan, D., Laocharoensuk, R., Zimmerman, M. et al. (2010). Rapid delivery of drug carriers propelled and navigated by catalytic nanoshuttles. *Small* 6: 2741–2747.

49 Wu, Z., Lin, X., Zou, X. et al. (2015). Biodegradable protein-based rockets for drug transportation and light-triggered release. *ACS Applied Materials and Interfaces* 7: 250–255.

50 Fan, D.L., Cammarata, R.C., and Chien, C.L. (2008). Precision transport and assembling of nanowires in suspension by electric fields. *Applied Physics Letters* 92: 1–4.

51 Wu, Z., Si, T., Gao, W. et al. (2016). Superfast near-infrared light-driven polymer multilayer rockets. *Small* 12: 577–582.

52 Díez, P., Esteban-Fernández De Ávila, B., Ramírez-Herrera, D.E. et al. (2017). Biomedical nanomotors: efficient glucose-mediated insulin release. *Nanoscale* 9: 14307–14311.

53 Hansen-Bruhn, M., de Ávila, B.E.F., Beltrán-Gastélum, M. et al. (2018). Active intracellular delivery of a Cas9/sgRNA complex using ultrasound-propelled nanomotors. *Angewandte Chemie – International Edition* 57: 2657–2661.

54 Wang, Y., Hernandez, R.M., Bartlett, D.J. et al. (2006). Bipolar electrochemical mechanism for the propulsion of catalytic nanomotors in hydrogen peroxide solutions. *Langmuir* 22: 10451–10456.

55 Gao, W., Uygun, A., and Wang, J. (2012). Hydrogen-bubble-propelled zinc-based microrockets in strongly acidic media. *Journal of the American Chemical Society* 134: 897–900.

56 Manjare, M., Yang, B., and Zhao, Y.P. (2013). Bubble-propelled microjets: model and experiment. *Journal of Physical Chemistry C* 117: 4657–4665.

9

Molecular Dynamics Simulations of the Network Strand Dynamics and Nanoparticle Diffusion in Elastomers

Yulong Chen[1] *and Jun Liu*[2]

[1] *Zhejiang Key Laboratory of Plastic Modification and Processing Technology, College of Materials Science and Engineering, Zhejiang University of Technology, Hangzhou 310014, China*
[2] *State Key Laboratory of Organic-Inorganic Composites, College of Materials Science and Engineering, Beijing University of Chemical Technology, Beijing 100029, China*

9.1 Introduction

Elastomers are polymer networks that are used at temperatures above their glass transition temperature. They are typically composed of covalently bonded macromolecules, though they can also be constructed from noncovalent interactions (the latter are referred to as "thermoplastic elastomers"). As a result, they are soft but nonfluid materials (Young's moduli of $\approx 10^6$ Pa or lower) that can be reversibly stretched to high extensions without breaking [1]. Despite their excellent elasticity and reversible deformability, the original form of elastomers is too weak and has limited usage. In the rubber industry, with the exception of vulcanization, no process is more important or as widely practiced as the reinforcement of elastomers by particulate fillers. Adding fillers, especially nanosized fillers, to an elastomer matrix, even at a relatively small volume fraction, has considerable impact on mechanical strength, stiffness, and functional properties of the materials [2]. Elastomers filled with carbon black or silica particles may have an elastic modulus up to a few hundred times higher than that of the pure elastomer, exhibit high dissipative efficiency, and may be extremely resistant to tear and wear. These characteristics have been widely used for materials such as tires, shock absorbers, and impact modifiers.

Comprehensive knowledge of the mechanical and other properties of filled elastomers and deep understanding of their microscopic origin are of great fundamental and practical importance. However, their microscopic structures and dynamic properties remain largely unknown and uncontrolled. Hence, systematic investigations of structure–dynamic–property relations face difficulties, since experimentally it is nearly impossible to accurately and independently characterize the chemical details, topological or morphological structures, and relaxational or diffusional dynamics of the network or embedded nanoparticles (NPs) of the studied samples [3, 4]. Furthermore, different cross-linking chemistries give rise to

Dynamics and Transport in Macromolecular Networks: Theory, Modeling, and Experiments,
First Edition. Edited by Li-Tang Yan.

different types of network defects such as dangling loops and dangling ends that acerbate the complexity of characterizing their structures and dynamics [5, 6].

Computer simulations have proved invaluable for the study of molecular systems, with the main advantage to link macroscopic and microscopic phenomena, as well as to test predictions and assumptions of theoretical models. A key advantage of computer simulation techniques in the context of polymer science is their ability to conclusively determine the effect of a limited set of parameters on the physical behavior, while constraining all others. Therefore, computational approaches have garnered a lot of attention in studying the structures, dynamics, mechanics, and others of pure or NP-filled elastomer networks [7–14].

This chapter reports some representative studies of our group based on molecular dynamics (MD) simulations on the structures and dynamics of elastomer networks as well as the NP diffusion in the elastomer melts and networks [15–20]. The topological structure of networks is related to the connectivity of network subcomponents, i.e. it is determined by a synthetic methodology. To construct more realistic networks that include defects, in our simulations, we followed routes similar to those employed experimentally, including (i) random cross-linking of an equilibrated melt, mimicking vulcanization, and (ii) end-linking of polymer chain ends by multifunctional junctions. Topological features of these two classes of networks are analyzed. For the randomly cross-linked networks, the effect of cross-link density, cross-link distribution, and temperature on network dynamics is discussed. For the end-linked networks, the influences of network strand length and network junction functionality are investigated. Then, the diffusion dynamics of NPs in elastomers either in their melt state or after cross-linking are studied. For particle diffusing in elastomer melts, the influences of particle size, shape, surface grafting, and the chain architecture of the melt (i.e. linear and bottlebrush) are considered. For particles in end-linked elastomer networks, the network confinement effect, determined by the size of particle relative to network mesh, and the polymer–particle attraction are the main factors concerned. Finally, we give conclusions and outlook in the last part.

9.2 Structures and Dynamics of Model Elastomer Networks

Elastomer networks can be formed from a mixture of bi- and multifunctional monomers or by cross-linking precursor chains [21]. These precursor strands, also known as prepolymers, are polymer chains containing reactive functional groups that can form macroscopic networks under appropriate conditions, by random cross-linking (i.e. vulcanization) [22] or by end-linking with a multifunctional junction [23]. Therefore, in this section, simulation studies related to the structures and dynamics of randomly cross-linked networks [15, 16] and end-linked elastomer networks [20] are discussed.

9.2.1 Randomly Cross-linked Elastomer Networks

Sulfur vulcanization of rubbers is a well-known example to obtain randomly cross-linked elastomer networks where the functional groups present on the backbone or side chains of prepolymers are cross-linked by reaction with sulfur. In this subsection, we will introduce simulation studies on the structures and dynamics of elastomer networks based on a randomly cross-linking technique similar to the rubber vulcanization [15, 16]. The effects of cross-link density, cross-link distribution, and temperature on the network dynamics were comprehensively examined.

9.2.1.1 Network Models and Simulation Methodology

Computer simulation techniques have been extensively used to study polymer solutions, melts, brushes, and networks [24]. In classical simulations, all-atom (AA) simulations are the most detailed models, and as such they provide details on the atomic and molecular levels. Coarse graining (CG) of the molecular model enables increased computational efficiency, which allows longer time and length scales to be explored, at the expense of limited chemical detail. Several CG representations for polymers are commonly used in simulating polymer networks. Herein, the well-known bead–spring model proposed by Kremer and Grest [25] was employed to represent the network precursor chains. Each chain consists of N beads with mass equal to m and diameter equal to σ. The adjacent beads in each chain are connected via a stiff finite extensible nonlinear elastic (FENE) potential with standard parameters. Nonbonded interactions between polymer beads, both intra- and intermolecular, are described by the truncated and shifted Lennard–Jones (LJ) potential with a cutoff distance of $r_c = 2^{1/6}\sigma$ or $r_c = 2.5\sigma$ to introduce purely repulsive or weakly attractive interaction between polymer monomers depending on the situation we investigated. The cross-linker molecules are also modeled as beads with radius and mass the same with the polymer monomers. The pairwise interactions between cross-linkers and between cross-linkers and polymer monomers are also represented by the truncated and shifted LJ potential.

In our MD simulations, the NVT ensemble was adopted, where the temperature in most cases is fixed at $T^* = 1.0$ by using the Nose–Hoover thermostat. The velocity–Verlet algorithm was used to integrate the equations of motion with a time step $\delta t = 0.001$, where the time is reduced by τ (τ is the LJ time unit of the simulation). The periodic boundary conditions were used in all three directions. The systems were equilibrated as polymer melts at a number density $\rho = 0.85\,\sigma^{-3}$. After sufficient equilibrium, the polymer chains were dynamically cross-linked under the same NVT conditions by randomly selecting the cross-linkers and polymer monomers within a distance of 1.25σ and tethering them together with a probability of 0.1 by a chemical bond also using FENE potential. Such a setting of low probability aims to slow down the rate of network formation and reduce the bonding heterogeneity. It is noteworthy that one cross-linker can only bond to two other beads (polymer monomers or other cross-linkers), and no polymer monomers are permitted to have more than two cross-links. These constraints allow polymer chains to construct a better three-dimensional framework. In our simulation, there inevitably exist a certain number ($\sim 20 \pm 5\%$) of self-linkages, i.e. the cross-linkages connecting polymer segments of the same chain, in a similar manner to that described by Grest and Kremer [26]. Conceivably, the self-linkages will generate unfavorable network defects such as dangling loops and consequently decrease the cross-link efficiency. From an experimental point of view, such a situation should be legitimate enough to mimic the real cross-linking or vulcanization process. The obtained network structures were further equilibrated under NVT ensemble with $T^* = 1.0$. The apparent cross-link density n is specified as the overall density of network junctions (either elastically active or inactive) in simulation systems, which is explicitly given by $n = N_{cl}/M$, where N_{cl} is the total number of cross-linkers and M is the total number of precursor linear chains.

To examine the effects of cross-link distributions, the cross-linker/cross-linker interaction parameters are varied in the modeling and initialization processes (before the cross-linking procedure) to regulate the spatial distribution of cross-linkers. Explicitly, at the cross-link density of $n = 2.4$, three polymer systems, i.e. aggregated dispersion system, random dispersion system, and uniform dispersion system, are chosen as representative samples of different class of spatial distribution of cross-linkers, as schematically shown in Figure 9.1. The three dispersion states of cross-links are verified by both the simulation snapshots and the probability distribution of smallest distance between neighboring cross-links, as shown in Figure 9.2.

9.2.1.2 Network Topology

The macroscopic networks created in our simulations have lots of topological imperfections and are far from perfect tree-like network structure. Therefore, we commence with an examination of the topological features of polymer networks with various cross-link densities and distributions. The connectivity of network junctions is determined using a similar analysis method to that of Kroll and Croll [27] on the basis of burning algorithm, thus allowing the elastically active strands and cross-links, the dangling loops, and the dangling ends of the network to be identified. Simulation results for the topological features of the networks are tabulated in Table 9.1. The number values of network topological structures are normalized by the total number of precursor chains in the largest cluster or the network, and the length values are expressed in terms of the number of polymer beads. It is worth stressing that only these topological structures belonging to the largest cluster (the elastic backbone) that forms the network are quantified, which means that all the structures in the elastically inactive sol part are not taken into account. Because

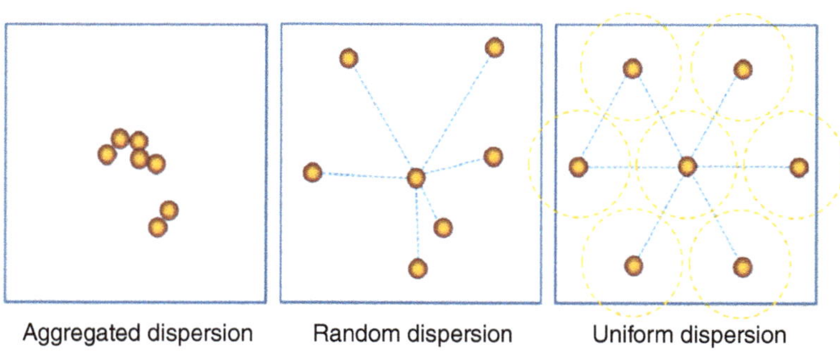

Figure 9.1 Schematic illustration of the three cross-link distribution states examined.

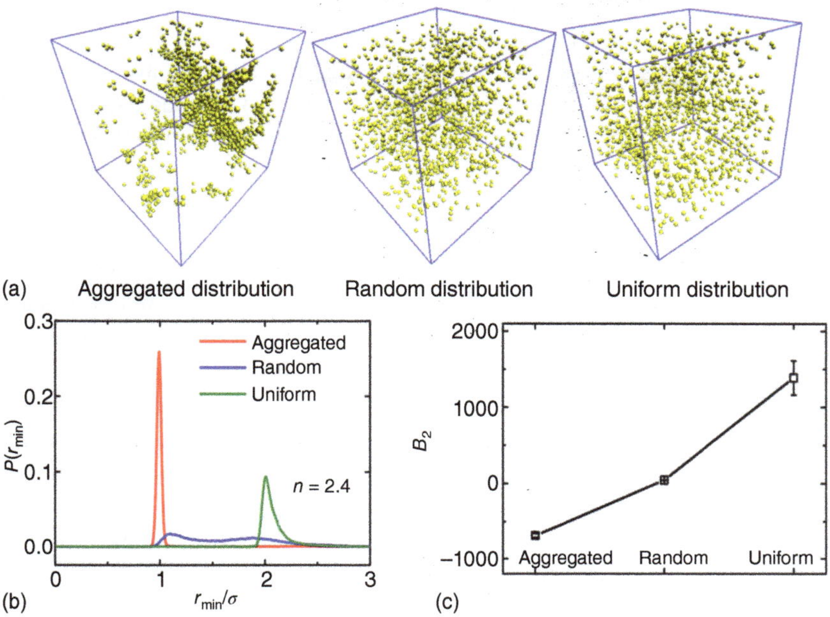

Figure 9.2 Comparison of three representative systems with different cross-link distributions: (a) the snapshots of simulation systems; (b) the probability distribution of the smallest distance between neighboring cross-links; (c) the second virial coefficient for cross-linkers. The small yellow spheres denote the cross-linkers, and polymer chains are not shown for visual clarity.

Table 9.1 Topological features of the randomly cross-linked elastomer networks.

System[a)]	$\langle L_{dl} \rangle$[b)]	$\langle L_{de} \rangle$[c)]	$\langle L_{st} \rangle$[d)]	ν_{st}[e)]	μ_e[f)]	ξ[g)]	ξ/ν_{st}[h)]
			Effect of cross-link density				
1.2/R	11.33	15.06	10.51	0.49	0.43	0.06	0.12
2.4/R	4.94	10.70	8.71	1.66	1.22	0.44	0.27
3.0/R	3.28	9.33	7.22	2.62	1.75	0.87	0.33
			Effect of cross-link distribution				
2.4/R	4.94	10.70	8.71	1.66	1.22	0.44	0.27
2.4/U	3.48	10.38	5.72	1.64	1.18	0.46	0.28

a) Apparent cross-link density/dispersion state of cross-linkers. Average length of b) dangling loops, c) dangling ends, and d) network strands between cross-links. Number of e) network strands and f) effective cross-links that are connected by at least three paths to the network. g) Cycle rank of the network and h) cycle rank per strand. Note that the number values of topological structures are expressed as concentrations with respect to the total number of precursor chains in the largest cluster or the network, and data for the lengths are expressed in terms of the number of polymer beads.

higher-order loops take very few portions in the network cyclic defects, only the primary loops that consist only one chain are taken into account herein.

From the results of random distribution systems, not only the average strand length $\langle L_{st} \rangle$ but also the average dangling loop length $\langle L_{dl} \rangle$ and the dangling end length $\langle L_{de} \rangle$ are seen to be shortened dramatically by raising the cross-link density. In particular, the network strand length $\langle L_{st} \rangle$ is a bit larger than the dangling end length $\langle L_{de} \rangle$; both of them approach to the expected value of $L/(2n + 1)$, which is in agreement with that found by Grest and Kremer [26]. Further analysis shows that the network strands and the elastically active cross-links are considerably increased in quantity at higher cross-link density, manifesting a better-connected network structure. Such a behavior of network structure is expected to lower down the dynamics of polymer segments and lead to an increase of glass transition temperature T_g (Section 9.2.1.3). In addition, compared with the random dispersion system of $n = 2.4$, the uniform dispersion system is found to have much shorter network strands between cross-links, while the number of network strands and effective cross-links seems to be slightly dependent on the cross-link distribution. The dangling loops and dangling ends are also shortened somewhat in the uniform dispersion system. In such a picture, polymer networks with higher degree of cross-link dispersion behave akin to those with higher cross-link density and imaginably will show similar network dynamics, which will be discussed in Section 9.2.1.4.

In the framework of the phantom network model [28], the elastic free energy from the connectivity of polymer network is proportional to the cycle rank (ξ) of the network, i.e. the number of scissions necessary to reduce it to a spanning tree. For any (even imperfect) network consisting of ν_{st} chains (strands) and μ_e junctions (effective cross-links), the cycle rank is given by $\xi = \nu_{st} - \mu_e$. In the presence of elastically inactive sol part, we adopt the notion that an effective cross-link is a vertex of a graph that is connected by at least three paths to the network. As can be seen from Table 9.1, the cycle rank ξ of the network increases monotonically with the cross-link density and distribution. For an ideal network devoid of any imperfections, the cycle rank can also be expressed in terms of the number of effective strands and the functionality of network cross-links $\xi = \nu_{st}(1 - 2/f)$, where f is the average functionality of network junctions. The cycle rank for the real network ($f = 4$) in our simulations, however, is proven to be much smaller than that of perfect network. Specifically, the cycle rank per strand (ξ/ν_{st}) of highly cross-linked networks ($n = 3.0$) is seen to be close to 1/3, which is less than that of the perfect tetrafunctional network (0.5). This can be ascribed to the formation of the dangling loops as well as the presence of dangling ends. The value of ξ/ν_{st} tends to be larger at higher cross-link density or dispersion degree, which confirms the decreasing network loop population. For the particular case of low cross-link density $n = 1.2$, not only the formation of loops but also the presence of sol causes the low efficiency of network connection.

9.2.1.3 Effect of Cross-link Density on Network Dynamics

A broad range of cross-link densities was explored, ranging from the uncross-linked system with $n = 0$ to the highly cross-linked system with $n = 12$ [15]. First, the total pair radial distribution functions (RDFs) were calculated, and the results showed that the static structure factor $S(q)$ is nearly independent of cross-link density. Additionally, we monitored the potential energy of the system, including the bonded and nonbonded energies, and found that the potential energy of the system varies linearly as a function of cross-link density. The effect of cross-link density on the network glass transition temperature T_g was also investigated. The results showed that T_g increases from approximately 0.42 to 0.50 when the system changes from a uncross-linked to a cross-linked state at a density $n = 12$, which is in agreement with the experimental results [29].

We then investigated the effect of cross-link density on the mobility and relaxation of the network strands. The monomer mobility on short time scales is almost the same when the cross-link density is small; while when the cross-link density is higher than $n = 2$, the short-time monomer mobility is also retarded. On long time scales, the monomers experience the constrained effects imposed by the cross-links, so the monomer mobility is significantly reduced with the increase of cross-link density. For $n = 12$, the long-time mean-square displacement (MSD) of the network strand monomers even becomes nearly unchanged.

In order to extract the effect of cross-link density on the relaxation behavior of polymer chains, a dynamic quantity, the incoherent intermediate dynamic structure (IIDS) factor $\phi_q^s(t)$ is calculated from the relationship: [30]

$$\phi_q^s(t) = \frac{1}{N}\sum_{k=1}^{N}\langle \exp(i\mathbf{q}\cdot[\mathbf{r}_k(t) - \mathbf{r}_k(0)])\rangle \tag{9.1}$$

where N stands for the total number of polymer segments in the rubbery state, $\mathbf{r}_k(t) - \mathbf{r}_k(0)$ indicates the displacement of the scattering center $\mathbf{k}$ over a period of time t, and q denotes the scalar wave vector transfer defining the spatial scale of the measurement. We choose two representative q values of 3 and 6.9. The smaller q value ($q = 3$) probes the length scale of a chain, whereas $q = 6.9$ probes the segmental or local relaxation, approximately corresponding to the maximum of the structure factor $S(q)$. The results are presented in Figure 9.3.

From Figure 9.3a, one can see that all $\phi_q^s(t)$ decay to zero within approximately $t = 10\tau$ except for $n = 12$ and shows little difference among different cross-link density, again suggesting that the relaxation dynamics of network strands at segmental scales almost not affected by cross-link density. From Figure 9.3b, we find that the relaxation dynamics at chain length scale shows a pronounce decrease with the increase of cross-link density. We define the time when $\phi_q^s(t)$ decays to $1/e$ as the characteristic relaxation time τ_q^s of network strands. We find τ_q^s increases linearly with cross-link density, as shown in Figure 9.3c, indicating a linear dependence of the characteristic relaxation time of network strands on cross-link density. The terminal relaxation dynamics of the network strands was also evaluated through the time decay of the autocorrelation function of the strand end-to-end vector correlation, $\phi_{\mathrm{ete}}^s(t)$. Similar behavior as that in Figure 9.3b was also observed. Meanwhile, the $\phi_{\mathrm{ete}}^s(t)$ curves can be well fitted by the Kohlrausch–William–Watts (KWW) function [31] when $n \leq 3$

$$\phi_{\mathrm{ete}}^s(t) = \exp\left[-\left(t/\tau_q^s\right)^{\beta}\right] \tag{9.2}$$

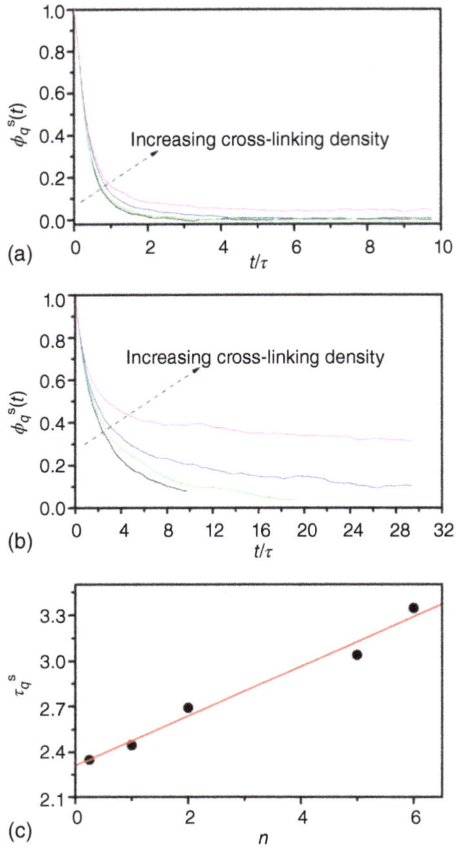

Figure 9.3 The decay of $\phi_q^s(t)$ with (a) $q = 6.9$ and (b) $q = 3$ at cross-link density $n = 0.25$, 2, 6, and 12. (c) The characteristic relaxation time at chain length scale τ_q^s as a function of cross-link density.

or the modified KWW function when $n > 3$

$$\phi_{\text{ete}}^s(t) = (1 - a)\exp\left[-\left(t/\tau_q^s\right)^\beta\right] + a \tag{9.3}$$

The stretching exponent parameter β extracted from the KWW function, which measures the distribution of the characteristic time, was found to decease with the increase of cross-link density due to the enhanced intermolecular coupling or cooperativity.

9.2.1.4 Effect of Cross-link Distribution on Network Dynamics

Now we turn our attention to the effect of cross-link distribution on the dynamic properties of network chains. Figure 9.4 shows the IIDS factor $\phi_q^s(t)$ for $q = 2$ and $q = 6.9$, propping the dynamics of network strands at chain length and segmental scales, respectively. From Figure 9.4, we see that all $\phi_q^s(t)$ curves with different cross-link densities and distributions almost coincide in short times of $t < 0.2\tau$, corresponding to the ballistic motion of monomers [32]. In contrast to the short-time regime, the $\phi_q^s(t)$ curves separate from one another with the increase of time. The $\phi_q^s(t)$ with higher degree of cross-link dispersion is found to decay more slowly at both the segment and the chain length scale. Particularly, the influence upon the

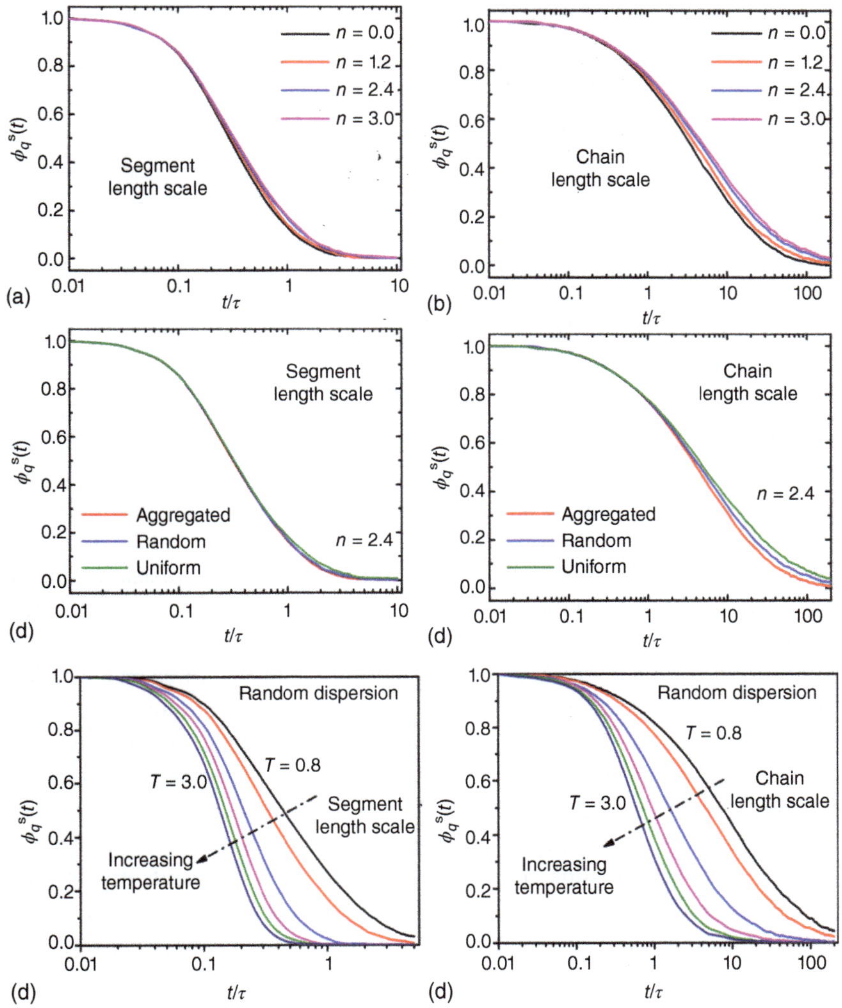

Figure 9.4 Decay of $\phi_q^s(t)$ with (a, c) $q = 6.9$ and (b, d) $q = 2$. (e, f) Decay of $\phi_q^s(t)$ for the random dispersion system with the cross-link density equal to $n = 2.4$ at different temperatures of $T = 0.8$, 1.0, 1.5, 2.0, 2.5, and 3.0, respectively.

relaxation behavior on the segment length scale is not noticeable, whereas for the larger length scale than the strand length, the effects of cross-linking become more pronounced. The depression of chain dynamics can be ascribed to the constraint effect of permanent cross-links, which increases appreciably with the distribution degree.

To probe the α-relaxation behavior in detail, we also use the empirical KWW function (Eq. (9.2)) to fit the data. The stretching exponent parameter β is then extracted from the KWW function, as tabulated in Table 9.2. It is found that β decreases monotonically with an increase in the degree of cross-link dispersion, suggesting the wider distribution of the relaxation time of either polymer segments or polymer chains.

Table 9.2 Dynamic properties of the randomly cross-linked elastomer networks.

	Segmental length scale			Chain length scale	
System	β[a)]	τ_q[b)]	κ[c)]	β	τ_q
2.4/A	0.866	0.468	1.101	0.634	7.587
2.4/R	0.847	0.475	5.133	0.575	8.756
2.4/U	0.794	0.485	7.980	0.525	10.420

a) Best-fitting parameter for $\phi_q^s(t)$ to the stretched exponential KWW function.
b) The characteristic α-relaxation time.
c) The cross-link constraint.

Furthermore, the characteristic α-relaxation time τ_q is evaluated through the time decay of $\phi_q^s(t)$. As displayed in Table 9.2, the characteristic α-relaxation time τ_q at either length scale is seen to increase with the cross-link distribution. Such a change can be rationalized by the more uniform network junctions and the resulting more intense cross-link constraints imposed on the polymer network, which will indeed slow down the chain dynamics. For our typical Rouse chains, the relaxation time is expected to be cut down with the decrease of chain length, given by $\tau_R \sim \xi N^2$, where ξ is the bead friction factor and N is the number of polymer beads in the chain. The Rouse friction coefficient, ξN, is inversely proportional to the diffusion coefficient describing the chain dynamics. Therefore, the cross-link constraint strength can be approximated by the value of τ_q/N. Considering the length of dangling ends is actually close to that of network strands, the cross-link constraint thus can be roughly measured by $\kappa \equiv \tau_q^{\text{net}}/\langle L_{\text{st}}\rangle$, where τ_q^{net} is the characteristic relaxation time for the polymer network at length scales less than $\langle L_{\text{st}}\rangle$. From Table 9.2, it can be clearly seen that the cross-link constraint upon the chain fragments dramatically changes with the cross-link distribution state. The uniform dispersion system shows the most significant cross-link constraint effect to the polymer network.

9.2.1.5 Effect of Temperature on Network Dynamics

The effect of temperature on the relaxation dynamics of the model elastomer networks was also investigated. The results showed that with increasing the temperature, the relaxation of network strands becomes faster. When rescaling the time with respect to the characteristic relaxation time, the $\phi_q^s(t)$ at segmental length scale reduce to a single master curve while this phenomenon is not observed for the case at chain length scale, implying that the time–temperature superposition principle (TTSP) is valid at segmental scale but beaks down at chain length scale. This is because at segmental length scale, the monomers do not experience the cross-link constraints, exhibiting the similar behavior to the uncross-linked system. However, at chain length scale, the movement of monomers is greatly influenced by the network junctions and the network heterogeneous causes the monomers to show different behaviors in response to the change in temperature.

The characteristic α-relaxation time τ_q for each decay curve of $\phi_q^s(t)$ is further identified to examine the TTSP of cross-linked elastomers. For the characteristic

α-relaxation time, the mode coupling theory (MCT) [32, 33] predicts a power law divergence:

$$\tau_q \propto (T - T_c)^{-\gamma} \tag{9.4}$$

where T_c is the critical temperature when the dynamics of polymer material qualitatively changes from a liquid-like to a solid-like behavior. The value of T_c is determined by fitting the data by Eq. (9.4). The quality of the fits can be judged in Figure 9.5a, where the relaxation time τ_q is plotted as a function of $T - T_c$ in a log–log representation where Eq. (9.4) should appear as a straight line. For our simulated polymer, T_c is found to be ~0.52 ± 0.04, slightly above T_g, and shows little dependence on the cross-link distribution at a given cross-link density.

For the large q ($q = 6.9$), the α-relaxation time τ_q agrees fairly well with the MCT prediction over the whole range of temperature we examined, whereas for the small q ($q = 2.0$) the τ_q fails to exhibit the predicted power law globally. Such a behavior of α-relaxation time is roughly independent of the cross-link distribution. These results also imply that the TTSP holds on the segment length scale but breaks down on the chain length scale. To be more specific, the TTSP on the chain length scale can be valid at sufficiently high temperature above the critical temperature.

It should be mentioned that the α-relaxation time can in principle verify the celebrated Vogel–Fulcher–Tammann (VFT) equation [34] with an Arrhenius-like temperature dependence, as given by:

$$\tau_q \propto \tau_0 \exp\left(\frac{E}{T - T_0}\right) \tag{9.5}$$

In this regard, the relaxation time τ_q is plotted as a function of $1/(T - T_0)$ in a log–linear representation where Eq. (9.5) should appear as a straight line. As can be seen in Figure 9.5b, the VFT prediction equally provides a good description of τ_q for the large q but fails as well for the small q, which is consistent with the conclusion from Figure 9.5a.

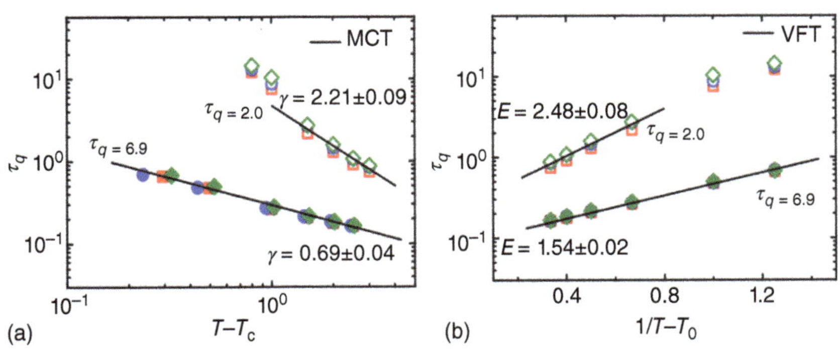

Figure 9.5 (a) MCT and (b) VFT plot of the relaxation times. Squares, circles, and diamonds correspond to the aggregated, the random, and the uniform dispersion systems, respectively. Open scatters are for the wave vector $q = 2.0$, and closed scatters are for the wave vector $q = 6.9$.

9.2.2 End-linked Elastomer Networks

Cross-linked polymer networks can be formed by various techniques, leading to a diverse and complex set of structures and properties. Of these network types, considerable attention has been paid to those formed by a process termed "end-linking." End-linked networks, elastomers in which polymer chains are connected to the network only by their ends, are highly examined by researchers due to their simplicity as well as their well-defined properties, which render them suitable candidates to be described theoretically and replicated computationally. Arguably, one of the most well-studied end-linked networks is polydimethylsiloxane (PDMS) [35, 36]. Usually, bifunctional PDMS precursor chains are mixed with cross-linker molecules of certain functionality (the number of available active groups per cross-linker molecule), f, at a specified stoichiometric imbalance between the two participating kinds of active groups, r =(cross-linkers active groups)/(precursors active groups), which, in the presence of a catalyst to initiate the reaction, form the end-linked network. PDMS rubbers not only are the most widely used silicon-based polymers, but also their chemistry allows for a variety of network topologies that can be used to tune their mechanical properties. Therefore, MD simulations were also performed to investigate the topological features and dynamics properties of this type of elastomer networks.

9.2.2.1 Network Models and Simulation Methodology

The precursor linear polymers that composed of N_x monomers are also modeled as bead–spring chains [25]. The two beads on the polymer ends are defined to be able to react with multifunctional junctions to form permanent cross-links. Nonbonded interaction between polymer beads is described by the truncated and shifted LJ potential, i.e. Weeks–Chandler–Andersen (WCA) potential [37]. The bonded interaction between any two connected monomers along a chain is also controlled by the FENE potential. A bending potential, $U_{\text{bend}}(r) = k_\theta(1 + \cos\theta)$, with $k_\theta = 1.5\varepsilon$, is included to reduce the entanglement chain length, N_e, from 85 to 28 [38]. The multifunctional junctions, each can join f ends of precursor polymers (f = 3–8), are also modeled as beads the same with the polymer monomers. The pairwise interactions between junctions and between junctions and polymer monomers are also treated using the WCA potential.

The simulations are started from a nonoverlapped configuration of precursor polymers and multifunctional junctions in a large simulation box implementing periodic boundary conditions. The simulation box was first compressed under isothermal–isobaric (NPT) ensemble to a volume V with desired bead number density. Then, canonical (NVT) simulations were performed to equilibrate the structures. For systems with long chains ($N_x > 30$), Monte Carlo bond swapping technique was also employed to accelerate the equilibrium process [39]. The end-to-end vector autocorrelation function $\langle \mathbf{u}(t) \cdot \mathbf{u}(0) \rangle$, with $\mathbf{u}(t)$ being the unit vector along the end-to-end vector of a chain, was employed to confirm the equilibration of the polymer chains. Then, the cross-linking process was carried out under the same NVT condition by randomly selecting the junctions and the end beads of the precursor polymers within 1.25σ and tethering them together by a permanent bond using the FENE potential. One junction permits to link f polymer ends, while one end bead only allows to react with one junction. During this process, the nonbonded LJ interaction between junctions and polymer end beads was switched to be attractive by setting $r_c = 2.5\sigma$. Once the network is completely developed (required about $10^5\tau$), i.e. all polymer ends are connected by the multifunctional junctions, the aforementioned LJ attraction is turned off. Afterward, the obtained network structures were further equilibrated for at least $10^5\tau$, ensuring that the network structures are fully relaxed. Subsequently, much longer production runs, up to $10^6\tau$, were carried out for data acquisition.

9.2.2.2 Network Topology

The topological features of the networks with various strand lengths and junction functionalities are tabulated in Table 9.3. Note that here the contents of junction monomers are slightly overdosed. The use of an excess number of junctions is based on the consideration that if only $n_j = 2n_{chain}/f$ of junctions are added, fluctuations in the spatial density of junctions create regions with unsaturated active ends and regions with free ends [40], which makes it difficult to prepare model networks with low dangling ends especially for long chains. Although dangling ends and

Table 9.3 Topological features of the end-linked elastomer networks.

N_x a)	n_{chain} b)	f c)	n_j d)	ϕ_{de} e) (%)	ϕ_{dl} f) (%)	ϕ_{st} g) (%)	$\bar{f}$ h)	d_x i) (σ)	P j)	d_{fluct} k) (σ)
5	4800	4	2520	0.54	4.12	93.56	3.81	2.78	1.90	2.31
10	2400	4	1260	0.46	3.38	93.88	3.81	4.58	4.22	3.11
20	1200	4	630	0.75	1.83	95.67	3.83	6.96	7.33	3.77
30	800	4	420	0.62	2.12	95.25	3.81	8.59	8.99	4.40
100	240	4	126	0.83	1.25	96.25	3.81	15.51	16.08	6.19
200	120	4	63	0	0.83	96.67	3.81	22.22	23.65	7.27
5	4800	3	3360	1.65	3.06	89.02	2.87	2.74	1.69	3.06
5	4800	6	1680	0.31	6.27	93.10	5.71	2.82	2.04	2.00
5	4800	8	1260	0.21	8.90	90.77	7.61	2.83	2.07	1.95

a) Monomer number per precursor chain.
b) Number of precursor chains.
c) Functionality of junctions.
d) Number of junctions.
e) Fraction of dangling ends.
f) Fraction of dangling loops, where both ends of the chain are attached to the same junction.
g) Fraction of the active network strands.
h) Average functionality of the active junctions.
i) Average size (end-to-end distance) of the active network strands.
j) Overlap parameter, i.e. the parameter that is defined as the number of network strands within the volume pervaded by one network strand.
k) Average fluctuation distance of the active network strands due to the restrictions of the adjacent network junctions.

heterogeneity in cross-link density can be largely reduced by this end-linking method, dangling loops and some unsaturated junctions inevitably exist in all polymer networks. Such topological defects would weaken the mechanical strength of the polymer networks to some extent [4]. Despite this fact, the networks produced by the present method are of active network strands over $\phi_{st} > 89\%$, indicating the successful construction of high-quality networks with monodispersed strands and low topological defects.

A typical polymer network has both chemical cross-links and topological entanglements. The relative length scales determine which type of network dominates the system properties. By implementing the same primitive path analysis (PPA) proposed by Everaers et al. [41], the entanglement length of the present networks was calculated and found to be about $N_e = 25 \pm 2$, comparable to the estimation by Hsu and Kremer [38] for linear chains ($N_e \approx 28$). The slightly lower N_e values are thought to be due to the chain end effect [42] of the relatively short precursor chains used in the present study. The change of N_e with the increase of f or the inclusion of a low loading of NPs is very limited and is almost within the statistical error.

The size of a network strand d_x is calculated as the average distance between two neighboring network junctions along the chain. In a network, there are many network strands overlapping within the volume pervaded by a network strand. We calculate the overlap parameter P by:

$$P = \pi d_x^3 n_{\text{chain}} \phi_{\text{st}}/(6V) \approx \pi d_x^3 \rho_P \phi_{\text{st}}/(6N_x) \tag{9.6}$$

which is defined as the number of network strands within the volume $\pi d_x^3/6$ pervaded by one network strand. We observe that the P values are larger than 1, and the variation trend of P with N_x or f is in parallel with that of d_x (Table 9.3), coinciding with the prediction by the scaling theory where $P \sim d_x^3/N_x \sim d_x$.

9.2.2.3 Network Dynamics

The IIDS factor $\phi_q^s(t)$ of the central monomers of the network strands for systems at various N_x and f was also analyzed, as given by Eq. (9.1). The results show that the segment-scale dynamics of network strands is almost unaffected by the strand length N_x (Figure 9.6a), while increasing the junction functionality f would slightly delay the tail of the relaxation process, though the shifts are nearly negligible (Figure 9.6b). The strand dynamics over larger length scales were also investigated by quantifying the IIDS factor at lower q values ($q = 1$, 2, 3, and 5) and find that with decreasing q value, the decay behavior becomes slower. This trend becomes more significant for the networks at lower N_x or higher f. For some cases, the low-q IIDS factor even cannot decay to zero due to the constraint effect of the permanent cross-links which enhances appreciably with decreasing N_x or increasing f.

The strand dynamics on longer length scale and time scale were also investigated by monitoring their MSD $\langle \Delta r^2 \rangle$. For MSD of the strand central monomers, there is a ballistic motion regime with MSD $\sim t^2$ at very short times ($t < 0.06\tau$), which is independent of N_x and f, as shown in Figure 9.6c,d, respectively. Beyond this ballistic motion regime, the slope of the MSD in logarithmic coordinates continuously decreases. When the time scale is larger than the characteristic time $\tau_0 \approx 0.39\tau$,

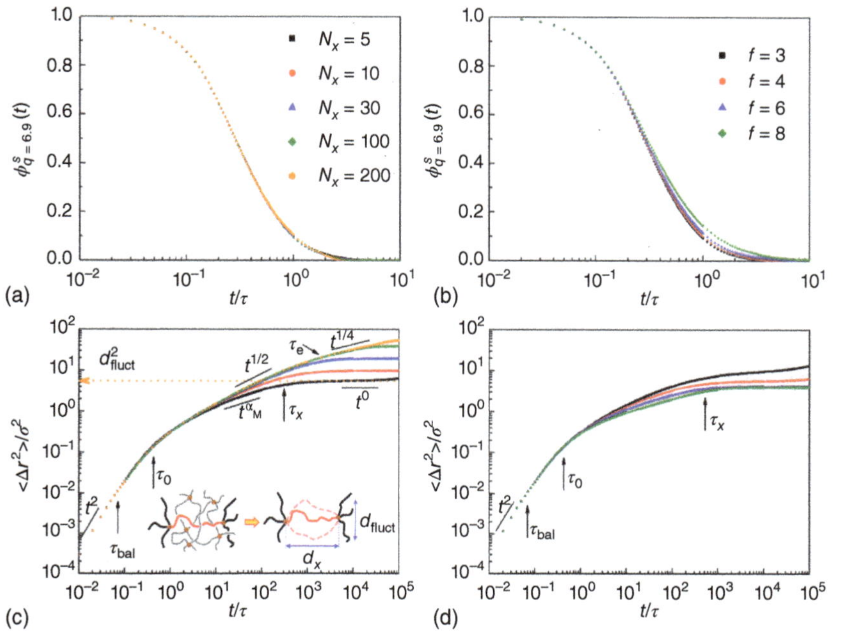

Figure 9.6 Decay of IIDS factor $\phi^s_{q=6.9}(t)$ of the strand central monomers for systems at different (a) strand lengths N_x and (b) junction functionalities f. MSD $\langle \Delta r^2 \rangle$ of the strand central monomers for systems at different (c) N_x and (d) f. In the left column, the junction functionality is fixed at $f = 4$, and in the right column, the strand length is fixed at $N_x = 5$. The inset in (c) shows the schematic view of the size d_x and fluctuation distance d_{fluct} of a strand in a cross-linked network.

corresponding to the segment-scale relaxation time τ_q (defined as the time when the $\phi^s_{q=6.9}(t)$ decays to $1/e$) [33], the motion of the strand central monomers becomes subdiffusive with MSD $\sim t^{\alpha_M}$, where $0 < \alpha_M < 1$.

For long network strands (i.e. $N_x = 100$ and 200), we find they exhibit a similar subdiffusive behavior with the heavily entangled linear polymers [38] before being restricted by the permanent network junctions at time τ_x (Figures 9.6c and 9.7a). Such dynamics behavior is almost unaffected by increasing f (from 3 to 8) or introducing a few NPs (Figure 9.7b). There is a transition in α_M at $\tau_e \approx 2 \times 10^3 \tau$ from 1/2 to 1/4 (Figures 9.6c and 9.7a) due to the starting of entanglement effect at this critical time, though the time window for the slope 1/4 regime is short ($\tau_e < t < \tau_x < 10^2 \tau_e$). According to the previous theory [43, 44], the entanglement tube diameter can be measured from the intersection of the slope 1/2 and 1/4 lines $\langle \Delta r^2 \rangle_{\tau_e} \sim d_T^2 \approx 2 \langle R_g^2(N_e) \rangle = (4.34 \pm 0.12\sigma)^2$, which is consistent with the result obtained by Hsu and Kremer [38] ($d_T \approx 5\sigma$). After reaching the Rouse relaxation time τ_x, monomers in a strand that are restricted within an imaginary tube have moved a distance determined from $\langle \Delta r^2 \rangle_{\tau_x} \sim d^2_{\text{fluct}} \approx d_T^2 (N_x/N_e)^{1/2} \approx 2 \langle R_g^2(N_e) \rangle (N_x/N_e)^{1/2}$. However, since the strand ends are fully linked, the subsequent chain reptative dynamics channel is closed, and hence the MSD reaches a plateau at $t > \tau_x$. We call this maximum displacement that the network strands can achieve as strand

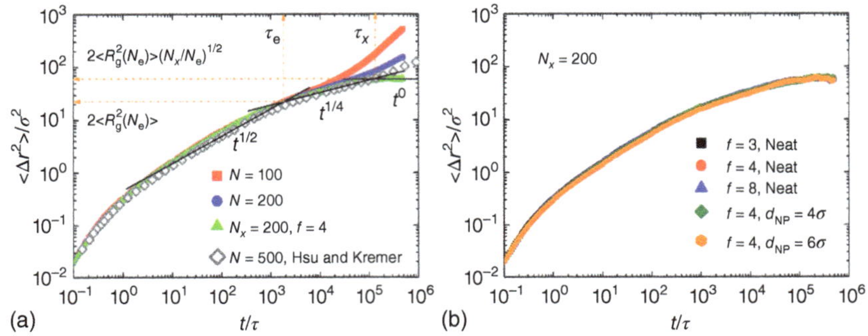

Figure 9.7 (a) MSD $\langle \Delta r^2 \rangle$ of the central monomers of the linear polymers with chain lengths $N = 100$, 200, and 500 (from Ref. 38) and that of the strands of the end-linked network with $N_x = 200$ and $f = 4$. (b) MSD $\langle \Delta r^2 \rangle$ of the strand central monomers for neat networks at fixed $N_x = 200$ but at varied f as indicated and that for filled networks at $N_x = 200$ and $f = 4$ with 0.94 vol.% and 1.57 vol.% of NPs of sizes $d_{NP} = 4$ and 6σ, respectively.

fluctuation distance d_{fluct} (Table 9.3). If we estimate the entanglement length from $d_{fluct}^2 \approx d_T^2(N_x/N_e)^{1/2}$, we get $N_e \approx 24.2$ and 25.4 for $N_x = 100$ and 200, respectively, which are consistent with the results from PPA [41].

When the network strand length is comparable to or lower than the entanglement length $N_e \approx 25$, the cross-linked network dominates the system properties. At time scales within the regime of $\tau_0 < t < \tau_x$, the exponent α_M decreases to a value lower than 1/2 due to the slowdown of the strand Rouse dynamics imposed by network junctions (Figure 9.6c). Increasing the junction functionality f would further enhance this constraint effect until to an extreme level (Figure 9.6d). Accordingly, the fluctuation distance d_{fluct} of the network stands is also reduced with decreasing N_x or increasing f (Table 9.3). Overall, the d_{fluct} values are lower than the network strand sizes d_x due to the restrictions of the overlapping "elementary" networks and/or the chain entanglements (see the inset in Figure 9.6c).

9.3 Diffusion Dynamics of Nanoparticles in Elastomers: Melts and Networks

The introduction of nanosized particles, such as carbon black, silica, carbon nanotube, and graphene, into elastomer matrices is an important strategy to endow new functions and/or improve their properties. After incorporating NPs within an elastomeric host, the relaxation dynamics of elastomer chains and thereby the corresponding dynamical and rheological properties change with the diffusivity of embedded NPs. It is imperative to control the diffusion of NPs in polymers to achieve elastomer-based nanomaterials with desired dynamics-related characteristics. In this section, we focused on the diffusion dynamics of NPs in elastomer matrices, either in their melt states or those after cross-linking.

9.3.1 Diffusion of Nanoparticles in Elastomer Melts

The motion of NPs is significantly influenced by their surrounding polymer environments. In melts of elastomers, various factors are important to the dynamics of NPs, resulting in the breakdown of the classic continuum Stokes–Einstein (SE) relation [45]. In this subsection, we will introduce some of our recent MD simulation studies related to the diffusion of NPs in elastomer melts, where many important factors, including the shape, size, surface modification of NPs, molecular weight, and topological architecture of polymers were considered [17–20].

9.3.1.1 Models and Simulation Methodology

In this part, the standard bead–spring model [25] was also adopted to represent the elastomer melt chains. For spherical NPs in linear melts, the NPs were modeled as LJ spheres of various diameters. The interactions between polymer monomers and particles and between particles and particles were modeled using modified LJ functions, which account for the excluded volume of the beads and particles by offsetting the interaction range by R_{EV}. The polymer–particle and particle–particle interactions were turned to weakly attractive and purely repulsive, respectively, to ensure a good dispersion of NPs in the elastomer melts.

To consider the effect of surface grafting on NP dynamics, model systems containing polymer-grafted NPs in linear elastomer melts were also constructed in which both the matrix and graft chains are simulated by the standard bead–spring model [25]. The particle core was modeled as an icosahedron composed of 12 vertex beads of diameter and mass the same to the polymer monomers. The same FENE bond potential as used for polymers was also applied to connect the beads between adjacent vertices. A harmonic angle potential in the form of $U_\theta = k_\theta(\theta - \theta_0)^2$ with $k_\theta = 200\,\varepsilon/\text{rad}^2$ and $\theta_0 = 60°$ was used to control the 60 angles of the 20 equilateral triangle faces on an icosahedron. Such treatments ensure that the particle can maintain its icosahedral structure and hence can be regarded approximately as a spherical rigid body. Since the edges of the icosahedron are controlled by FENE bonds of equilibrium length $l_b \approx 0.96\sigma$, the diameter of the particle core is about 2.83σ. The pair interactions between the NP vertex beads as well as those between the NP vertex beads and the polymer monomer beads were also treated using the WCA potential, and hence the NPs can be considered as athermal and nonsticky particles without enthalpic interaction. Each NP is uniformly grafted with $z = 6$ to 42 polymer chains, as shown in Figure 9.8a. Accordingly, the grafting density, which is calculated as $\Sigma_g = z/(\pi d_{NP}{}^2)$, is about 0.24 to 1.67 chain/σ^2. Note that one polymer end permits to graft with only one NP vertex bead, while one NP vertex site allows to link with 1–4 grafted chains by FENE bonds.

We also considered the influences of particle shape and chain architecture on the NP diffusion, and hence model systems of spherical or rod-like NPs in bottlebrush melts were also built. The bottlebrush polymer model is shown in Figure 9.8b, which shows several important structural parameters, such as the backbone length (N_{bb}), the side chain length (N_{sc}), and the gap of side chains (N_g). We set $N_{bb} = 100$, $N_{sc} = 10$, and $N_g = 1$ ($N_g = 1$ implies that every backbone bead is grafted with a side

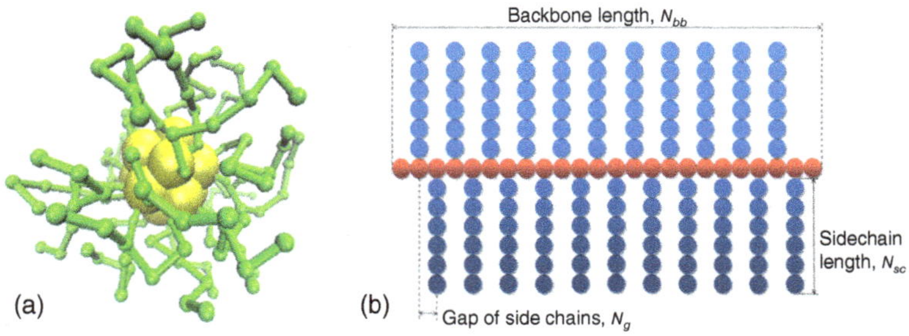

Figure 9.8 (a) Coarse-grained model of a polymer-grafted nanoparticle (PGNP) with 24 grafted chains each with 6 beads. (b) Diagram of the bottlebrush model displaying architectural parameters.

chain). This setup allows us to build a typical bottlebrush polymer and significantly differentiates it from the comb polymer. The rod-like particles consist of 10 beads arranged along a straight line. The bending angles between three adjacent beads are represented by the harmonic potential also in the form of $U_\theta = k_\theta(\theta - \theta_0)^2$. For the polymer chains, the rigidity $k_\theta = 20\,\varepsilon/\text{rad}^2$ and the bond angle is $\theta_0 = 180°$ or $\theta_0 = 90°$, corresponding to the bending angle on a straight line or formed by the beads of the backbone and side chains together. For the rod-like particles, $k_\theta = 20\,\varepsilon/\text{rad}^2$ and $\theta_0 = 180°$.

The simulations are started from a nonoverlapped configuration of the above-constructed model polymers and particles. Similar procedure as that described in Section 2.1.1 and 2.2.1 was also adopted to compress and equilibrate the simulation systems. After equilibration, production runs were carried out for data acquisition. All simulations were also performed using the LAMMPS software package [46].

9.3.1.2 Size Effect on Nanoparticle Diffusion

The diffusion coefficient D of NPs was obtained using

$$D = \lim_{t\to\infty} \langle [r(t) - r(0)]^2 \rangle / 6t \tag{9.7}$$

First, we investigated the effect of NP size on its dynamics in the dilute limit. The results are shown in Figure 9.9a. Note that the mass of the NP is proportional to its volume, so the mass density of the particle remains unchanged and is equal to that of polymer monomers. In this simulation, we consider elastomer chains of length $N = 60$ with their radius of gyration $R_\text{g} \approx 4.0\sigma$, which is unperturbed by the addition of NPs. It is found that the diffusion coefficient of the NP decreases with increasing the particle size. To test the validity of the classic SE law, the results calculated using the SE formula, as given by [47]

$$D = \frac{k_\text{B}T}{f\pi\eta R_\text{h}} \tag{9.8}$$

is also presented in Figure 9.9a. Here, R_h is the hydrodynamic radius of the particle and is approximately the actual radius of the particle R plus the radius of the adsorbing monomers $\sigma/2$, since the polymer–particle interaction used here is weakly

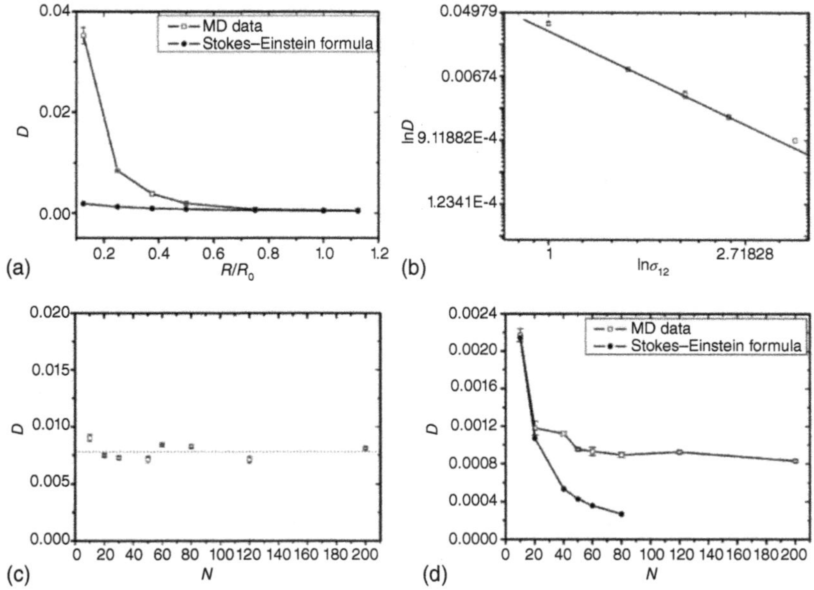

Figure 9.9 (a) Diffusion coefficient D of the particle as a function of the size ratio R/R_g. (b) The plot of ln D vs. ln σ_{12}, where σ_{12} is the hydrodynamics radius. Diffusion coefficient D of the particle of radius (c) $R = 1\sigma$ and (d) $R = 3\sigma$ with respect to different chain length N.

attractive. η is the viscosity of the surrounding polymer melt and can be got from literature using the same simulation condition, i.e. $\eta \approx 42.5$ [48].

It is found that, with the increase of the size ratio of R/R_g, the SE prediction gradually approximates the MD data under the slip boundary condition. As the size ratio of R/R_g increases to 1, the SE prediction and MD diffusion coefficient are nearly the same. Obviously, the smaller the size ratio of R/R_g, the larger is the deviation between the SE and MD diffusion coefficients, qualitatively agrees with the experimental results by Tuteja et al. [45] where the actual diffusion coefficient is 200 times faster than the SE prediction when the size ratio of R/R_g is as low as 0.25. However, in the experiment, the polymer chains are entangled, while the polymer length used here is shorter than the entanglement length. The reason may be that, in the SE formula, the diffusion coefficient is obtained on the basis of the macroscopic viscosity of the polymer melt. However, the actual viscosity that the NP can experience is determined by only the local subsections of polymer chains, which is much smaller than the macroscopic viscosity, as speculated by Wyart and de Gennes [49], which therefore leads to underestimation of the diffusion coefficient of the NP in the polymer melt by the SE equation. The mechanism for the small NP to sense the local viscosity is that the small NP can move through the polymer melts without necessarily waiting for the fully relaxation of the whole chain. With the increase of R/R_g, the chain-like solvent becomes a continuum on the length scale of the chain size. Certainly, the big particle experiences the macroscopic viscosity of polymer melts and recovers to the Brownian behavior.

To further investigate the regime of particle size smaller than the chain gyration radius of the melt, we present in Figure 9.9b the diffusion coefficient of the particle as a function of the hydrodynamic radius of the particle. The results showed that the diffusion coefficient of the particle is inversely proportional to the cube of its hydrodynamic radius when $R < R_g$, which is quite different from the SE prediction where the diffusion coefficient is inversely proportional to the hydrodynamic radius. For small NP, the friction between the polymer and the particle is caused by the monomer rubbing the NP surface. Therefore, the resulting friction is proportional to the particle surface. The local viscosity attributed to this friction is supposed to scale as R^2. Applying the local viscosity rather than the macroscopic viscosity into the SE formula, we can get the result illustrated in Figure 9.9b.

We now keep the particle size fixed at $R = 1\sigma$ or 3σ and vary the chain length of the polymer from $N = 10$ to 200, considering both the unentangled and entangled cases. For particles with radius $R = 1\sigma$, which is smaller than the gyration radius of all the simulated chains, the diffusion coefficient shows no polymer weight or polymer chain length dependence within the statistical uncertainty (Figure 9.9c), because the particles only experience local viscosity related to the monomeric friction coefficient and the surface area of the particle. For the particle of radius $R = 3.0\sigma$, when the melt chain length is shorter than 20, the simulated diffusion coefficient agrees well with the SE prediction, because in this case, the particle radius is larger than the gyration radius of the polymer chains (Figure 9.9d). When the chain length is higher than 20, the SE equation underestimates the NP diffusion coefficient, since the gyration radius of the polymer becomes comparable to or even larger than the particle size. When the melt chain length is within the range from 20 to 50, weak but finite molecular weight dependence of NP diffusion still exists because the motion of the entire chain is still correlated with the particle motion to some extent. Further increasing the chain length from $N = 50$, accompanying with the fact that the particle size becomes much smaller than the gyration radius of the chain, the particle motion becomes decoupled with the motion of the entire polymer chain, only being dependent on the local viscosity of the polymer melts. So, the particle diffusion becomes independent of the chain length and remains nearly unchanged with increasing the chain length N.

9.3.1.3 Effect of Surface Grating on Nanoparticle Diffusion

Grafting polymers to NP surface is a common approach to modify the surrounding environments of and the interactions between NPs for enhancing dispersion or controlling self-assembled structures [50, 51]. As polymers are grafted onto the NP surface, they are forced to depart from their free chain, random-walk isotropic conformation, causing them to stretch in the direction perpendicular to the NP surface [52]. Consequently, the dynamics of the grafted chains are strongly altered [53–56], showing a mobility gradient, i.e. relax slower in the vicinity of the NP surface but faster in the corona region [18]. This dynamical heterogeneity in the grafted polymer shell makes the dynamics of the polymer-grafted nanoparticles (PGNPs) quite different from that of bare NPs.

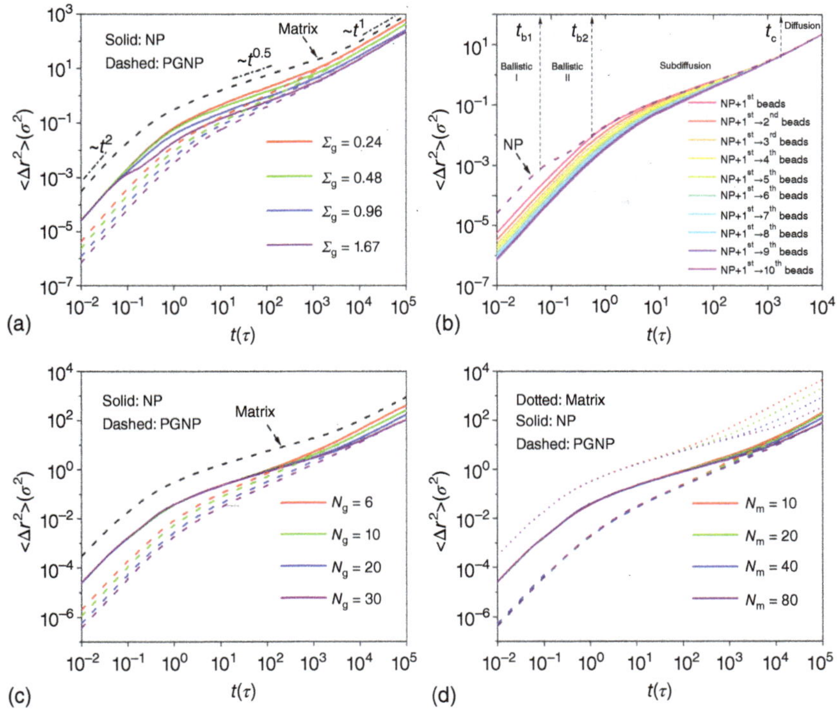

Figure 9.10 Time dependence of the MSDs $\langle \Delta r^2 \rangle$ of the center of mass of only the particle (solid lines), the whole PGNP (dashed lines) and the matrix monomers in chain middle for systems (a) at fixed $N_g = 10$ and $N_m = 40$ but at different Σ_g, (c) at fixed $\Sigma_g = 0.96$ chain/σ^2 and $N_m = 40$ but at different N_g and (d) at fixed $\Sigma_g = 0.96$ chain/σ^2 and $N_g = 30$ but at different N_m, as indicated. (b) Time dependence of the MSDs $\langle \Delta r^2 \rangle$ of the center of mass of only the particle core (dashed line) and that of the grafted NP including progressively larger numbers of graft monomers (solid lines) for the system with $\Sigma_g = 1.67$ chain/σ^2, $N_g = 10$ and $N_m = 40$.

Figure 9.10a shows the MSD data for PGNPs with a fixed graft length $N_g = 10$ but varying Σ_g in melt of $N_m = 40$, where the MSD of the matrix monomers in chain middle is also included for comparison. We find that at short times ($t < t_{b1} \approx 0.06\tau$), there is a ballistic motion regime where $\langle \Delta r^2 \rangle \sim t^2$ for both the particle core and matrix monomers. Since the size of the NP is larger than that of polymer segment, the NP's MSD shows slower motion. Beyond that time, we find a sudden drop in the slope x of the NP's MSD in logarithmic coordinates, especially for systems at high Σ_g. To gain insight into the micro-source of this behavior, the MSDs of the grafted NPs that including progressively larger numbers of graft monomers were calculated, as shown in Figure 9.10b. Clearly, the reduction in the slope of the NP's MSD is due to the participation of the permanently grafted (1st) monomers to the NP motion which makes the particle of effectively larger size. Since these grafted monomers move rigidly with the NP, the NP together with the permanently grafted monomers can be considered as an effectively bigger particle and still moves ballistically, extending the ballistic regime to $t_{b2} \approx 0.6\tau$. Therefore, the scaling exponent x is gradually recovered to be higher than 1 (though still lower than 2).

With the passing of time, the grafted NP would enter a subdiffusive regime prior to the further participation of outer graft segments to the particle motion, since these graft segments do not move rigidly with the NP. When regarding the NP and the graft segments that have already participated in the particle motion at $t < \tau_R \approx \tau_0 N_m{}^2$ as an entity, it is locally trapped by the polymer matrix and can only move according to the Rouse relaxations of the surrounding matrix (the first subdiffusive regime). If the overall PGNP size is larger than the matrix coil size, the particle dynamics can enter the second subdiffusive regime when $t > \tau_R$ where the inner part of the grafted NP tends to move diffusively while it is dragged down by the outer corona layer of graft segments. This may result in a slight increase of the subdiffusive exponent from $x \approx 1/2$ based on the Rouse model of unentangled polymer dynamics [49]. However, this regime usually overlaps with the ending of the first subdiffusive stage and the crossover to the final Fickian diffusive regime [57], since the studied PGNP size is comparable to that of matrix polymer chains. Therefore, the second subdiffusive regime cannot be well separated from the first subdiffusive and the terminal Fickian diffusive regimes here. Ge et al. [58] have examined that the apparent second subdiffusive regime can only be observed when the grafted chains are sufficiently long and/or the grafting density is low.

As the NP dynamics is dragged down by the grafted polymer and the surrounding coupled matrix chains, the overall dynamics of the PGNP is slower than the NP core from the ballistic to the subdiffusive regime (Figure 9.10a,c,d). Different from the impact of grafting density, changes in graft and matrix chain lengths do not affect the dynamics of the particle at short and intermediate time scales, i.e. the MSD curves for the NP core are nearly coincidence with each other before Fickian diffusion (Figure 9.10c,d). This is because the participation of graft segments in the NP motion is space- and time-dependent, initially those close to the NP surface and then those farther away from the core, while the interior monomer packing state and dynamics are nearly unaffected by N_g and N_m. The PGNP would diffuse when all graft monomers participate in the NP motion. Therefore, the curves in the terminal diffusion regime separate from one another as N_g and N_m are varied (Figure 9.10c,d) because the grafted polymers extend farther at higher N_g and the dynamics of the outer graft segments is suppressed by the matrix chains of increasing N_m.

From Figure 9.10, we find that finally, all the grafted chain segments move cooperatively with the NP, showing a Fickian diffusive behavior, i.e. the MSD increases linearly with the elapsed time. The long-time diffusion coefficients D, calculated according to Eq. (9.7), are shown in Figures 9.11 and 9.12. From Figure 9.11, we find that with increasing Σ_g, the D continuously decreases. According to the SE relation [47], the diffusivity D is related to the friction coefficient ζ of the particle with the surrounding environment, i.e. $D = k_B T/\zeta$. For loosely grafted NPs, Ge and Rubinstein [59] argued that the friction coefficient can be estimated as $\zeta = \zeta_{core} + \zeta_0 z(N_g - 1)$, where ζ_{core} is the friction of the particle core (consisting of the NP and the permanently grafted monomers) and $\zeta_0 z(N_g - 1)$ is the friction of

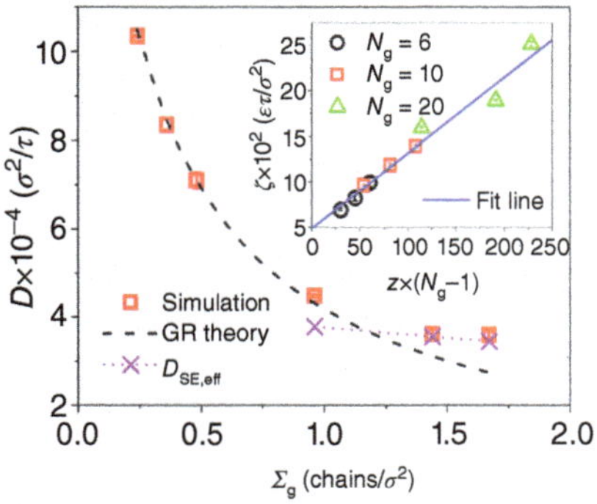

Figure 9.11 Diffusion coefficient D of the PGNPs as a function of Σ_g. Red open squares are the simulation data. Black dashed line is the prediction from the scaling theory by Ge and Rubinstein (GR) [59] for loosely grafted NPs. Purple cross (×) symbols and dotted line are the diameter-renormalized Stokes–Einstein prediction for densely grafted NPs considering the PGNPs as effectively larger particles. The inset shows the friction coefficients ζ for loosely grafted NPs with different combinations of z and N_g (open symbols). The blue line shows the best linear fit. These results are for $N_g = 10$ and $N_m = 40$.

the grafted chains, i.e. the monomeric friction coefficient ζ_0 times the number of monomers in the grafted chains $z(N_g - 1)$ (do not include the permanently grafted monomers that move rigidly with the NP). As expected, the ζ can be well fitted by $\zeta = \zeta_{core} + \zeta_0 z(N_g - 1)$, as shown in the inset of Figure 9.11. According to the relation $\zeta_0 = f\pi\eta_0\sigma$ with the full-slip condition $f = 2$ [49], the fitted local viscosity of polymer monomers $\eta_0 \approx 1.33\tau\varepsilon/\sigma^3$, which is comparable to the measured $\eta_0 \approx 1.4\tau\varepsilon/\sigma^3$ by nonequilibrium MD simulations. In addition, for the particle core with size smaller than the surrounding matrix chains, the friction can be estimated as $\zeta_{core} = f\pi\eta d_{core}$ with $\eta \approx \eta_0(d_{core}/\sigma)^2$, corresponding to the viscous drag associated with Rouse motion of a matrix polymer subsection of coil size $\approx d_{core}$. The fitted d_{core} is about

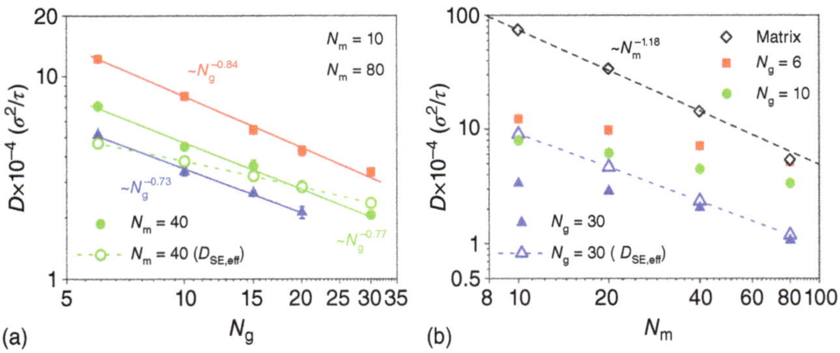

Figure 9.12 (a) Diffusion coefficients D of PGNPs as a function of N_g for systems at fixed $\Sigma_g = 0.96$ chain/σ^2 but varying N_m and (b) the diffusivity as a function of N_m for systems at fixed $\Sigma_g = 0.96$ chain/σ^2 but varying N_g as indicated. The green open cycles in panel a and the blue open triangles in panel b are the diameter-renormalized SE predictions for the systems indicated in Figure 9.10c,d, respectively. The black open diamonds in panel b are the D of the pure matrix chains.

$3.92\,\sigma \approx d_{NP} + 1.1\sigma$. Based on these fitted η_0 and d_{core}, the diffusivity D of the PGNPs in loosely grafted regime can be well described by [58, 59]

$$D = \frac{k_B T}{2\pi\eta_0\sigma\left((d_{core}/\sigma)^3 + z(N_g - 1)\right)} \tag{9.9}$$

as shown in Figure 9.11. However, this relation fails to estimate the diffusivity at $\Sigma_g > 1$ chain/σ^2 because many matrix monomers are expelled from the inner part of the grafted polymer layer that the friction coefficient ζ calculated using $\zeta = \zeta_{core} + \zeta_0 z(N_g - 1)$ is overestimated. Many previous theoretical and experimental studies [58–62] argued that the NP densely grafted with or strongly adsorbed with polymers can be treated as a "core–shell" object with an effective diameter d_{eff} and the diffusivity can be described by the diameter-renormalized SE relation:

$$D_{SE,eff} = \frac{k_B T}{f\pi\eta_{full} d_{eff}} \tag{9.10}$$

when d_{eff} is larger than the matrix polymer coil size, where $f = 2$ or 3 depending on whether the particle-medium boundaries are slip or stick and η_{full} is the macroscopic viscosity of the pure polymer liquid. Here, we use h_{eff} measured according to [63]

$$\left\langle h_{eff}^2 \right\rangle^{1/2} = \left[\frac{\int_{R_{NP}}^{\infty} 4\pi r^2 dr (r - R_{NP})^2 \rho_g(r)}{\int_{R_{NP}}^{\infty} 4\pi r^2 dr \rho_g(r)}\right]^{1/2} \tag{9.11}$$

to calculate the effective diameter $d_{eff} = d_{NP} + 2h_{eff}$. Here, $r - R_{NP}$ denotes the distance of graft monomer from the NP surface and $\rho_g(r)$ is the number density profiles for graft monomers from the NP surface. Since the calculated effective particle radii $d_{eff}/2$ are smaller than the estimated slip length $L_s \approx (\eta_{full}/\eta_0)\sigma \geq 10\sigma$ [58, 64], the full-slip condition is satisfied in all simulated systems examined in the present study. Therefore, $f = 2$ is used. We find that when $\Sigma_g > 1.2$ chain/σ^2, the estimation by Eq. (9.10) is in good agreement with the simulation results (Figure 9.11). However, for case with $\Sigma_g < 1.2$ chain/σ^2, the prediction by Eq. (9.10) is questionable due to the increased significance of the interpenetration between graft and matrix polymers, termed grafted layer "wetting."

In athermal systems, the relative length between the graft and matrix chains plays a vital role in controlling the grafted layer wetting [53]. We therefore also investigated the influence of the graft and matrix lengths on the PGNP diffusion. From Figure 9.12a, we see that the D decreases with increasing N_g and can be scaled by $D \sim N_g^{-0.77}$. As aforementioned, the diffusivity of PGNPs in loosely grafted regime can be estimated by Eq. (9.9) that $D \sim N_g^{-1}$, while in densely grafted regime, it can be estimated by Eq. (9.10) that $D \sim (d_{NP} + 2h_{eff})^{-1} \sim N_g^{-0.5}$, if the particle core is small enough or the graft length is long enough. Here, the NP size $d_{NP} = 2.83\sigma$ is comparable to or slightly lower than that of the grafted polymer $R_e \approx 3$–8σ, and meanwhile the grafting density $\Sigma_g = 0.96$ chain/σ^2 is located around the crossover from mushroom to brush; therefore, the scaling exponent of D by N_g is about 0.8.

From Figure 9.12b, we find that the D show large violations from the diameter-renormalized SE prediction (Eq. (9.10)), suggesting that the PGNPs cannot be simply treated as hard spheres due to grafted layer wetting, especially at low Σ_g and at high

N_g/N_m [65]. Here, we propose that the diffusivity of a PGNP in a melt is controlled by the viscous drags of an effective nondraining core and a free-draining layer outside the core:

$$D = \frac{k_B T}{\zeta_{core} + \zeta_0 z N_f} \quad (9.12)$$

where ζ_{core} is the fraction of the core consisting of the NP and the nondraining portion of the grafted chains and N_f is the length of the free-draining grafted polymer segments in the corona region. When Σ_g is very low that the NP surface is exposed to the matrix, Eq. (9.12) is equivalent to Eq. (9.9) since $N_f = N_g - 1$. While when Σ_g is extremely high that the grafted layer significantly "dewetting," i.e. N_f is ignorable, Equation 9.12 reduces to Eq. (9.10). At moderate Σ_g, with increasing N_g, both ζ_{core} and $\zeta_0 z N_f$ increase that D reduces with $D \sim N_g^{-\gamma}$, where γ is located within 0.5–1 depending on d_{NP}, Σ_g, N_g, and N_m, as discussed in Figure 9.12a. Increasing N_m would enlarge the core size but at the same time would reduce N_f due to grafted layer dewetting. The counteraction between these two opposite effects leads to the fact that the D only slightly decreases with N_m (Figure 9.10b), significantly different from the typical SE behavior for bare NPs in unentangled melts [66] where $D \sim N_m^{-1}$ for $d_{NP} > R_e$. From Figure 9.12, we also see that at high enough N_g/N_m, the PGNPs even show diffusivities lower than that predicted by the diameter-renormalized SE relation (Eq. (9.10)) due to the extremely significant grafted layer wetting which produces viscous drags higher than that estimated as effective spheres.

9.3.1.4 Nanoparticle Diffusion in Bottlebrush Elastomers

Bottlebrush polymers are promising grafted polymers consisting of a linear backbone with densely grafted side chains. High grafting density leads to strong spatial exclusion between side chains, resulting in a worm-like conformation of the backbone extension. The unique conformation reduces overlap and entanglement between the molecular chains of the bottlebrush polymer [67]. We now investigate how such architectural feature of bottlebrush polymer melts affects the diffusion dynamics of the embedded NPs of different geometries [19].

Figure 9.13a shows the MSDs of spherical NPs in a bottlebrush elastomer melt, from which we find, not surprisingly, that the smaller the radius of the particle, the more motile it is. The diffusion coefficients obtained from the long time limit of the MSDs using Eq. (9.7) are shown in Figure 9.13b. For comparison, the diffusion coefficients predicted by the SE relation (Eq. (9.8)) are also provided in Figure 9.13b, where the melt bulk viscosity $\eta \approx 47.5$ is obtained from reverse nonequilibrium MD simulations. We find that when the particle size is small, the simulated diffusion coefficients deviate from the SE predictions; however, the deviations are very low, i.e. only one to two times the SE predicted values. This is because, the bottlebrush polymer we established has a high grafting density, a stretched backbone, and less entanglements between chains that the difference between the local viscosity and the macroscopic viscosity felt by the small particle is not too obvious. With increasing the particle size, the diffusion gradually recovers to the SE behavior.

The diffusion behavior of rod-like NPs in bottlebrush elastomers was also investigated. Figure 9.13c shows that the MSD values of nanorods decrease with increasing

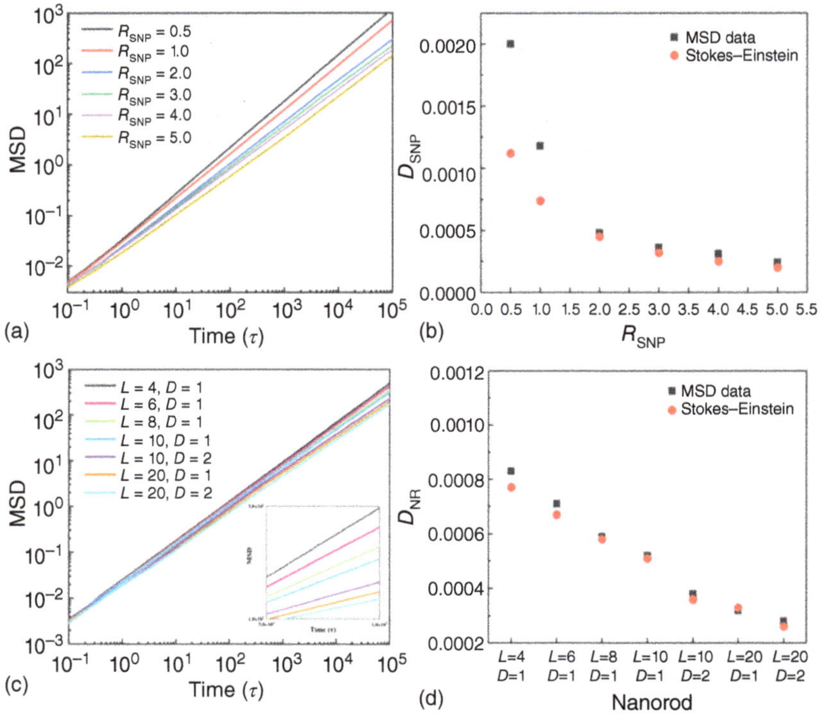

Figure 9.13 MSDs of (a) spherical NPs and (b) nanorods of various sizes and diffusion coefficients of (c) spherical NPs and (d) nanorods obtained from simulations and those from SE predictions in a bottlebrush polymer.

the rod length (L) and diameter (D). Figure 9.13d shows the overall translational diffusion coefficients of nanorods obtained from the MSD curves. Based on the continuum SE theory, the diffusivity of nanorods can be predicted by: [68]

$$D_{\mathrm{SE}} = \frac{k_{\mathrm{B}}T}{3\pi\eta L}\ln(L/D) \tag{9.13}$$

which is also put into Figure 9.13d. From the figure, we find that when the rod length $L \geq 8$, the diffusion coefficients obtained from simulations largely match with those predicted by the SE equation, whereas, for $L < 8$, the simulated values become deviate from the SE predictions. Here, the R_g of the bottlebrush polymer chain is about 8.8σ, so for rods with $D = 1\sigma$, when $L < R_g$, the rods do not feel the viscosity of the entire size bottlebrush molecule, and the motion of nanorods is decoupled from the motion of the whole bottlebrush chain. However, when $L > R_g$, the SE relation becomes applicable because the rods would feel the macroscopic viscosity completely at such large size. It is well known that for particles of anisotropic shape, different diffusion coefficients in the directions parallel to and perpendicular to their major axis are predicted [69, 70]. We further investigate the anisotropic dynamics of nanorods by calculating the diffusion coefficients along the rod axis ($D_{\parallel}$) and the diffusion coefficients perpendicular to the rod axis ($D_{\perp}$). Both $D_{\parallel}$ and

$D_\perp$ decrease with increasing L and D, as nanorods experience more drag from the surrounding polymers.

In addition to translational diffusion, nanorods in a polymer matrix also undergo rotational diffusion. We analyzed the rotational diffusion of nanorods by calculating end-to-end vector autocorrelation function:

$$C(t) = \langle R_{ee}(t) - R_{ee}(0) \rangle / \left\langle R_{ee}^2 \right\rangle \tag{9.14}$$

where $R_{ee}(t)$ is the end-to-end vector of the nanorod at time t. As shown in Figure 9.14a, rotational relaxation of nanorods progressively decreases with increasing L and D. The rotation relaxation time of rods with larger L is significantly longer, suggesting that the L plays a major role in the coupling of motion between nanorods and polymer chains.

To get an insight into the diffusion mechanism of nanorods, we calculate the anisotropic diffusion parameter A(t) based on MSD, as given by: [71]

$$A(t) = \frac{3\langle \mathrm{MSD}_{\parallel}(t) \rangle}{\langle \mathrm{MSD}(t) \rangle} - 1 \tag{9.15}$$

where $\mathrm{MSD}_{\parallel}(t) = \langle [(r(t_0 + t) - r(t_0)) \cdot e_{\parallel}(t_0)]^2 \rangle$ and $e_{\parallel}$ is the unit vector parallel to the rod major axis. $A(t) = 2$ if the rod diffuses totally along its end-to-end vector, whereas when $A(t)$ is closer to 0, its diffusion anisotropy is weaker. The results are presented in Figure 9.14b from which we find that all rods show strong anisotropic diffusion on

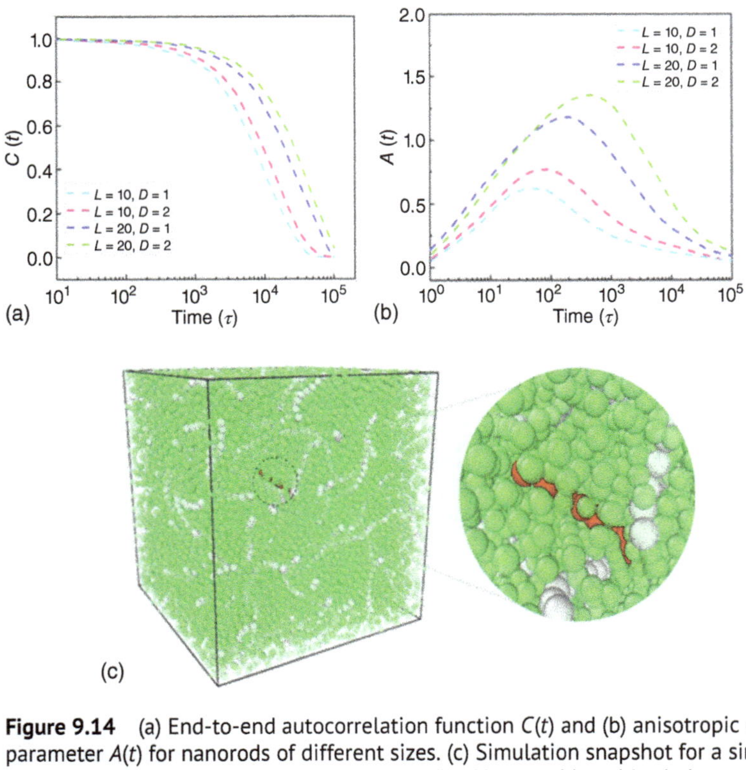

Figure 9.14 (a) End-to-end autocorrelation function $C(t)$ and (b) anisotropic diffusion parameter $A(t)$ for nanorods of different sizes. (c) Simulation snapshot for a single rod (red, $L = 10$, $D = 1$) in a bottlebrush polymer melt (backbone: white; side chain: green).

relatively short time scales while weak anisotropic diffusion on relatively long time scales. This is because there are almost no polymer entanglements between different bottlebrush polymer chains and the backbone is wrapped by dense long side chains, so the motion of nanorods is mainly limited by the steric effects generated by the side chains, especially for rotational motion. According to the snapshot shown in Figure 9.14c, we can see that it is mainly the side chain beads of the bottlebrush polymer that wrap the nanorod. From Figure 9.14b, we also find that the time for the $A(t)$ value reaches maximum decreases with reducing L and D. This is because the small-sized rods are less restricted by the surrounding medium and hence can rotate much faster, resulting in a more rapid transition from anisotropic to isotropic diffusion. Moreover, the maximum value of $A(t)$ also increases with increasing L because long NRs undergo faster translational motion along the parallel direction but much slower translation in the direction perpendicular to the rod major axis as they feel higher restriction in their normal direction by the surrounding polymer.

In addition, we also investigated the diffusion of spherical and rod-like NPs in bottlebrush polymers with various backbone lengths. As shown in Figure 9.15a, the motility of the polymer chains gradually decreases as N_{bb} increases. Also, with

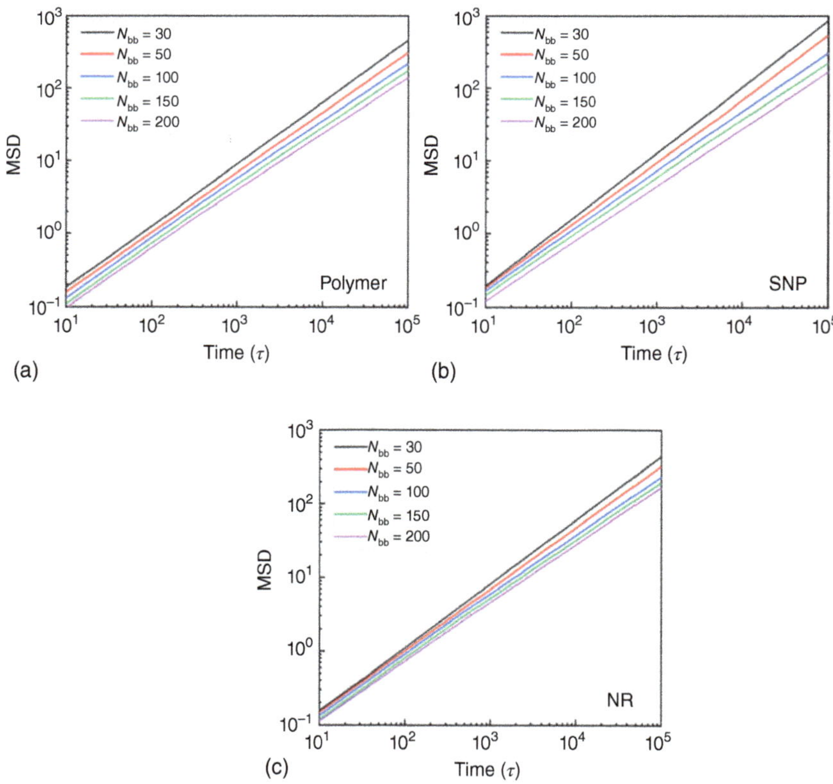

Figure 9.15 (a) MSD of bottlebrush polymers with various backbone length N_{bb} and that of the embedded (b) spherical NP and (c) nanorod.

the increase of N_{bb}, the MSD values of spherical NPs and nanorods decrease, as shown in Figure 9.15b,c, respectively. We expect that when the backbone length of the bottlebrush polymer exceeds a certain value (multiple times the entanglement length), the MSD of NPs would no longer change with the bottlebrush polymer backbone length and couples with only the local dynamics of the bottlebrush polymer. However, the reproduction of such a phenomenon by simulation is quite hard, because the side chains of the bottlebrush polymer is such dense that the interpenetration between macromolecules is significantly suppressed by the steric repulsion between side chains, resulting in a more extended chain conformation and much longer entanglement length.

9.3.2 Diffusion of Nanoparticles in Elastomer Networks

Understanding the diffusion dynamics of nanosized particles in elastomer networks is crucial for the design and fabrication of new NP-filled elastomer materials with targeted properties. Unlike particles in melts of linear polymers where the entanglements can slide (constraint release) along the polymer chains, in cross-linked elastomers, the network junctions are permanent, significantly restricting the motion of the embedded NPs. Previous theoretical studies by Cai et al. [72] and Schweizer et al. [73] both argued that the only way for a particle with size exceeding the network mesh size to leave a confinement cage is by hopping–waiting for the fluctuation of a gate (loop) between two neighboring confinement cages to become large enough to slip around the particle. Though the network confinement effect has been fully considered in their theories, the impact of the permanent cross-links on the network dynamics and thereby the coupling of the network dynamics to the mobility of the embedded NPs is less understanding. In this section, we will introduce one of our resent MD simulation studies on the diffusion of NPs in permanently cross-linked elastomer networks whether the network strands are entangled with each other or not [20].

9.3.2.1 Models and Simulation Methodology

Using the same procedure described in Section 2.2.1, the polymer networks were constructed by end-linking of monodisperse precursor linear polymers with multifunctional junctions. The length of the precursors (N_x) relative to the length of entanglements (N_e) determines whether the resulting cross-linked network is entangled ($N_x/N_e > 1$) or not ($N_x/N_e < 1$). Once the networks are completely developed, NPs (modeled as LJ spheres) are inserted to obtain the particle-filled elastomer networks. The insertion of NPs started with creating well-separated small voids (on a rectangular lattice that spans the simulation box) with volumes $<\sigma^3$. This corresponds to a well-dispersed initial configuration. Then the volumes of the voids are gradually increased, by changing the offset value R_{EV} of the modified LJ potential between particles and polymer monomers, until each void is large enough to accommodate a NP. Through such procedure, NPs of different sizes can be successfully incorporated into the already constructed networks. To ensure a good dispersion of NPs, the interaction between NPs is tuned to be purely repulsive while the particle–polymer interaction is adjusted to be attractive. The obtained particle-filled elastomer networks are then further equilibrated for at least $10^5\tau$, ensuring that the simulation systems are fully equilibrium. All the simulations are also performed using LAMMPS package [46].

9.3.2.2 Size Effect on Nanoparticle Diffusion

As discussed in Section 2.2.3, the conformational freedom and the dynamics of strands of the elastomer networks are restricted due to the introduction of the permanent network junctions, which would significantly suppress the diffusion of the embedded NPs. Figure 9.16 shows MSD data for NPs in cross-linked networks at different values of NP diameter d_{NP} and network strand length N_x. Note that to ensure good dispersion, the interactions between NPs and polymers were tuned to be attractive enough with a cutoff at $r_c = 2.5\sigma$ and attraction strength at $\varepsilon_{\text{np}} = 2.0\varepsilon$. This attraction strength is only slightly higher than the critical adsorbing strength at $\varepsilon_{\text{np}}^{\text{ads}} \approx 1.6\varepsilon$[74], where the depletion of polymer chains on the NP surface is just completely released, allowing the absolute dispersion of NPs while without too significant polymer adsorption. From Figure 9.16, we find that the curves share some common features. At very short times ($t < \tau_{\text{bal}}$), there is a ballistic motion of NPs where MSD $\sim t^2$. Beyond that regime, the NPs show a subdiffusive behavior at intermediate times, wherein MSD $\sim t^{\alpha_{\text{NP}}}$ with $0 < \alpha_{\text{NP}} < 1$, and become Fickian diffusive at long times where MSD $\sim t$. The minimum apparent exponents $\alpha_{\text{NP,min}}$ in the subdiffusive regime and the times at the crossover from subdiffusive to Fickian-diffusive τ_c extracted from Figure 9.16 are plotted as a function of N_x in Figure 9.17a,b, respectively. The long-time NP diffusion coefficients, calculated according to Eq. (9.7) are shown in Figure 9.17c. For $d_{\text{NP}} = 6\sigma$ and $N_x \leq 10$, the NP MSD does not reach the Fickian diffusion regime as τ_c is longer than the simulated time window (Figure 9.16e).

From Figures 9.16 and 9.17a–c, we see that the intermediate- and long-time behaviors are qualitatively different, depending on d_{NP} and N_x. For $d_{\text{NP}} = 1\sigma$, the behaviors show a negligible dependence on N_x, since the dynamics of the coupled polymer on segment length scale is almost unaffected by N_x (Figure 9.6a,c). With increasing the NP size, the coupling effect becomes more remarkable that stronger and longer subdiffusive behaviors are observed (Figures 9.16b–e, 9.17a,b). For NP with size $\xi < d_{\text{NP}} < d_{\text{T}}$ in an entangled linear polymer, where ξ is the polymer correlation length (about 1σ in melt condition), coupling theory [75] predicts the subdiffusive exponent to be about $\alpha_{\text{NP,min}} = 0.5$. However, many recent experiments and simulations [57, 66, 76, 77] showed that the actual subdiffusive exponent is often larger than predicted ($0.5 < \alpha_{\text{NP,min}} < 1$). Chen et al. [77] attributed this discrepancy to the presence of coupling between NP dynamics and polymer center-of-mass (COM) motion (NPs move with the polymer COM) in addition to that predicted by the traditional theory based on the coupling between NP motion and polymer segmental relaxations (NPs move relative to the polymer COM). Such combined coupling mode is also existed in the present simulation systems, as schematically shown in Figure 9.17d. For the cases with long network strands where $N_x > N_e$, we observe the $\alpha_{\text{NP,min}}$ is independent of N_x (Figure 9.17a), because the Rouse dynamics of the

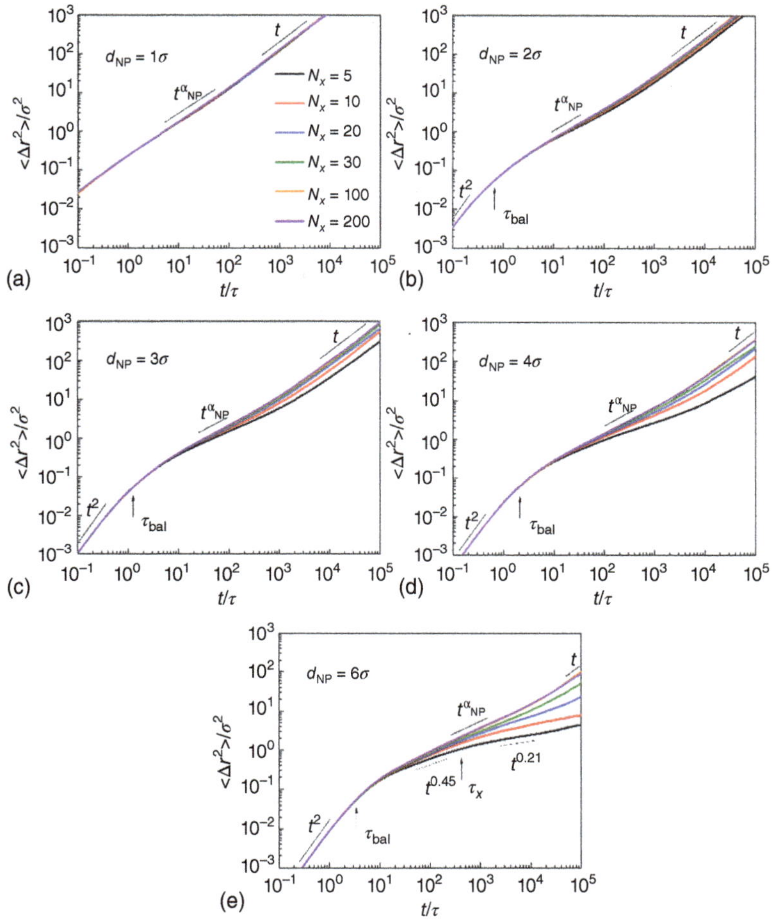

Figure 9.16 MSD $\langle \Delta r^2 \rangle$ of NPs with sizes (a) $d_{NP} = 1\sigma$, (b) 2σ, (c) 3σ, (d) 4σ, and (e) 6σ in cross-linked networks at a fixed junction functionality $f = 4$ while with varying strand length N_x as indicated.

coupled polymer subsections of sizes $\approx d_{NP} < d_T$ is unaffected by the strand length (Figure 9.6c). However, there is a decay in $\alpha_{NP,min}$ from 0.86 to 0.58 with increasing the NP size (Figure 9.17a), indicating that the coupling of NP dynamics to strand COM motion is reduced as the NP size is increased. Similar trend was also shown in entangled polymer solutions where the recovery of $\alpha_{NP,min}$ to 0.5 is only observed for large NPs with $d_{NP} \gtrsim 10\xi$ [76].

As the subdiffusive behavior is suppressed with increasing the NP size, the time crosses over to the Fickian diffusive regime becomes longer (Figure 9.17b). When $d_{NP} \leq 4\sigma < d_T$, the crossover time τ_c is smaller than the polymer entanglement time $\tau_e \approx 2 \times 10^3 \tau$. While for NPs with size comparable to the tube diameter ($d_{NP} = 6\sigma$), the crossover occurs at $\tau_c > \tau_e$ due to the confinement of the entanglement network. The theory by Brochard-Wyart and de Gennes [49] predicts that $\alpha_{NP,min}$ decreases

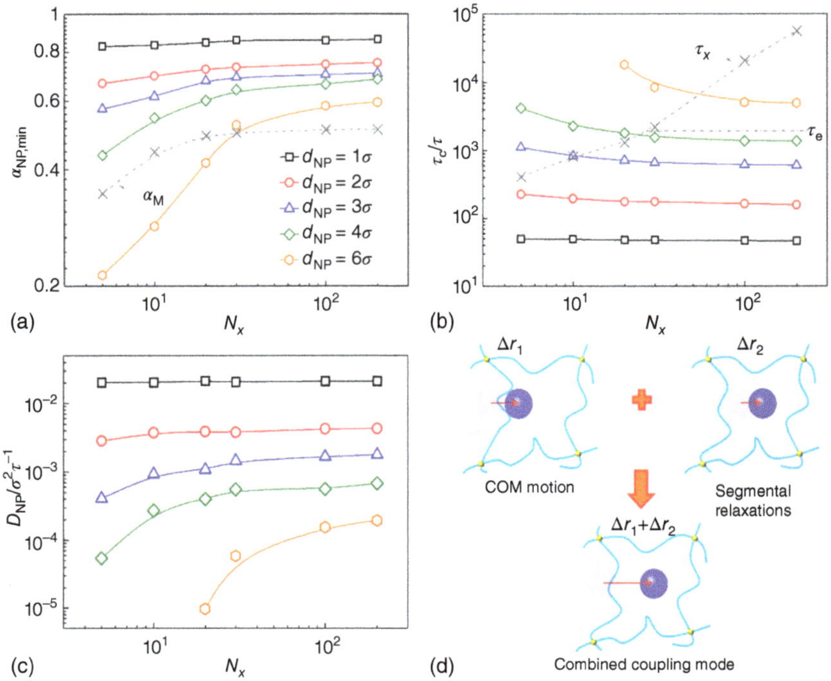

Figure 9.17 (a) Minimum subdiffusive exponents $\alpha_{NP,min}$, (b) crossover times (from subdiffusive to Fickian diffusive) τ_c, and (c) diffusive coefficients D_{NP} extracted from Figure 9.16. For comparison, subdiffusive exponents α_M and Rouse relaxation times τ_x of network strands extracted from Figure 9.6c are also shown in (a) and (b), respectively. (d) Schematic illustrating the coupling modes of the NP dynamics in a cross-linked polymer network: COM motion coupling, segmental relaxation coupling, and the combined coupling that includes both.

to 0 sharply at $d_{NP} \approx d_T$. While in the present study, the $\alpha_{NP,min}$ for $d_{NP} = 6\sigma \gtrsim d_T$ in the networks of $N_x \geq 100$ is about 0.6 (Figure 9.17a), because the entanglements can partially relax through longitudinal Rouse modes in a confining tube, tube length fluctuation, and constraint release. Meanwhile, the additional coupling mode of NP dynamics to polymer COM motion (Figure 9.17d) would also lead to the increase in $\alpha_{NP,min}$.

Similar to the trends of $\alpha_{NP,min}$, the long-time diffusivity D_{NP} for NPs with size $d_{NP} \leq 4\sigma < d_T$ in entanglement dominant networks ($N_x > N_e$) is also independent of N_x (Figure 9.17c). While such plateau is not presented for NPs with size comparable to or larger than d_T due to entanglement confinement. In this situation, the NPs can diffuse only through hopping or due to the constraint release of the entanglement mesh [72, 73, 75, 78, 79]. However, since the network strands are end-linked by multifunctional junctions, the constraint release effect is restricted to some extent. With increasing N_x, faster constraint release is allowed. Therefore, for NPs with $d_{NP} = 6\sigma$, which is comparable to d_T, the diffusivity D_{NP} increases as N_x is increased (Figure 9.17c).

For NPs within the network dominated by permanent cross-links ($N_x < N_e$), both the intermediate- and long-time motion behaviors are suppressed (Figures 9.16 and 9.17a–c) due to the reduction of network dynamics and the enhancement of network confinement. The restrictions arising from network dynamics would become more significant with decreasing N_x (Figure 9.6c). While the network confinement effect is determined by the relative sizes between the NPs and network meshes. Each mesh is composed of several connected strands. For a tetrafunctional network with short precursor chains, Lang et al. [80] estimated that the average strand number per mesh is about 7. These meshes are significantly overlapped and interconnected with each other, since the strand overlap parameters P are larger than 1 (Table 9.3). Therefore, Cai et al. [72] inferred that the actual network mesh size is about d_x. While through comparing the behavior in Figure 9.17b with the parameters in Table 9.3, we find that the actual network confinement that the NPs can feel is correlated with the strand fluctuation distance d_{fluct} rather than the strand size d_x. When $d_{\text{NP}} < d_{\text{fluct}}$, the crossover times τ_c are smaller than the strand Rouse relaxation times τ_x because the NPs are coupled with only the local subsections of strands. While when $d_{\text{NP}} > d_{\text{fluct}}$, the τ_c becomes larger than τ_x, because the NPs should wait for fully relaxation of at least one network strand to move. Note that in the present study, the network confinement is still not so high as to extremely stretch the strands around the NPs even for the case with $d_{\text{NP}} = 6\sigma$ and $N_x = 5$, where only the strand dimension is slightly extended, but the bond length is nearly unaffected.

Theoretical study by Cai et al. [72] argued that hopping is the only motion mode for NPs with size exceeding the network mesh size. At times shorter than the strand relaxation time τ_x, the NP motion is coupled to the segmental dynamics of the network strands. At time scales longer than τ_x but shorter than τ_c, the NPs are fluctuating within the network cages without leaving them. To move between cages, the NPs have to wait for time τ_w (determined by the hopping free energy barrier $\Delta U_{\text{hop}} \sim d_{\text{NP}}^2/d_{\text{fluct}}^2$, at which the fluctuation of a network mesh become large enough to allow the particles to hop between cages. Therefore, Cai et al. [72] expected a transient MSD plateau of $\alpha_{\text{NP,min}} \approx 0$ at times $\tau_x < t < \tau_c$. However, such transient plateau is never observed in the present study (Figure 9.16). Despite this fact, we observe an apparent transition in α_{NP} from 0.45 to 0.21 and from 0.48 to 0.28 at $t > \tau_x$ for the two systems with $d_{\text{NP}} = 6\sigma$ and $N_x = 5$ and 10, respectively (Figure 9.16e). The existence of an extended power law scaling regime at $\tau_x < t < \tau_c$ for these two systems indicates that the NPs are trapped within the network cages and would hop when $t > \tau_w$, where $\tau_x < \tau_w \leq \tau_c$. A value of $\alpha_{\text{NP,min}} \approx 0.21$ for the largest $d_{\text{NP}}/d_{\text{fluct}} \approx 2.6$ implies that the NPs are far from the idealized localization situation, owing to the superposition of the NP fluctuations around the center of a network cage and the hopping between neighboring cages at $t > \tau_w$. As the NP hopping events are partially masked by the NP thermal fluctuations, the extended power law scaling regime at $\tau_x < t < \tau_c$ is even unobservable for some systems although with $d_{\text{NP}}/d_{\text{fluct}} > 1$ (Table 9.3), for example, the system with $d_{\text{NP}} = 4\sigma$ and $N_x = 5$ (Figure 9.16d).

To quantify the interesting NP hopping behavior, the displacement distribution of NPs at different times for systems at a fixed $N_x = 5$ but with varying d_{NP} is monitored using the self-part of the van Hove function (VHF): [81]

$$G_s(r, \Delta t) = \langle \delta[r - |\mathbf{r}_i(t_0 + \Delta t) - \mathbf{r}_i(t_0)|] \rangle_{t_0} \tag{9.16}$$

which gives the probability distribution to find a NP at a distance r from the initial position (at time t_0) after a time interval Δt. The results are multiplied by a factor $4\pi r^2$ and normalized as $\int 4\pi r^2 G_s(r, \Delta t)dr = 1$. For the system with $d_{NP} = 2\sigma$, since $d_{NP} < d_{fluct}$, the data at all time intervals can be well fitted by a Gaussian function with a peak value corresponding to the most probable traveled distance during that time interval Δt, as can be seen from Figure 9.18a1–a4. For the system with $d_{NP} = 4\sigma$, the curves at small time intervals $\Delta t = 100\tau$ and 500τ can still be fitted by a single Gaussian peak (Figure 9.18b1,b2), indicating that the NPs at that time intervals are fluctuating within the network cages ($t < \tau_w$). However, at longer time intervals ($\Delta t = 2000\tau$ and 5000τ), though only one peak is observed in each curve, they can no longer be well fitted by a single Gaussian function, but instead a bimodal function is necessary to model the NP displacement (Figure 9.18b3,b4). This behavior suggests the existence of a jump-like motion of NPs with size slightly larger than the cross-linked network mesh size, though it is not so observable due to the masking of the NP thermal fluctuations. For NPs with much larger size $d_{NP} = 6\sigma$ ($d_{NP}/d_{fluct} \approx 2.6$), two distinguishable peaks can be seen at long time intervals ($\Delta t = 10\,000\tau$, $20\,000\tau$, and $100\,000\tau$), as shown in Figure 9.18c2–c4, indicating that the hopping-type motion of NPs becomes directly observable from the NP trajectories.

Considering that a hopping fashion of NP motion is existed in networks with $d_{fluct} < d_{NP}$, we are also curious to know whether such behavior can be affected by changing the functionality of the network junctions f. For this purpose, NP behaviors in systems at a fixed NP size $d_{NP} = 4\sigma$ and a fixed network strand length $N_x = 5$ but with varying f were examined. We find that both the intermediate- and long-time dynamics of the NPs are suppressed with increasing f (Figure 9.19a). Accordingly, the NP hopping behavior, though being delayed, becomes more apparent (Figure 9.19b–d) as the network confinement ratio d_{NP}/d_{fluct} is increased (Table 9.3) and the network dynamics is reduced (Figure 9.6b,d).

9.3.2.3 Nanoparticle Diffusion in Attractive Networks

Creating an attractive interfacial potential between NPs and polymers is important to promote NP dispersion and the corresponding mechanical properties of the nanocomposites. We now investigate the effect of particle–polymer attraction on the NP mobility ($d_{NP} = 4\sigma$) within cross-link dominant networks ($N_x = 5$ and 30). Figure 9.20a shows the representative results of the site–site pair correlation function between a NP and a polymer monomer, $g_{np}(r)$. We note that $g_{np}(r)$ peaks at $r \approx 2.5\sigma$ with an amplitude that grows with ε_{np}, thereby demonstrating the strong adsorption of polymer monomers on NPs. The $g_{np}(r)$ is essentially independent of N_x, which is consistent with the behavior in dense polymer melts [82].

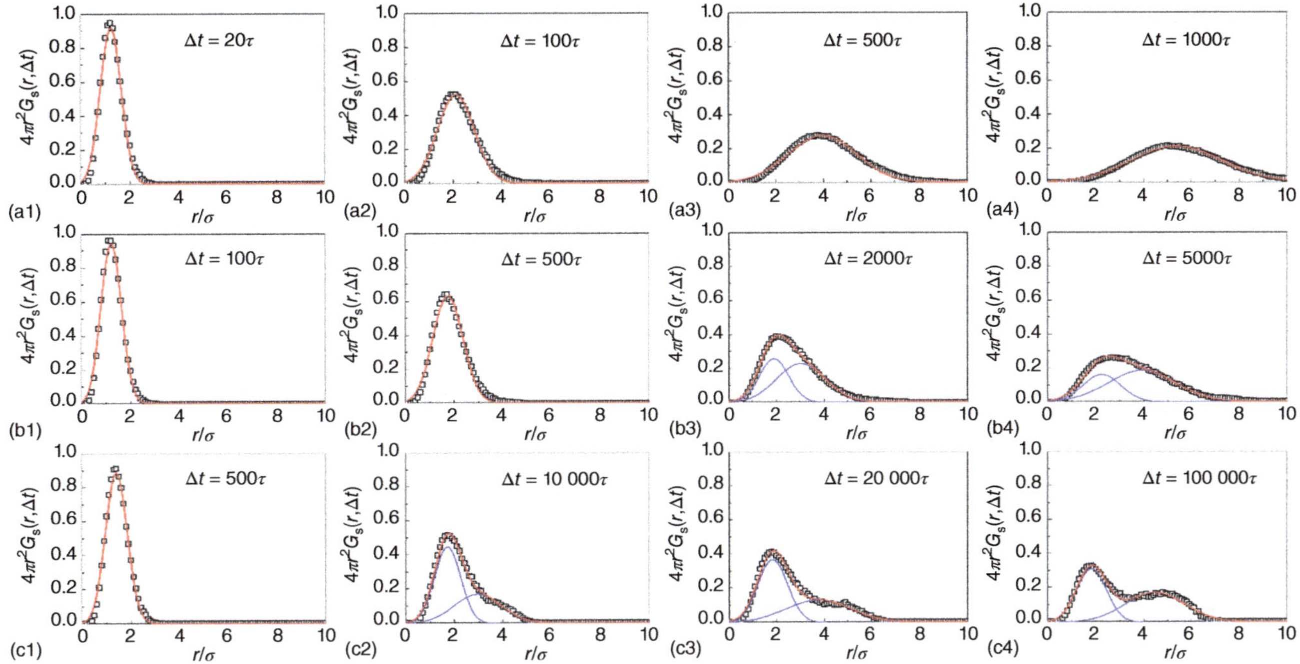

Figure 9.18 van Hove function $G_s(r,\Delta t)$ for NPs with sizes (a1–a4) $d_{NP} = 2\sigma$, (b1–b4) $d_{NP} = 4\sigma$, and (c1–c4) $d_{NP} = 6\sigma$ in a cross-linked network with strand length fixed at $N_x = 5$ at different time intervals Δt as indicated. Symbols are MD data, red lines are the fitting of single or bimodal Gaussian functions, and blue lines represent the individual contributions of the bimodal Gaussian curves.

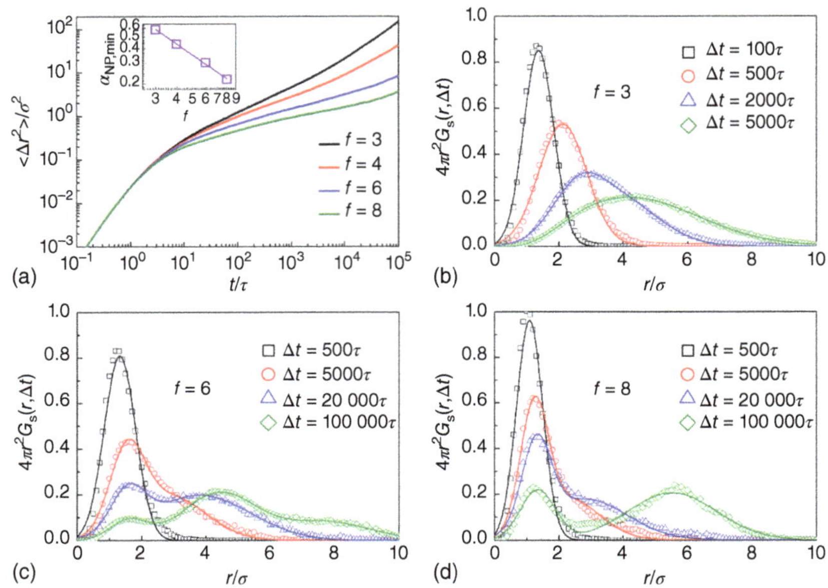

Figure 9.19 (a) MSD $\langle \Delta r^2 \rangle$ of NPs with size $d_{NP} = 4\sigma$ in cross-linked networks at a fixed $N_x = 5$ but with varying f as indicated. The inset shows $\alpha_{NP,min}$ vs. f. The corresponding van Hove functions $G_s(r,\Delta t)$ during different time intervals Δt as indicated: (b) $f = 3$, (c) 6, and (d) 8.

The autocorrelation of monomers distributed inside the adsorption layer surrounding a NP: [61, 74]

$$Q_D(t) = \frac{1}{\langle N_P^{ads} \rangle} \left\langle \sum_{ID(0)} \sum_{ID(t)} \delta_{ID(0),ID(t)} \right\rangle \tag{9.17}$$

was calculated to describe the dynamical relaxation of polymer adsorption, where N_P^{ads} is the number of monomers inside the adsorption layer and ID(0) and ID(t) are the adsorbed monomer IDs at the times 0 and t, respectively. The adsorption layer is defined as the polymer segments adsorbed around a NP within a cutoff distance of $r \approx 3.0\sigma$ (the location of the first minimum of $g_{np}(r)$). We see that for $N_x = 30$, the curves follow the normal decay behavior as that in linear polymer melts since $d_{NP} < d_{fluct}$ (Figure 9.20b). While for $N_x = 5$ where $d_{NP} > d_{fluct}$, the $Q_D(t)$ shows a two-step relaxation behavior. The first step relaxation (off the plateau from 1.0) is fully coincide with that for $N_x = 30$, indicating that the polymer desorption behavior initially is N_x independent. However, beyond that initial time regime, the decay behavior is slowed down since the polymer desorption is partially retarded by the confined network. The desorption time τ_{des} of the adsorbed polymer monomers is then estimated as the time when the $Q_D(t)$ decays to $1/e$. We find that the τ_{des} increases with ε_{np} and follows a simple Arrhenius form of $\tau_{des} \propto \exp(A\varepsilon_{np}/k_B T)$, where $A = 0.40 \pm 0.03$ and 0.33 ± 0.01 for $N_x = 5$ and 30, respectively. This result is presented as $\ln(\tau_{des}/\tau)$ vs. $\varepsilon_{np}/k_B T$ in the inset of Figure 9.20b.

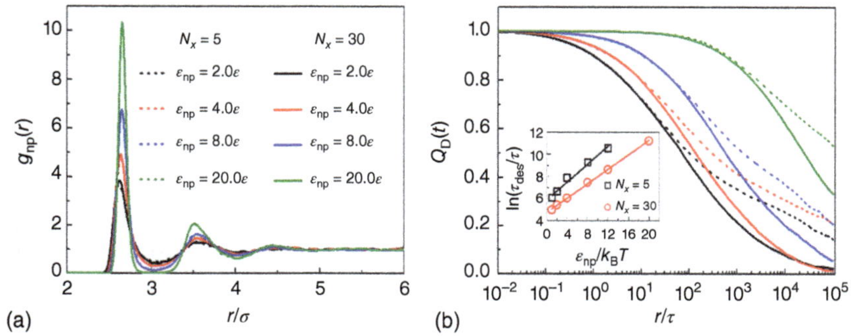

Figure 9.20 (a) Site–site pair correlation function between a NP and a polymer monomer as a function of separation, $g_{np}(r)$, for NPs of size $d_{NP} = 4\sigma$ in networks with $N_x = 5$ (dashed lines) and 30 (solid lines) at different NP–polymer attraction strengths ε_{np} as indicated. (b) The corresponding adsorption autocorrelation function of monomers distributed inside the adsorption layer $Q_D(t)$. The inset gives the monomer desorption time τ_{des} as a function of ε_{np} plotted via $\ln(\tau_{des}/\tau)$ vs. ε_{np}/k_BT, where the solid lines are the best Arrhenius function fit to the simulation data (open symbols).

The MSD data are presented in Figure 9.21a from which we observe that the NPs are more difficult to move as the NP–polymer attraction is increased. The short-time ballistic behavior is unaffected by N_x but becomes slower with increasing ε_{np}, because at high ε_{np}, some polymer segments are so tightly bound to the NP that they effectively move ballistically with the NP on these time scales, increasing the NP effective mass and hence the NP ballistic times. For $N_x = 30$, when $\varepsilon_{np} \leq 4.0\varepsilon$, the minimum subdiffusive exponent $\alpha_{NP,min}$ is almost unchanged with ε_{np} (Figure 9.21b). In this regime, since the monomer desorption times $\tau_{des} < 10^3\tau$ (Figure 9.20b) are lower than the network strand relaxation time $\tau_x \approx 2\times10^3\tau$, the NP subdiffusive behavior is still controlled by the strand Rouse dynamics. While the higher effective viscosity η_{eff} "felt" by the NPs at stronger ε_{np} results in the slower

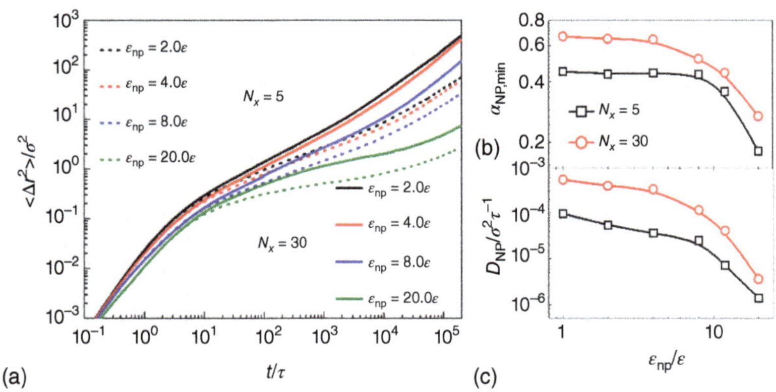

Figure 9.21 (a) MSD $\langle\Delta r^2\rangle$ of NPs of size $d_{NP} = 4\sigma$ in networks with $N_x = 5$ (dashed lines) and 30 (solid lines) at different ε_{np} as indicated. The corresponding (b) $\alpha_{NP,min}$ and (c) D_{NP} as a function of ε_{np}.

Fickian diffusivity (Figure 9.21c). Since $\tau_{des} < \tau_x$, the NPs move through the network using the local subsections of network strands as "carriers" or "vehicles." [61] When $\varepsilon_{np} > 4.0\varepsilon$, as the desorption times τ_{des} become longer than the strand relaxation time τ_x (Figure 9.20b), the NPs are significantly trapped within the network (cannot move before polymer desorption). Therefore, both the intermediate- and long-time dynamics are dropped dramatically with increasing ε_{np} (Figure 9.21b,c). For $N_x = 5$ where $d_{NP} > d_{fluct}$, the variation trends of $\alpha_{NP,min}$ and D_{NP} as a function of ε_{np} are similar to those for $N_x = 30$ but with relatively lower values due to the network confinement effect (Figure 9.21b,c). The critical ε_{np} at which the $\alpha_{NP,min}$ and D_{NP} significantly drop is determined by the relative values between the polymer desorption time τ_{des} and the waiting time τ_w of NP hop rather than τ_x since $d_{NP} > d_{fluct}$. The waiting time estimated by Cai's theory $\tau_w \simeq \tau_x(d_{NP}/d_{fluct}) \exp\left(d_{NP}^2/d_{fluct}^2\right)$ is about $6 \times 10^3\tau$, which is on the same order of magnitude with the desorption time at $\varepsilon_{np} = 8.0\varepsilon$ ($\tau_{des} \approx 10^4\tau$, Figure 9.20b). Therefore, the critical ε_{np} occurs at about 8.0ε. When $\varepsilon_{np} \geq 8.0\varepsilon$, the degree of non-Gaussianity of NP motion is decreased with increasing ε_{np} as the NP hopping due to strand fluctuations is significantly suppressed by the strong polymer adsorption.

9.4 Conclusions

In this chapter, we present MD simulation studies on the topological structures and dynamics of two classes of elastomer networks and the diffusion of NPs in both elastomer melts (from linear to bottlebrush) and networks. For the randomly cross-linked networks, with increasing the cross-link density and distribution, the numbers of elastically active strands, network junctions, and cycle rank are increased, while the lengths of active strands, dangling loops, and dangling chain ends are shortened. Improving the cross-link dispersion state is found to be essentially equivalent to increasing the effective cross-link density. The structural relaxation behavior is analyzed in terms of the IIDS factor. The results indicate that the relaxations of the network strands at different length scales are slowed down with the increase of cross-link density and cross-link dispersion degree, which is ascribed to stronger intermolecular coupling imposed by the introduced cross-linkers. The relaxation behavior is further examined by MCT and VFT equation, and the results indicate that the TTSP holds on the segmental length scale but fails on the chain length scale.

For the end-linked networks, the relative length scales of the network strands N_x and entanglements N_e determine which type of constraints (chemical cross-links or topological entanglements) govern the network properties. When $N_x/N_e > 1$, the network properties are dominated by the topological entanglements between network strands. As permanent network junctions are introduced to the strand ends, the strand dynamics exhibits a heavily entangled subdiffusive feature before being significantly restricted within the fluctuation distance d_{fluct}. When $N_x/N_e < 1$, the network is dominant by chemical cross-links. The constraints arising from the permanent cross-links would restrict the movement of the network strands within distance d_{fluct}. Overall, the d_{fluct} values are lower than the network strand sizes d_x

due to the restrictions of the overlapping "elementary" networks. The dynamics of the strands within such distance follows a subdiffusive behavior with an exponent α_M lower than 1/2. Such restrictions imposed by network junctions would become more significant with reducing the network strand length N_x or increasing the junction functionality f.

For spherical NPs diffusing in linear elastomer melts, we find only when the particle size is larger than the size of the polymer chain, the SE formula can predict well the diffusion of the particle. When the particle radius is smaller than the radius of gyration of the melt chains, the SE relation underestimates the particle diffusivity because the particle can feel only the local viscosity of the polymer melt rather than the macroscopic viscosity used in the SE formula.

Grafting polymer chains to the NP surface would change the surrounding environment of the particle, being dynamically heterogeneous, namely, the graft monomers closer to the particle surface relax more slowly than those farther away due to the constraint of the grafted surface and the confinement of the neighboring chains. Such heterogeneous relaxations of the surrounding environment would perturb the particle motion, making them fluctuating around their centers before they can diffuse through the melt. For the long-time Fickian diffusion, we found that the diffusivity D decreases with increasing grafting density Σ_g, grafted chain length N_g, and matrix chain length N_m. This is due to the fact that the diffusivity is controlled by the viscous drag of an effective core, consisting of the NP and the hydrodynamically coupled graft segments, and that of the free-draining graft layer outside the "core." With increasing Σ_g, the PGNPs become harder with greater effective size and thinner free draining layer, resulting in a reduction in D. At extremely high Σ_g, the diffusivity can even be estimated by the diameter-renormalized SE relation. With increasing N_g, both the effective core size and the thickness of the free-draining layer increase, leading to a reduction in diffusivity by $D \sim N_g^{-\gamma}$ with $0.5 < \gamma < 1$. Increasing N_m would lead to the enlargement of the effective core size but meanwhile result in the reduction of the free-draining layer thickness due to autophobic dewetting. The counteraction between these two opposite effects leads to only a slight reduction in the diffusivity, significantly different from the typical SE behavior where $D \sim N_m^{-1}$.

We also evaluated the validity of the SE relation in predicting the diffusion coefficients of nanofillers in bottlebrush elastomer melts. The results demonstrate that the simulated diffusion coefficients deviate from the SE predictions when the particle is small, but the deviation magnitude is low because the local viscosity felt by the particle in the bottlebrush polymer does not differ much from the macroscopic viscosity. For rod-like particles in bottlebrush polymers, they also exhibit a pronounced anisotropic translation, namely, move faster along their major axis but slower in the normal direction, due to the hindrance of the surrounding bottlebrush polymer which also restricts the rotation of the rod, particularly when the rod aspect ratio is high.

The cross-linking of elastomer chains would further suppress the diffusion of the particles. In an end-linked network, when the network strands are entangled with each other, i.e. $N_x/N_e > 1$, the entanglements dominate the network properties. For NPs with size comparable to the entanglement tube diameter d_T, their motion is suppressed by the entanglements. The constraint release of the entanglements would allow the particles to pass through but is slightly restricted by the permanent network junctions. While for the unentangled networks, where $N_x/N_e < 1$, the particle with size larger than the strand fluctuation distance d_{fluct} are trapped by the network cages and can diffuse at long times through hopping. Such hopping fashion of NP motion becomes more apparent with increasing the network confinement ratio before being permanently localized within the network. In either entangled or unentangled networks, when the particle size is smaller than the mesh size of the chemically cross-linked network or topological entanglements, their diffusion is controlled by only the local subsections of network strands of sizes comparable to d_{NP} and is independence of N_x.

Increasing NP–polymer attraction ε_{np} would make the NPs become more difficult to move. There exists a critical ε_{np} beyond which the NP mobility dramatically drops. For NPs with $d_{NP} < d_{fluct}$, the critical ε_{np} is determined by the relative values between the polymer desorption time τ_{des} and the strand Rouse relaxation time τ_x. If $\tau_{des} < \tau_x$, the NP motion is still controlled by the strand Rouse relaxation, while when $\tau_{des} > \tau_x$, the NP mobility significantly decreases as the NPs are significantly trapped within the network due to the strong polymer adsorption. For NPs confined by the cross-linked network ($d_{NP} > d_{fluct}$), the critical ε_{np} is then decided by the relative scales between the polymer desorption time τ_{des} and the waiting time τ_w of NP hop rather than τ_x. When ε_{np} is beyond this critical value, as the NP hopping due to strand fluctuations is suppressed by the strong polymer adsorption, the degree of non-Gaussianity of the NP motion is decreased.

In general, the progress provided by this chapter verifies lots of experimental findings and makes a good connection with many previous theories and is anticipated to provide some insights into understanding various dynamical issues in elastomer nanocomposites. However, there are still unresolved questions that require future investigation. Developing a unified picture connecting the behavior from MD of different components to the full range of performance and properties of the entire elastomer-based materials should be the focus of future studies. This will include understanding the effect of network heterogeneity, particle size polydispersion and shape irregularity, the complex nature of the interaction between polymers and particles on the dynamics of the particle-filled elastomers as well as the establishment of the connections between the microscopic dynamics and the macroscopic properties. We keenly anticipate that the development of advanced techniques, improvements in computational power, and informatics will enable the solution of these important issues with the effort from many scientists and researchers.

Acknowledgments

This work was supported by the National Natural Science Foundation for Excellent Young Scholars (52122311), the National Natural Science Foundation of China (51703198), and the Zhejiang Provincial Natural Science Foundation of China (LY20E030007).

References

1 Varnava, C.K. and Patrickios, C.S. (2021). Polymer networks one hundred years after the macromolecular hypothesis: a tutorial review. *Polymer* 215: 123322.

2 Hamed, G.R. (2000). Reinforcement of rubber. *Rubber Chemistry and Technology* 73 (3): 524–533.

3 Danielsen, S.P.O., Beech, H.K., Wang, S. et al. (2021). Molecular characterization of polymer networks. *Chemical Reviews* 121 (8): 5042–5092.

4 Zhong, M., Wang, R., Kawamoto, K. et al. (2016). Quantifying the impact of molecular defects on polymer network elasticity. *Science* 353 (6305): 1264–1268.

5 Di Lorenzo, F. and Seiffert, S. (2015). Nanostructural heterogeneity in polymer networks and gels. *Polymer Chemistry* 6 (31): 5515–5528.

6 Seiffert, S. (2017). Origin of nanostructural inhomogeneity in polymer-network gels. *Polymer Chemistry* 8 (31): 4472–4487.

7 Escobedo, F.A. and De Pablo, J.J. (1999). Molecular simulation of polymeric networks and gels: phase behavior and swelling. *Physics Reports* 318 (3): 85–112.

8 Schneider, J., Fleck, F., Karimi-Varzaneh, H.A. et al. (2021). Simulation of elastomers by slip-spring dissipative particle dynamics. *Macromolecules* 54 (11): 5155–5166.

9 Li, K., Li, Z., Liu, J. et al. (2022). Designing the cross-linked network to tailor the mechanical fracture of elastomeric polymer materials. *Polymer* 252: 124931.

10 Zhang, Z., Liu, J., Li, S. et al. (2019). Constructing sacrificial multiple networks to toughen elastomer. *Macromolecules* 52 (11): 4154–4168.

11 Zhang, Z., Hou, G., Shen, J. et al. (2018). Designing the slide-ring polymer network with both good mechanical and damping properties via molecular dynamics simulation. *Polymers* 10 (9): 964.

12 He, J., Chen, Q., Qu, J. et al. (2023). Mechanically robust and self-healing elastomers based on dynamic oxime–carbamate bonds: a combined experiment and all-atom simulation study. *ACS Applied Polymer Materials* 5 (4): 3161–3172.

13 Liu, J., Zheng, Z., Li, F. et al. (2016). Nanoparticle chemically end-linking elastomer network with super-low hysteresis loss for fuel-saving automobile. *Nano Energy* 28: 87–96.

14 Chen, Y., Li, Z., Wen, S. et al. (2014). Molecular simulation study of role of polymer–particle interactions in the strain-dependent viscoelasticity of elastomers (Payne effect). *The Journal of Chemical Physics* 141 (10): 104901.

15 Liu, J., Cao, D., and Zhang, L. (2009). Static and dynamic properties of model elastomer with various cross-linking densities: A molecular dynamics study. *The Journal of Chemical Physics* 131 (3): 034903.

16 Shen, J., Lin, X., Liu, J. et al. (2018). Effects of cross-link density and distribution on static and dynamic properties of chemically cross-linked polymers. *Macromolecules* 52 (1): 121–134.

17 Liu, J., Cao, D., and Zhang, L. (2008). Molecular dynamics study on nanoparticle diffusion in polymer melts: a test of the Stokes–Einstein law. *Journal of Physical Chemistry C* 112 (17): 6653–6661.

18 Chen, Y., Xu, H., Ma, Y. et al. (2022). Diffusion of polymer-grafted nanoparticles with dynamical fluctuations in unentangled polymer melts. *Physical Chemistry Chemical Physics* 24 (18): 11322–11335.

19 Qu, J., Chen, Q., Huang, W. et al. (2022). Dispersion and diffusion mechanism of nanofillers with different geometries in bottlebrush polymers: insights from molecular dynamics simulation. *The Journal of Physical Chemistry. B* 126 (39): 7761–7770.

20 Chen, Y., Ma, R., Qian, X. et al. (2020). Nanoparticle mobility within permanently cross-linked polymer networks. *Macromolecules* 53 (11): 4172–4184.

21 Gu, Y., Zhao, J., and Johnson, J.A. (2020). Polymer networks: from plastics and gels to porous frameworks. *Angewandte Chemie, International Edition* 59 (13): 5022–5049.

22 Coran, A.Y. (2013). Chapter 7 – Vulcanization. In: *The Science and Technology of Rubber*, 4e (ed. J.E. Mark, B. Erman, and M. Roland), 337–381. Waltham: Academic Press.

23 Norris, S.C.P., Kasko, A.M., Chou, T. et al. (2020). Stochastic model of randomly end-linked polymer network microregions. *Macromolecules* 54 (1): 126–142.

24 Binder, K. (1995). *Monte Carlo and Molecular Dynamics Simulations in Polymer Science*. Oxford University Press.

25 Kremer, K. and Grest, G.S. (1990). Dynamics of entangled linear polymer melts: a molecular dynamics simulation. *The Journal of Chemical Physics* 92 (8): 5057–5086.

26 Grest, G.S. and Kremer, K. (1990). Statistical properties of random cross-linked rubbers. *Macromolecules* 23 (23): 4994–5000.

27 Kroll, D.M. and Croll, S.G. (2015). Influence of crosslinking functionality, temperature and conversion on heterogeneities in polymer networks. *Polymer* 79: 82–90.

28 Flory, P.J. (1985). Network topology and the theory of rubber elasticity. *British Polymer Journa* 17 (2): 96–102.

29 Sasaki, T., Uchida, T., and Sakurai, K. (2006). Effect of crosslink on the characteristic length of glass transition of network polymers. *Journal of Polymer Science Part B: Polymer Physics* 44 (14): 1958–1966.

30 Bennemann, C., Baschnagel, J., and Paul, W. (1999). Molecular-dynamics simulation of a glassy polymer melt: Incoherent scattering function. *European Physical Journal B* 10: 323–334.

31 Liu, J., Wu, S., Cao, D. et al. (2008). Effects of pressure on structure and dynamics of model elastomers: a molecular dynamics study. *The Journal of Chemical Physics* 129 (15): 154905.

32 Paul, W. (2004). Molecular dynamics simulations of the glass transition in polymer melts. *Polymer* 45 (11): 3901–3905.

33 Bennemann, C., Paul, W., Binder, K. et al. (1998). Molecular-dynamics simulations of the thermal glass transition in polymer melts: α-relaxation behavior. *Physical Review E* 57 (1): 843.

34 Rault, J. (2000). Origin of the Vogel–Fulcher–Tammann law in glass-forming materials: the α–β bifurcation. *Journal of Non-Crystalline Solids* 271 (3): 177–217.

35 Urayama, K., Miki, T., Takigawa, T. et al. (2004). Damping elastomer based on model irregular networks of end-linked poly (dimethylsiloxane). *Chemistry of Materials* 16 (1): 173–178.

36 Esteves, A.C.C., Brokken-Zijp, J., Laven, J. et al. (2009). Influence of cross-linker concentration on the cross-linking of PDMS and the network structures formed. *Polymer* 50 (16): 3955–3966.

37 Weeks, J.D., Chandler, D., and Andersen, H.C. (1971). Role of repulsive forces in determining the equilibrium structure of simple liquids. *The Journal of Chemical Physics* 54 (12): 5237–5247.

38 Hsu, H.P. and Kremer, K. (2016). Static and dynamic properties of large polymer melts in equilibrium. *The Journal of Chemical Physics* 144 (15): 154907.

39 Auhl, R., Everaers, R., Grest, G.S. et al. (2003). Equilibration of long chain polymer melts in computer simulations. *The Journal of Chemical Physics* 119 (24): 12718–12728.

40 Duering, E.R., Kremer, K., and Grest, G.S. (1994). Structure and relaxation of end-linked polymer networks. *The Journal of Chemical Physics* 101 (9): 8169–8192.

41 Everaers, R., Sukumaran, S.K., Grest, G.S. et al. (2004). Rheology and microscopic topology of entangled polymeric liquids. *Science* 303 (5659): 823–826.

42 Moreira, L.A., Zhang, G., Müller, F. et al. (2015). Direct equilibration and characterization of polymer melts for computer simulations. *Macromolecular Theory and Simulations* 24 (5): 419–431.

43 Likhtman, A.E. and McLeish, T.C.B. (2002). Quantitative theory for linear dynamics of linear entangled polymers. *Macromolecules* 35 (16): 6332–6343.

44 Doi, M., Edwards, S.F., and Edwards, S.F. (1988). *The Theory of Polymer Dynamics*. Oxford University Press.

45 Tuteja, A., Mackay, M.E., Narayanan, S. et al. (2007). Breakdown of the continuum Stokes– Einstein relation for nanoparticle diffusion. *Nano Letters* 7 (5): 1276–1281.

46 Thompson, A.P., Aktulga, H.M., Berger, R. et al. (2022). LAMMPS-a flexible simulation tool for particle-based materials modeling at the atomic, meso, and continuum scales. *Computer Physics Communications* 271: 108171.

47 Einstein, A. (1906). Eine neue Bestimmung der Molekldimensionen. *Annalen der Physik* 324 (2): 289–306.

48 Kröger, M., Loose, W., and Hess, S. (1993). Rheology and structural changes of polymer melts via nonequilibrium molecular dynamics. *Journal of Rheology* 37 (6): 1057–1079.

49 Brochard Wyart, F. and de Gennes, P.G. (2000). Viscosity at small scales in polymer melts. *European Physical Journal E: Soft Matter and Biological Physics* 1 (1): 93–97.

50 Kumar, S.K., Jouault, N., Benicewicz, B. et al. (2013). Nanocomposites with polymer grafted nanoparticles. *Macromolecules* 46 (9): 3199–3214.

51 Chen, Y., Xu, Q., Jin, Y. et al. (2018). Design of end-to-end assembly of side-grafted nanorods in a homopolymer matrix. *Macromolecules* 51 (11): 4143–4157.

52 Midya, J., Rubinstein, M., Kumar, S.K. et al. (2020). Structure of polymer-grafted nanoparticle melts. *ACS Nano* 14 (11): 15505–15516.

53 Hore, M.J.A. (2019). Polymers on nanoparticles: structure & dynamics. *Soft Matter* 15 (6): 1120–1134.

54 Holt, A.P., Bocharova, V., Cheng, S. et al. (2016). Controlling interfacial dynamics: covalent bonding versus physical adsorption in polymer nanocomposites. *ACS Nano* 10 (7): 6843–6852.

55 Wei, Y., Xu, Y., Faraone, A. et al. (2018). Local structure and relaxation dynamics in the brush of polymer-grafted silica nanoparticles. *ACS Macro Letters* 7 (6): 699–704.

56 Mapesa, E.U., Street, D.P., Heres, M.F. et al. (2020). Wetting and chain packing across interfacial zones affect distribution of relaxations in polymer and polymer-grafted nanocomposites. *Macromolecules* 53 (13): 5315–5325.

57 Ge, T., Kalathi, J.T., Halverson, J.D. et al. (2017). Nanoparticle motion in entangled melts of linear and nonconcatenated ring polymers. *Macromolecules* 50 (4): 1749–1754.

58 Ge, T., Rubinstein, M., and Grest, G.S. (2020). Effects of tethered polymers on dynamics of nanoparticles in unentangled polymer melts. *Macromolecules* 53 (16): 6898–6906.

59 Ge, T. and Rubinstein, M. (2019). Mobility of polymer-tethered nanoparticles in unentangled polymer melts. *Macromolecules* 52 (4): 1536–1545.

60 Carroll, B., Bocharova, V., Carrillo, J.M.Y. et al. (2018). Diffusion of sticky nanoparticles in a polymer melt: crossover from suppressed to enhanced transport. *Macromolecules* 51 (6): 2268–2275.

61 Yamamoto, U., Carrillo, J.M.Y., Bocharova, V. et al. (2018). Theory and simulation of attractive nanoparticle transport in polymer melts. *Macromolecules* 51 (6): 2258–2267.

62 Lin, C.C., Griffin, P.J., Chao, H. et al. (2017). Grafted polymer chains suppress nanoparticle diffusion in athermal polymer melts. *The Journal of Chemical Physics* 146 (20): 203332.

63 Vogiatzis, G.G. and Theodorou, D.N. (2013). Structure of polymer layers grafted to nanoparticles in silica–polystyrene nanocomposites. *Macromolecules* 46 (11): 4670–4683.

64 Ge, T., Grest, G.S., and Rubinstein, M. (2018). Nanorheology of entangled polymer melts. *Physical Review Letters* 120 (5): 057801.

65 Poling-Skutvik, R., Slim, A.H., Narayanan, S. et al. (2019). Soft interactions modify the diffusive dynamics of polymer-grafted nanoparticles in solutions of free polymer. *ACS Macro Letters* 8 (8): 917–922.

66 Kalathi, J.T., Yamamoto, U., Schweizer, K.S. et al. (2014). Nanoparticle diffusion in polymer nanocomposites. *Physical Review Letters* 112 (10): 108301.

67 Paturej, J., Sheiko, S.S., Panyukov, S. et al. (2016). Molecular structure of bottlebrush polymers in melts. *Science Advances* 2 (11): e1601478.

68 Riseman, J. and Kirkwood, J.G. (1950). The intrinsic viscosity, translational and rotatory diffusion constants of rod-like macromolecules in solution. *The Journal of Chemical Physics* 18 (4): 512–516.

69 Han, Y., Alsayed, A.M., Nobili, M. et al. (2006). Brownian motion of an ellipsoid. *Science* 314 (5799): 626–630.

70 Molaei, M., Atefi, E., and Crocker, J.C. (2018). Nanoscale rheology and anisotropic diffusion using single gold nanorod probes. *Physical Review Letters* 120 (11): 118002.

71 Bao, C., Liu, B., Li, B. et al. (2020). Enhanced transport of shape and rigidity-tuned α-lactalbumin nanotubes across intestinal mucus and cellular barriers. *Nano Letters* 20 (2): 1352–1361.

72 Cai, L.H., Panyukov, S., and Rubinstein, M. (2015). Hopping diffusion of nanoparticles in polymer matrices. *Macromolecules* 48 (3): 847–862.

73 Dell, Z.E. and Schweizer, K.S. (2014). Theory of localization and activated hopping of nanoparticles in cross-linked networks and entangled polymer melts. *Macromolecules* 47 (1): 405–414.

74 Cao, X.Z., Merlitz, H., and Wu, C.X. (2017). Tuning adsorption duration to control the diffusion of a nanoparticle in adsorbing polymers. *Journal of Physical Chemistry Letters* 8 (12): 2629–2633.

75 Cai, L.H., Panyukov, S., and Rubinstein, M. (2011). Mobility of nonsticky nanoparticles in polymer liquids. *Macromolecules* 44 (19): 7853–7863.

76 Poling-Skutvik, R., Krishnamoorti, R., and Conrad, J.C. (2015). Size-dependent dynamics of nanoparticles in unentangled polyelectrolyte solutions. *ACS Macro Letters* 4 (10): 1169–1173.

77 Chen, R., Poling-Skutvik, R., Nikoubashman, A. et al. (2018). Coupling of nanoparticle dynamics to polymer center-of-mass motion in semidilute polymer solutions. *Macromolecules* 51 (5): 1865–1872.

78 Yamamoto, U. and Schweizer, K.S. (2015). Microscopic theory of the long-time diffusivity and intermediate-time anomalous transport of a nanoparticle in polymer melts. *Macromolecules* 48 (1): 152–163.

79 Yamamoto, U. and Schweizer, K.S. (2011). Theory of nanoparticle diffusion in unentangled and entangled polymer melts. *The Journal of Chemical Physics* 135 (22): 224902.

80 Lang, M., Michalke, W., and Kreitmeier, S. (2001). A statistical model for the length distribution of meshes in a polymer network. *The Journal of Chemical Physics* 114 (17): 7627–7632.

81 Volgin, I.V., Larin, S.V., Abad, E. et al. (2017). Molecular dynamics simulations of fullerene diffusion in polymer melts. *Macromolecules* 50 (5): 2207–2218.

82 Hooper, J.B. and Schweizer, K.S. (2005). Contact aggregation, bridging, and steric stabilization in dense polymer– particle mixtures. *Macromolecules* 38 (21): 8858–8869.

10

Experimental and Theoretical Studies of Transport of Nanoparticles in Mucosal Tissues

Falin Tian and Xinghua Shi

Laboratory of Theoretical and Computational Nanoscience, CAS Center for Excellence in Nanoscience, National Center for Nanoscience and Technology, Beijing 100190, P.R. China

10.1 Introduction

Biological hydrogels are found in various tissues and bodily fluids and are known to play important roles in various physiological processes. One of the largest hydrogels in the body is the mucus gel that is mainly presented on the apical epithelial surfaces of various organs in the body, such as oral cavity, the airway, the gastrointestinal (GI) and urogenital tracts, and so on [1–8] (Figure 10.1a). The most important component of mucus is water (approximately 95% by weight); other components include mucins, lipids, salts, DNA, cellular debris, and various nonmucin proteins, which form a heterogeneous 3-dimensional network with an average pore size of 10–500 nm and give mucus its viscoelastic nature and gel-forming properties (Figure 10.1b). The mucins are the primary structural and functional component of mucus as they are the largest molecules produced by the body. Most of them have a polypeptide backbone and some heavy weight of sugar residues attached to the polypeptide backbone, which gives them a brush-like structure. According to the component, the mucins can further be divided into two domains: regions rich in proline, serine, and threonine with sugar chains (PST domains) and regions linked to two PST domains (cysteine-rich domains). In addition to the complex network structure, another important characteristic of mucus is its special viscoelasticity [9]. Under low shear, such as the shear experienced during normal breathing or peristaltic movement in the gastrointestinal tract, mucus behaves like an elastic solid. It can recover its shape after deformation [5, 10, 11]. This adaptable behavior is due to the cross-linked mucin fibers that make up the mucus matrix. However, under high shear, such as the shear experienced during coughing or sneezing, mucus can behave like a viscous liquid and undergo irreversible deformation. This is because the shear forces are strong enough to disrupt the cross-linking between mucin fibers, causing the mucus matrix to break down and become more fluid-like. This viscous behavior is critical for removing mucus from the respiratory system.

Dynamics and Transport in Macromolecular Networks: Theory, Modeling, and Experiments, First Edition. Edited by Li-Tang Yan.

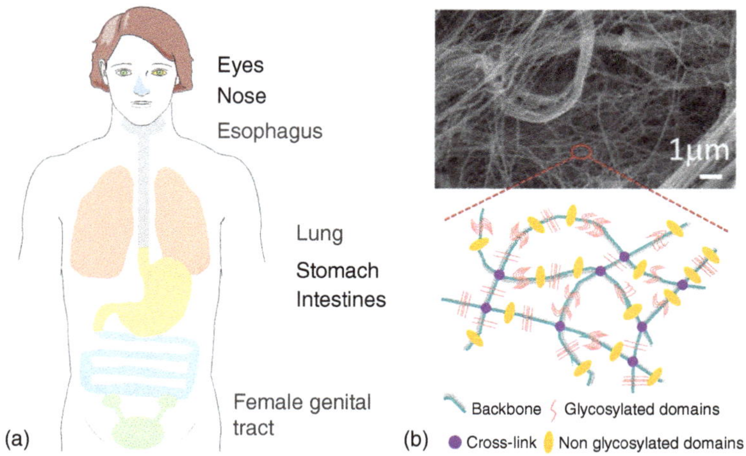

Figure 10.1 The scheme of the mucus. (a) The distribution of mucus throughout the body. (b) The typical structure and component of mucus.

These high complex structure and viscoelasticity give mucus its unique properties and acts as a protective barrier, preventing the spread of toxins, pollutants, and invading pathogens, and regulating the exchange of nutrients, gases, and waste products between cells and the surrounding environment [1, 12–17]. In addition, through continuous secretion and shedding, mucus can also quickly eliminate foreign particles trapped within it.

Unfortunately, this barrier function can also hinder the delivery of drug carriers [18–22]. Therefore, to enhance the delivery efficiency, it is required that the carrier must quickly penetrate the mucus layer in a short period of time. Therefore, understanding the complex properties of mucus and its interactions with nanoparticles is important in developing effective oral drug delivery strategies that can overcome the mucus barrier and improve drug efficacy. Currently, the permeability of mucus is believed to be achieved through two mechanisms [23–30]. The first mechanism suggests that the mesh size of the protein network in mucus plays a major role, where the protein network easily traps particles larger than the mesh size. In comparison, smaller particles can pass through the network structure (Figure 10.2). Indeed, lots of experiments support the idea of size effect, which refers to the selective retention and hindered transport of larger particles in mucus compared to smaller particles [31–34]. It was reported that the diffusivity of polystyrene nanoparticles densely coated with 5 kDa polyethylene glycols (PEG) (0.09 PEG/nm^2) in human respiratory mucus depended on their size [33]. The diffusion or mobility of nanoparticles decreases with increasing their size in sputum mucus, indicating that larger particles experience more obstruction and entanglement with the mucus fibers [35]. Moreover, the diffusivity of NPs will be enhanced in the tumor matrix, which also has a dense and complex network of proteins similar to the mucus, when its surface grafting the proteolytic enzyme rHuPH20 to destroy the network structure of protein and increases the pore size [36]. All these results indicated that size is a critical parameter for mucus permeation. However, in the case of cervicovaginal

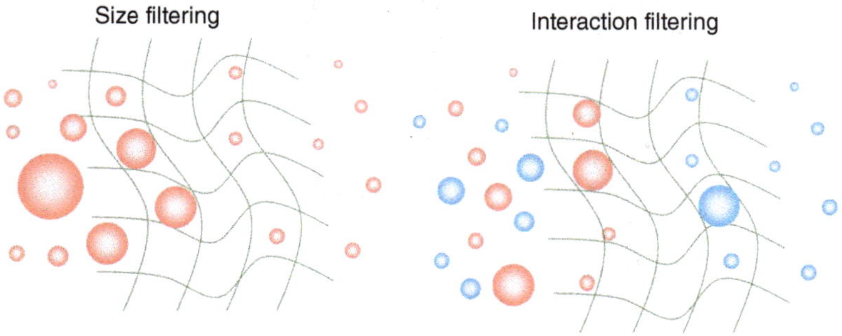

Figure 10.2 The scheme of two generic filtering mechanisms of mucus. Size filtering can restrict the passage of particles with sizes larger than the mesh size of the protein networks. Interaction filtering based on the surface properties (e.g., charge, hydrophobicity, or specific adhesive sites) of particles. Particles with a surface interacting strongly with the protein fibers can be hindered or trapped (red particle), while those with weak interactions can pass through the mucus (blue particle).

mucus, several studies have shown that small and highly negatively charged particles (<100 nm) are less mobile than larger, neutral particles (>500 nm), which is opposite to the size effect observed in other mucosal environments [37]. This suggests that alternative mechanisms, such as electrostatic interactions between the negatively charged particles and the negatively charged mucin fibers, may be at play. Indeed, lots of recent works have concluded that the surface properties of nanoparticles, such as their surface charge [38, 39], hydrophobicity [40], and functionalization [32, 35, 41–44], can have a significant effect on their adhesive interaction with mucin fibers, which in turn affects their mobility and permeation properties within mucus [25, 45]. Therefore, the interaction between NPs and protein fibers of mucus surface based on the properties of NPs-based filtering is the second and independent mechanism of serves as the second and independent filtering mechanism, i.e. interaction effect. In this scenario, with specific surface properties can engage in strong interactions with the protein fibers in mucus, leading to their trapping and reduced mobility within the mucus, while other particles with different surface properties may not interact strongly with the protein fibers and can diffuse efficiently through the mucus (Figure 10.2).

The above two filtering mechanisms can explain most of the phenomena in the mucus-particle system. In comparison, these two mechanisms may not be independent in real biological systems, and both size and interactions may simultaneously play a key role in some systems. For example, the diffusion of the polymeric NPs with sizes from 200 to 500 nm have the same diffusivity in cervicovaginal mucus when their surface is modified by the densely PEG chains [37]. Moreover, particles with a small size and negative charge can penetrate the mucus layer but fail to pass through the epithelial barrier because of the small size limits the internalization ability of epithelial cells [46]. Therefore, it is necessary to consider both the size and surface properties of the carrier for oral medications so that it can quickly penetrate through mucus and have a high ability to be internalized by epithelial cells. However, in addition to the size and surface properties, the shape and stiffness can also alter the interaction between particles and biological media and further the biological effects of NPs [47, 48]. Thus, the effect mechanism of the shape and stiffness on the diffusivity of NPs in mucus should be studied systemically.

As it is challenging to directly study the diffusion process of nanocarriers within the mucosal layers *in situ* in human gastrointestinal tract due to the technical limitations. Many model systems, such as *in vitro*, *ex vivo*, and *in vivo* models, can be employed to study the transport of nanocarriers across a mucosal barrier in a controlled environment. Experimental techniques such as multiple-particle tracking (MPT) can be used to track the diffusion of nanoparticles on short time and length scales in mucosal tissues, while fluorescence images can be used to evaluate the penetration and bulk diffusion, on the other hand, are conducted on longer time and length scales. Moreover, molecular simulation and mathematical models have also been employed to analyze the diffusion mechanism through the mucus and the regulatory role of key parameters. Indeed, a coarse-grained water-soluble regular polymer network has been used as a model system to study particle transport through mucus [49–51]. The properties of the polymer networks, such as the mesh size, the stiffness of the fibers, and the adhesive strength, can be tuned to mimic the properties of various types of mucus and study the effect of these properties on transport behavior. Although all these studying models have some deviations compared to the real system, we believe that these ideal models have recapitulated the composition and structure of native mucus and thus provided valuable insights into the transport of particles through mucus.

In this chapter, we mainly focus on introducing the mechanisms behind the impact of particle size and stiffness on their diffusion ability in mucus based on a combination of experiments, molecular simulations, and theoretical modeling.

10.2 Enhancing Diffusivity of Deformable Particles to Overcome Mucus Barriers Via Adjusting Their Rigidity

Rigidity is an important characteristic of the nanocarriers formed by the assembly of biological molecules, such as liposomes, peptides micelles, dendrimers, and protein nanotubes [52–54]. Previous studies suggest that the stiffness of the nanocarriers can influence their interactions with the cell membrane and the subsequent internalization mechanism. For instance, some studies have shown that nanocarriers with high rigidity can induce mechanical deformation of the cell membrane, leading to membrane invagination and subsequent endocytosis [55, 56]. In contrast, soft nanocarriers may be able to change their shape and conform to the cell membrane, entering the cell in different ways. Other studies suggest that the stiffness of nanocarriers can alter their ability to interact with specific receptors or trigger intracellular signaling pathways, ultimately affecting their cellular uptake [57, 58]. However, few works have reported the effect of rigidity on the diffusivity of NPs in the mucus, which is a main barrier to the delivery process of nanocarriers. In this section, we will review some recent works focused on how the rigidity of NPs affects their diffusivity in mucus.

10.2.1 The Preparation of the Hybrid NPs with Various Rigidities

Polymeric nanoparticles (PNPs) coated with lipids have been widely studied for their potential applications in drug delivery because of their unique advantages such as high efficiency and low risk of side effects. Usually, he polymeric matrix of PNPs can be made of various polymers such as poly(lactic-*co*-glycolic acid) (PLGA), polyethyleneimine (PEI), and chitosan. The lipids coated on the PNPs not only allow for improved targeting and increased cellular uptake but also provide stability to the PNPs and protect them from degradation. The lipid-coated PNPs can also serve as model systems for understanding the mechanisms of the biological effects of nanomaterials due to their special core-shell structure and tunable mechanical properties.

The lipid-coated PNPs with controllable rigidity can be generated using a two-stage microfluidic platform [56] (Figure 10.3a). The first stage consists of three inlets and a straight synthesis microchannel. The second stage comprises one centered inlet and a spiral mixing channel. The lipid solution is introduced in the first stage, while the PLGA solution is introduced in the second stage of the preparation process. At the first stage, the lipids self-assemble into a liposome in an aqueous solution and then re-assemble onto the surface of PLGA PNPs within a thin water layer between the lipid shell and PLGA core at the second stage. Using this chip,

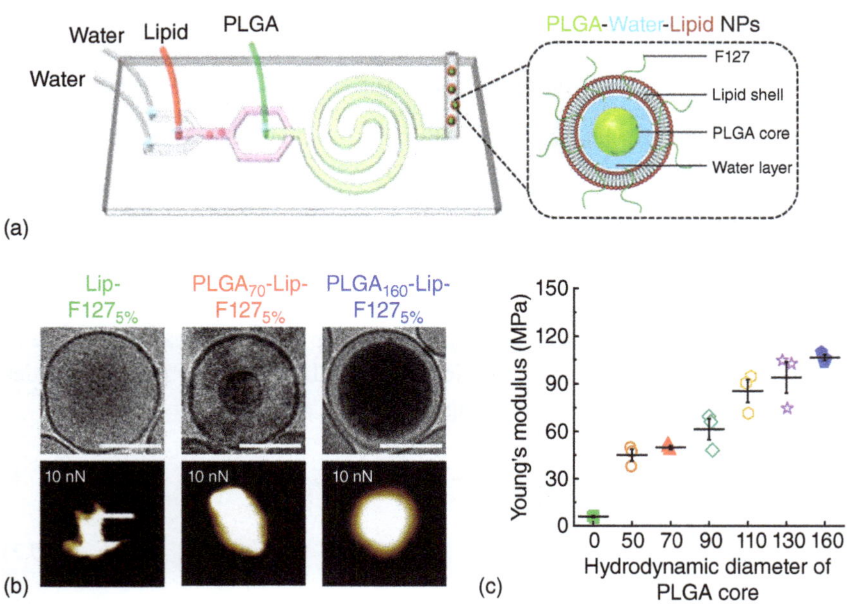

Figure 10.3 The characterizations of hybrid NPs of varying PLGA core size. (a) The scheme of two-stage microfluidic platforms for preparing the hybrid NPs. (b) Cryo-TEM and deformation images under the AFM compression of hybrid NPs with different PLGA core sizes. Lip represents the lipid shell molecules, PLGAx indicates a PLGA core of diameter x nm, and F1275% indicates 5 wt% content of Pluronic F127 (F127) on the lipid (Lip) shell. (c) Young's modulus values of hybrid NPs with different hydrodynamic diameters of PLGA core ($n = 3$).

the researchers synthesized PLGA-Water-Lipid hybrid NPs with varying rigidity by altering the order of introducing the reagents. For example, the core size of the PLGA can be controlled by changing the total flow rate (2 mL/min to 15 mL/min to produce NPs with hydrodynamic diameter from 50 nm to 160 nm). Moreover, using microfluidic chips to synthesize hybrid NPs offers several advantages over conventional methods, including improved control over particle size, composition, and surface properties, as well as higher throughput and reproducibility.

The prepared hybrid NPs have similar hydrodynamic diameters and surface charge properties but different PLGA core sizes [59]. The cryogenic transmission electron microscopy (cryo-TEM) results show no matter how the size of the PLGA core changes, the lipid shells will completely cover it (Figure 10.3b). The morphologies of the hybrid NPs with different PLGA core sizes are characterized using AFM-based nanomechanical mapping (Figure 10.3b). As the size of the PLGA core of the hybrid NPs decreases, their shape can undergo significant changes under the compression of AFM. Specifically, the nanoparticles may transform from a spherical shape to an irregularly shaped structure. This can be due to the small size of the PLGA core corresponds to the presence of a large amount of interfacial water, which make the NPs easier deformation under stress or external stimuli. The Young's modulus of the hybrid NPs increases with their PLGA core size (Figure 10.3c), which further confirms that NPs with large PLGA core size are too rigid to deform under the under the external stimuli.

10.2.2 The Diffusivity of Hybrid NPs with Different Rigidity in Mucus

The diffusivity of hybrid NPs with different rigidity in mucus is first studied through *ex vivo* tracking experiments. The fresh mucus was collected from the surface of the small intestine of rats that were euthanized. Next, the hybrid nanoparticles were pipetted onto the mucus and incubated for a short period of 0.5 hours prior to microscopy analysis. Then, the movies of the movement of NPs in the mucus are captured using the inverted fluorescence microscope, and each NP's positions and movement trajectory are analyzed using ImageJ for each experiment. Particle-averaged mean squared displacement (MSD) and effective diffusivities (D_{eff}) were calculated using the following equations:

$$\text{MSD}(t) = \frac{\sum\left[(x(t) - x_0)^2 + (y(t) - y_0)^2\right]}{N}, \tag{10.1}$$

$$D_{\text{eff}} = \frac{\sum[\text{MSD}(t)/(4t)]}{N}, \tag{10.2}$$

where x and y represent the coordinates of the particle, and t represents time scale or time lag.

To investigate how the rigidity of hybrid NPs influences their diffusivity in mucus, three types of hybrid NPs with Young's modulus of 5 MPa (Lip-F1275%, soft), 50 MPa ($PLGA_{70}$-Lip-F1275%, semi-elastic) and 110 MPa ($PLGA_{110}$-Lip-F1275%) are tracked using MPT technology. The results revealed that the semi-elastic NPs (PLGA70-Lip-F1275%) exhibited the fastest diffusion in the mucus and displayed

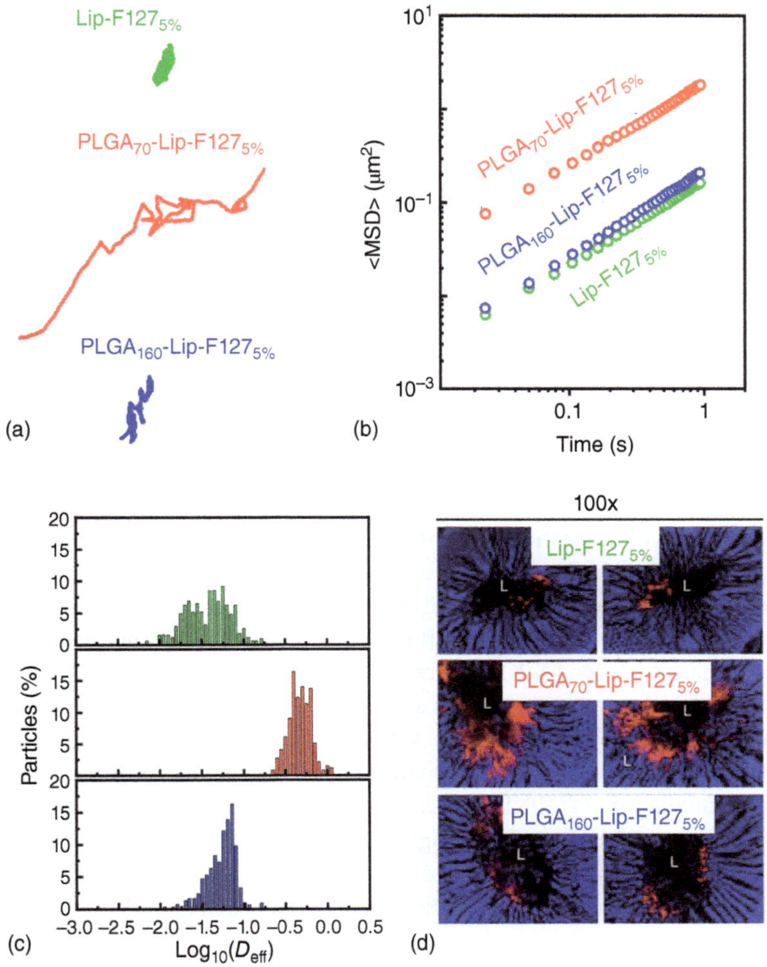

Figure 10.4 The transport of hybrid NPs with various rigidities in rat mucus. (a) Representative trajectories of NP motion in 1 s in fresh rat intestinal mucus. (b) MSD values as a function of time for NPs in mucus. (c) Distributions of the logarithms of individual particle effective diffusivity (D_{eff}) values in mucus on a time scale of 1 second. The data represent three independent experiments, with each experiment tracking 100 NPs. (d) Distribution of NPs in rat intestines. "*L*" indicates the lumen of the intestine. Blue: nuclei of the intestinal villi. Red: NPs.

the largest diffusion areas among the tested NPs (Figure 10.4a). Further analysis showed that the time scale-dependent ensemble mean squared displacement (MSD) of the semi-elastic NPs had a significantly higher value across all time scales than those of the soft NPs (Lip-F1275%) and rigid NPs (PLGA160-Lip-F1275%) (Figure 10.4b). The same result was also observed by tracking the motion of NPs with different rigidity in poly(ethylene oxide) (PEO) hydrogels, which have a network structure similar to mucus. Based on a previous study, it was expected that unusually fast-moving nanoparticles would be capable of penetrating mucus layers. The distribution of individual particle effective diffusivity (D_{eff}) values in mucus showed that all semi-elastic NPs diffused relatively quickly with a value exceeding 0.1 μm²/s, while only 4.7% of the soft NPs and 3.3% of the rigid NPs exceeded this diffusion speed (Figure 10.4c). These results indicated that NPs with moderate rigidity display superior mucus-penetrating capability to their soft and hard counterparts and can be used to deliver oral drugs.

The delivery of orally administered drugs requires the nanocarrier that can quickly penetrate the mucus layer on the surface of the gastrointestinal tract. *In vitro* tests have shown that semi-elastic NPs have the best ability to diffuse through the mucus. However, whether these types of NPs can effectively penetrate the mucus layer in the gastrointestinal tract *in vivo* remains to be seen. In order to verify whether semisoft NPs could quickly penetrate the real mucus layer in the gastrointestinal tract *in vivo*, an *in vivo* intestinal loop incubation method was used in rats to investigate the cross-sectional distribution of nanoparticles (NPs). The results showed that the semi-elastic NPs can penetrate throughout the mucus layer and have a largest distribution area in the intestine (Figure 10.4d). Interestingly, some semi-elastic NPs were also observed in both the epithelium and enterocytes, suggesting that this type of NP may be able to cross the epithelium via enterocytes. In contrast, both soft and rigid NPs remained clumped in the intestinal lumen and could not cross the epithelial barrier efficiently. To further investigate the distinct behavior of semi-elastic nanoparticles (NPs), different fluorophore-labeled particles were orally co-administered in rats. The results showed that the semi-elastic NPs were dispersed widely throughout the intestine, while the soft and hard NPs remained aggregated in the intestinal lumen. The *in vivo* results of the study confirmed that the semi-elastic nanoparticles (NPs) maintained their superiority in terms of mucus penetration compared with the soft and rigid NPs. This suggests that semi-elastic NPs could potentially be used to improve oral drug delivery and increase drug absorption in the intestines. However, further studies are needed to understand the underlying mechanisms involved in the superior behavior of semi-elastic NPs.

10.2.3 The Interaction Between NPs with Different Rigidity and Mucus Network

The mucus layer is a complex network of glycoproteins, which creates a physical barrier that prevents foreign particles from reaching the underlying cells and tissues. In addition, the interaction between nanoparticles (NPs) and the mucus network is an important factor in determining the ability of NPs to penetrate the mucus layer. To further elucidate the mechanism underlying the superior mucus-penetrating ability of semi-elastic NPs, a series of molecular dynamic simulations were used to investigate the diffusion behavior of NPs in mucus at the molecule level. The experimental test has been confirmed that the average size of the intestine mucus protein network pores of rat was around 240 nm, with the majority of pores of mucus ranging from 100 to 500 nm in diameter. The hydrodynamic diameter of the hybrid NPs used in the above experimental tests was about 200 nm, which is close to the size of the mucus pore. Since the scale of the system being studied is far beyond the size range of simulations using all-atomic models, a coarse-grained model is more suitable for studying the NP–mucus interaction. According to the characteristic features of the protein mesh structure in mucus and the main components of the network structure, a cross-linked regular polymer network was used to represent the protein network structure in mucus. The hybrid NPs were modeled as a lipid shell with tunable rigidities (Figure 10.5a). A classic Lennard–Jones (LJ) potential was used to describe the affinity between the polymer network and NPs, and the parameter was verified by a series of all-atom molecular dynamics simulations (details see the reference [51]). In the simulations, the Velocity–Verlet algorithm was utilized to perform time integration. The integration time step was $\Delta t = 0.01\tau$, and the Langevin thermostat was used to control the system temperature at $k_B T = 0.23\varepsilon$. All simulations were performed under the constant number-volume-energy (NVE) integration to update the position and velocity of the beads in each simulation.

The snapshots from the simulation showed that the soft NPs have a significant deformation and were able to conform to the corners of the polymer network, maintaining a stable state throughout the simulation (Figure 10.5b). In contrast, the rigid NPs do not undergo any significant shape change and almost trend to be trapped in one corner of the network during the simulation because of the high attraction density in that area (Figure 10.5b). Interestingly, the semi-elastic NPs exhibit a unique behavior compared to rigid NPs and soft NPs, they first attract to one corner of the network like the rigid NPs. However, the shape then deforms into an ellipsoid due to its attraction to another corner of the network. After that the NPs will detach from the first corner and move to toward the second corner (Figure 10.5b). The variation of the major axial length of NPs, which relates to the shape deformation of the NPs during the diffusion, indicates the major axial length of the soft and rigid NPs remains constant, while that for semi-elastic NPs fluctuates around its initial radius and exhibits two specific shape (Figure 10.5c). In addition, the average major axial length of the soft NPs is much higher than its initial value, while that of the rigid NPs is equal to its initial radius. These results confirmed that the shape of the soft NPs changes dramatically, while the rigid NPs remain a sphere during diffusion in the network. The semisoft changes its shape frequently during the diffusion process. The shape and state of NPs with different rigidities may result from the complex interactions between the NPs and the polymer network. Further analysis shows that rigidity remarkably affects the contact numbers between NPs and network fibers during the diffusion process (Figure 10.5d). Soft NPs have the highest contact number during diffusion because of their large deformation, resulting in more contact numbers with the network fibers. The high contact number corresponds to a strong attraction between the NPs and the network, which means that the soft NPs have a much stronger attraction for the network fibers than the semi-elastic and rigid NPs. Interestingly, the contact number of the semi-elastic NPs is not constant but constantly fluctuates over time because of the fluctuation of their shape during the diffusion.

Therefore, one can conclude that the soft NPs will be apt to undergo a large deformation during the diffusion, which increases its contact area with the hydrogel network, resulting in increased affinity and decreased diffusivity. The rigid NPs tend

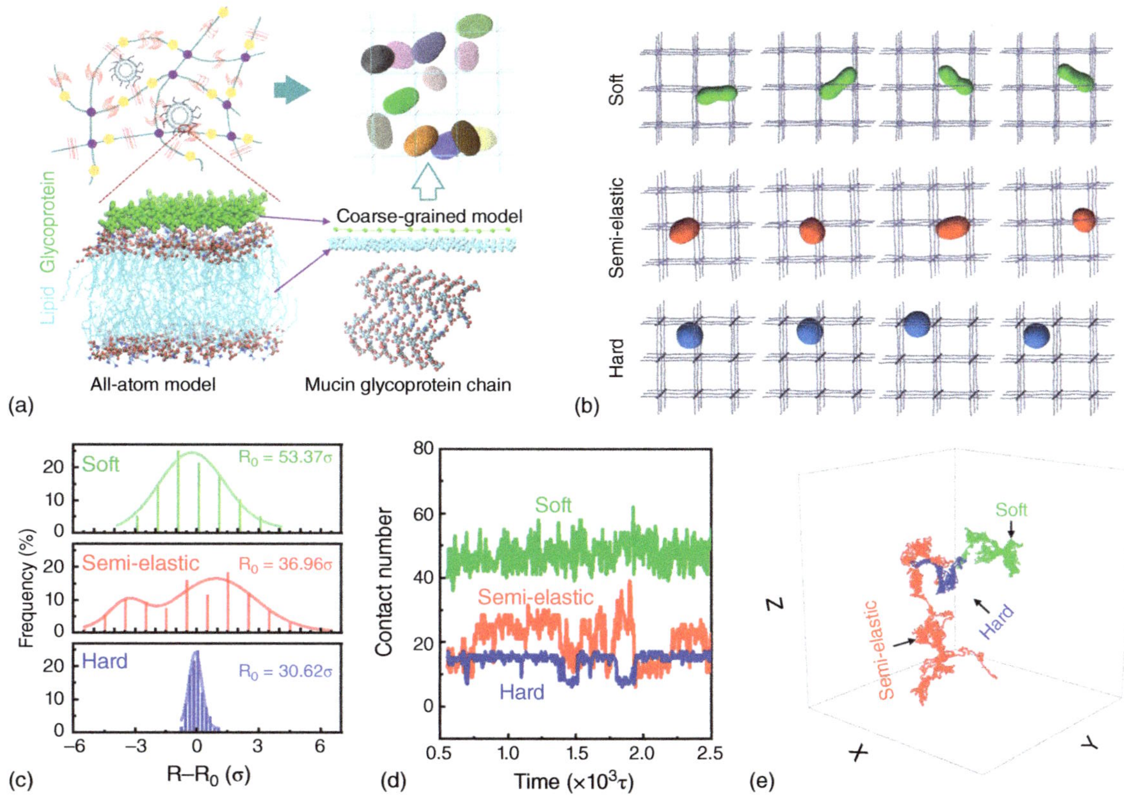

Figure 10.5 The diffusivity behavior of NPs with various rigidities from molecular simulations. (a) The scheme for constructing the coarse-grained model. The All-Atom system contained a lipid membrane, a mucin glycoprotein chain above the membrane. The mucin glycoprotein chain was modeled as a combination of 13 O-Glycans. The coarse-grained system is composed of a lipid membrane and a long polymer chain above the lipid membrane (Cyan Beads: lipid membrane; Green beads: the polymer chain). In the coarse-grained model, a single bead of polymer chain and lipid membrane represents one O-Glycan and lipid molecule, respectively. (b) Representative snapshots of NPs structures from the simulation. (c) The distribution of fluctuation of the length of the macroaxis of liposomes near their time average length R_0. (d) Number of beads in each NPs contacting the polymer network in mucus. (e) Typical centroid trajectories of NPs in simulations.

to be trapped in the corner of the hydrogel because of the high attraction density in that area, and these molecules oscillate around the corner without shape change. However, the semi-elastic NPs exhibit unique diffusion behavior due to their intermediate mechanical properties. They tend to attach and detach from the corners of the network due to the competition between the attractive forces between the NP and the two corners and the energy required to deform the NPs. Through this process, the semisoft liposomes exhibit superior diffusivity compared with that of the soft and hard liposomes (Figure 10.5e).

10.2.4 The Theoretical Model to Describe the Diffusion Behavior of Deformable Nanoparticles in Adhesion Network

Combining data from experiments with multiple coarse-grained molecular dynamics simulations, it was confirmed that the rigidity of the deformable nanoparticles (DNPs) is a key parameter for determining their diffusion behaviors in the protein network of mucus. The discoveries also implied that the rigidity of the DNPs and the interaction between DNPs and protein network could be coupled and possibly achieve a balance to maximize the diffusion rate. In addition, the diffusion of the DNPs in the protein network of mucus is a complex process that involves multiple physical and chemical processes, such as adhesion, deformation, and diffusion. The classic Stokes-Einstein equation, which assumes a constant diffusion coefficient, cannot accurately describe this process. Thus, there is a growing awareness that a theoretical model needs to be developed to reveal the key factors affecting the nanoscale interactions between the DNPs and adhesion network, identify key factors, and aid in designing optimized drug delivery systems.

Previous works have pointed out that the consumed time for the diffusion of a particle from one position to another could be obtained by calculating the mean first passage time (MFPT) [60, 61]. The particles have always been assumed to be rigid in those calculations. At the same time, the diffusion of the DNPs in the adhesion network always undergoes displacement accompanied by shape deformation, which would change the interaction energy and further affect the MFPT calculation. To accurately model the diffusion process, we first consider a system consisting of $M = \sum_{i=1}^{N} M_i$ NPs, where M_i represents the number of DNPs with the same shape, and N is the types of the NPs with different shapes. The proportion of each M_i is determined by the energy of this shape based on the assumption that all states follow Boltzmann distribution

$$p_i = \frac{\exp(-E_i/k_B T)}{\sum \exp(-E_i/k_B T)}, \tag{10.3}$$

where E_i is the energy of the ith-set of NPs, k_B is the Boltzmann constant, and T is the temperature. This way, the diffusion and deformation process of deformable NPs in an adhesion network can be decoupled, and the complex diffusion process can be considered a diffusion problem of a set of DNPs with different shapes in the same protein network.

10.2.4.1 Shape Distribution of NPs

Following the Helfrich framework [62], the total deformation energy of a DNP consists of membrane bending energy (E_{bending}), surface tension energy (E_{tension}), and work difference of osmotic pressure (E_{osmotic})

$$E_{\text{def}} \underbrace{\int \frac{k}{2}(c_1 + c_2)^2 \text{d}s}_{E_{\text{bending}}} + \underbrace{\lambda \Delta A}_{E_{\text{tension}}} + \underbrace{\Delta p \cdot \Delta v}_{E_{\text{osmotic}}}, \tag{10.4}$$

where c_1 and c_2 are two principal curvatures of the membrane surface, κ is the bending stiffness of the membrane, dS is the surface integral element, λ is the membrane surface stress, Δp is the pressure difference between outer and inner environment, and ΔA and ΔV are the area difference and volume difference between the current deformable and initial shape of DNP, respectively. To simplify, one can be assumed that liquid can be exchanged freely between outer and inner environment of NPs, which means that there is no osmotic pressure ($\Delta p = 0$). Using this deformation energy expression, the shape of DNP under different conditions (stiffness and surface membrane stress) can be easily identified. Because of the rotational symmetric property of the shape of DNP, the arc length s and angle θ have a relationship, as Figure 10.6a illustrates

$$\theta = \theta(s). \tag{10.5}$$

Thus, the deformation energy turns into

$$E_{\text{deformation}} = \pi\kappa \int_0^l \left(\frac{\text{d}\theta(s)}{\text{d}s} + \frac{\sin\theta(s)}{r(s)} \right)^2 r(s)\text{d}s + \lambda \int_0^l 2\pi r(s)\text{d}s, \tag{10.6}$$

with the boundary condition

$$\theta(0) = 0, \quad \theta(l) = \pi. \tag{10.7}$$

The Fourier series is suitable to describe the cross-section shape

$$\theta(s) = \frac{\pi}{l}s + \sum_{i=1}^{N} \eta_i \sin\left(\frac{\pi}{l} is\right). \tag{10.8}$$

Here, η_i is the coefficient of Fourier series as well as the decision variable of this optimization problem. Another requirement that should be satisfied is that the curve must be close

$$r(l) = \int_0^l \cos\theta(s)\text{d}s = 0. \tag{10.9}$$

Here, without loss of generality, if we select l_0 as the distance unit and κ the energy unit. Then, the dimensionless parameters can be derived as follows

$$\bar{s} = s/l_0, \bar{r} = r/l_0, \quad \bar{l} = l/l_0, \quad \overline{V_0} = V_0/l_0^3. \tag{10.10}$$

Therefore, the total deformation energy turns into

$$E_{\text{deformation}} = \kappa \left[\pi \int_0^{\bar{l}} \left(\frac{\text{d}\theta}{\text{d}\bar{s}} + \frac{\sin\theta}{\bar{r}} \right)^2 \bar{r}\text{d}\bar{s} + 2\pi \frac{\lambda l_0^2}{\kappa} \int_0^{\bar{l}} \bar{r}\text{d}\bar{s} \right]. \tag{10.11}$$

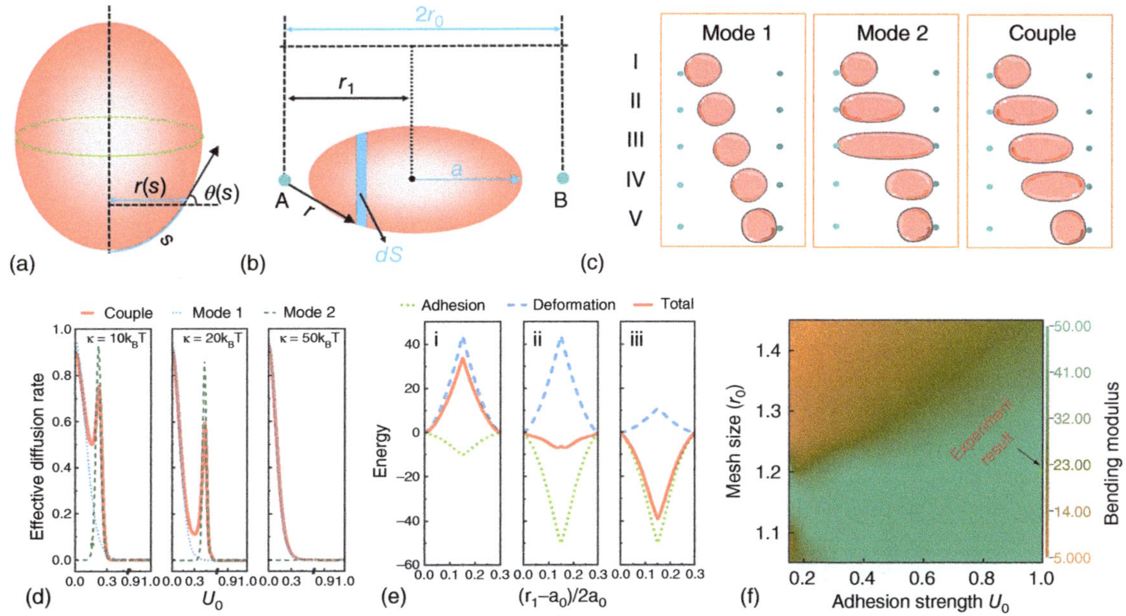

Figure 10.6 The diffusivity of deformable NPs predicted by the theoretical model. (a) The Schematic of a deformable NP can be modeled as a rotationally symmetric object. (b) Schematic for the interaction of a deformable NP with adhesion regions A and B. (c) The Schematic of the three diffusion modes between two adhesion regions. (d) Effective diffusion rate of deformable NPs with various rigidity. (e) Total energy profiles consisting of adhesion energy and deformation energy: (i) deformation energy is higher than adhesion energy, corresponding to the NPs with high rigidity; (ii) deformation energy is nearly equal to adhesion energy, corresponding to the NPs with intermediate rigidity; (iii) deformation energy is less than adhesion energy, corresponding to the NPs with low rigidity. (f) The phase diagram of predicted optimized stiffness κ to achieve maximum diffusion rate on the plan of adhesion strength U_0 and mesh size r_0. The red star is the experimental test value.

Since the ideal shape of the DNP should be determined by the deformation energy, so the problem is simplified into an optimization problem with nonlinear constraints, and the shape of the DNP is totally determined by η_i (Eq. (10.8)). In this problem, we need nonlinear constraints to obtain a unique solution. Here, we choose the length of the semi-major axis and the surface area to be constraints as follows

$$a = \frac{1}{2}\int_0^{\bar{l}} \sin\theta d\bar{s}, \tag{10.12}$$

$$S = \int_0^{\bar{l}} 2\pi\bar{r}d\bar{s}. \tag{10.13}$$

10.2.4.2 Diffusion Model

The diffusion process of a DNP in the polymer network without adhesion can be derived using Amsden's obstruction-scaling model [63]

$$D_g = \exp(-\overline{N})\frac{k_B T}{6\pi\mu \cdot R_H}, \tag{10.14}$$

where μ is the dynamics viscosity of the water, R_H is the hydrodynamic radius of the particle, $\overline{N}$ is the mathematical expectation of the number of fibers in contact with NP [64]. To model the diffusion process of deformable DNP in the adhesion polymer, we assume the adhesion zones within the network are discrete, and the affinity interaction between the DNP per unit length and the adhesion region is described by the Morse potential

$$V(r) = U_0(e^{-2r/\rho} - 2e^{-r/\rho}), \tag{10.15}$$

where U_0 is the adhesion strength per unit area, r_d is the distance between one point on the surface of DNP and the adhesion region, and ρ is the potential range, which is usually less than a dozen of nanometers (Figure 10.6b). Therefore, the adhesion energy of one adhesion region will be the integral on the whole surface of the DNP

$$U_{\text{adhesion}} = \int_{\frac{r_1}{2}-a}^{\frac{r_1}{2}+a} U_0\left(e^{-\frac{2r_d}{\rho}} - 2e^{-\frac{r_d}{\rho}}\right) dS, \tag{10.16}$$

where r_1 is the mass center of the particle, and dS is the integral surface element. Thus, the total adhesion energy will be the sum of adhesion energy with two adjacent region A and region B. The diffusion process of DNPs in this adhesive network depends on their rigidity and can be divided into three modes (Figure 10.6c). To state the model explicitly, two extreme cases in which the DNPs are extremely rigid and extremely soft are discussed first.

Mode 1: diffusion without shape deformation. The DNPs with high rigidity will keep their initial shape due to high deformation energy cost and undergo a hopping diffusion. In this case, the DNP is initially adsorbed at one adhesion region A, and then detach and adsorb to another adhesion region B (Figure 10.6c). The total energy is only contributed by the adhesive interaction:

$$U(r) = \int U_0\left(e^{-\frac{2r_A}{\rho}} - 2e^{-\frac{r_A}{\rho}}\right) dS + U_0\left(e^{-\frac{2r_B}{\rho}} - 2e^{-\frac{r_B}{\rho}}\right) dS, \tag{10.17}$$

where r_A and r_B is the distance from A and B to the point on the surface of NP.

Mode 2: diffusion with large deformation. For soft DNP, it is easy to change its shape due to the low deformation energy. Thus, the DNP will undergo a remarkable shape deformation during the diffusion. In this case, one end is attached to adhesive region A first; then it will lengthen, and another end will be attached to B due to the attraction of region B; after that, the DNP will detach from A and shorten until it reaches the initial shape near the region B (Figure 10.6c). In this process, the energy profile is composed of two parts: deformation energy and adhesion energy. Since there is always one end pinned, the shape of the DNP is fixed once the mass center of the DNP is fixed. And the total energy is only the function of its position

$$U(r) = U_{\text{deformation}} + U_{\text{adhesion}}. \tag{10.18}$$

The diffusion time for the above two modes can be calculated by MFPT theory as

$$t = \frac{1}{D_g} \cdot \int_0^{r_0} \exp\left(-\frac{U(r)}{k_B T}\right) \cdot \left[\int_0^{r_0} \exp\left(\frac{U(y)}{k_B T}\right) \mathrm{d}y\right] \cdot \mathrm{d}r, \tag{10.19}$$

where D_g is the diffusion coefficient of DNPs in the nonadhesion network. Finally, the effective diffusivity of the DNP in the adhesive polymer will be

$$D_e = \frac{r_0^2}{2t}. \tag{10.20}$$

Mode 3: diffusion with a couple of two modes. The DNP with intermediate rigidity first lengthens to a certain length (stage II) while its left end is pinned at adhesion region A (Figure 10.6c). But different from mode 2, its right end would not directly attach to adhesion region B. Alternatively, the left end of the DNP will detach from the adhesion region A, and it will move with this constant shape (stage III) until its right end attaches to the adhesion region B (stage IV). Eventually, the DNP bounces back to its original shape (stage V) while its right end is pinned at adhesion region B. In this process, the energy profile can be tuned by the adhesion energy and deformation energy simultaneously. It means there exists a critical point where these two energy barriers will cancel with each other, and the DNP will have a high diffusivity near this critical point because of the low energy barrier. As discussed above, this distribution follows the Boltzmann distribution as following:

$$p(x_b) = \frac{\exp(-E(x_b)/k_B T)}{\int_0^{r_0-a_0} \exp(-E(x)/k_B T)\mathrm{d}x}. \tag{10.21}$$

Here, $x_b = a - a_0$ represent the stretching length of the deformed DNPs, $a_0 = 1$ is the initial length of the semi-major axis. $E(x_b)$ is the total energy of the NP with the length of the semi-major axis a. Once the stretching length x_b is specified, we can calculate the effective diffusion rate on each x_b by MFPT theory. With the probability distribution of x_b and diffusion rate for each x_b obtained, there will be a total effective diffusion coefficient D_{eff} determined by the following three distributions

$$D_{\text{eff}} = \int \frac{D_e}{D_g(a,b)} \frac{D_g(a,b)}{D_0} p(x_b) D_0 \mathrm{d}x_b, \tag{10.22}$$

Actually, the effective diffusion rate of the coupled mode is equivalent to a kind of average of two extreme modes based on the probability of appearance of each shape.

With the derived theoretical model, we can analyze how the rigidity of the DNP influences their diffusivity in an adhesive network and further optimize the stiffness of the DNP to possess the highest diffusivity. For example, we can calculate and compare the effective diffusion rate of DNPs with different rigidity in networks with various adhesive strengths (Figure 10.6d). The results showed for the soft DNPs ($\kappa = 10k_BT$), the effective diffusion rate decreases in low-adhesion regions ($U_0 < 0.17$) and then increases to a peak when the affinity increases to $U_0 = 0.23$. After that, the effective diffusion rate drops remarkably with the increase of the adhesive strength U_0. The reason is that when the affinity strength is weak, the low adhesive energy cannot overcome the deformation energy of the DNPs; thus, the DNPs keep their initial spherical shape, and there exists a high energy barrier, which hinders the diffusion of the DNPs. If the affinity strength is too strong ($U_0 > 0.23$), the high adhesive energy is far higher than the deformation energy of the DNPs. Moreover, the high adhesive energy would induce a large deformation, further increasing the contact area between the DNP and the network and promoting the adhesive energy. When the affinity strength has a suitable value ($U_0 = 0.23$), the adhesive energy will be offset by the deformation energy, which results in a relatively low energy barrier and further leads to a high diffusion rate (Figure 10.6e). The variation of the effective diffusion rate for the DNP with intermediate stiffness ($\kappa = 20k_BT$) has the similar tendency like soft DNP. However, for the rigid DNP ($\kappa = 50k_BT$), the effective diffusion rate decreases with the adhesion strength monotonically. The reason is that the high rigidity induces a high deformation energy that makes it difficult for DNPs to undergo deformation in the adhesive zone, making it difficult to overcome the adhesion energy and resulting in a higher energy barrier (Figure 10.6e). Interestingly, to achieve the highest diffusion rate, the adhesive strength for the intermediate rigid DNP ($\kappa = 20k_BT$) is larger than that of the soft one ($\kappa = 10k_BT$).

To systematically understand how the rigidity of DNPs, the mesh size, and the adhesive strength of the network affect their diffusion, we conducted a series of calculations and derived the variation of the stiffness of those DNPs that have the highest effective diffusion rate in the adhesive network on the $U_0 - r_0$ plane (Figure 10.6f). Generally, the results show that the soft DNPs have the highest effective diffusion rate in the network with large mesh size and weak adhesion. In contrast, the rigid DNPs have the highest effective diffusion rate in the network, with a small mesh size and strong adhesive strength. Those with intermediate stiffness have the highest effective diffusion rate in the network with large mesh size and strong affinity. Once having confirmed the feature of the biological hydrogel, one can easily select DNPs with suitable rigidity that have the optimal diffusivity in this biological hydrogel based on this phase diagram.

10.2.5 Summary

Combining a series of experiments with systematical molecular dynamics simulations, the "rigidity effect" that influences the diffusion of deformable DNPs in the protein network of mucus was discovered. Specifically, under the premise of similar size, surface charge, and surface ligand modification, DNPs with moderate rigidity can quickly penetrate the intestinal mucus's protein network and have the optimal diffusion ability. The underlying mechanism by which rigidity regulates the diffusion of NPs in mucus was elucidated by examining the interface interaction between the DNPs and the protein network at the microscale using molecular dynamics simulations. DNPs with high rigidity exhibit a significant size effect due to their weak deformation ability, leading to low diffusivity. They can be trapped in the corner of the network for a long time. DNPs with low rigidity have excessive deformation ability, increasing the contact area and adhesion with the protein network that affects their diffusion capabilities. In contrast, liposomes with moderate rigidity continuously transform from spherical to elliptical during the diffusion process, allowing for effective adjustments to the interfacial adhesion and avoiding the influence of size effects. Both of these factors contribute to improving their diffusion capability. Furthermore, a theoretical model describing the diffusion behavior of deformable DNPs in the adhesion network was constructed, combining statistical and continuum mechanics. This model can accurately describe the diffusion process of DNPs with adjustable rigidity in adhesion networks and can precisely reflect the impact of three regulatory parameters on DNP diffusion, including particle rigidity, network mesh size, and adhesive strength. Moreover, this model can also predict the optimal stiffness of particles for diffusion in networks with different properties, providing a theoretical basis and design ideas for the rational design of efficient drug carriers.

10.3 The Effect of the Shape on the Diffusivity of NPs in Mucus

Pioneering works showed that the diffusion dynamics of gold nanorods (NRs) on an artificial lipid membrane, consists of two parts, namely rotation and translation [65, 66]. Indeed, the shape of NPs is another crucial factor that can significantly impact their interactions with biological medium and lot of works have studied the shape effect on the interaction between NPs and cell membrane [67, 68]. For example, some recent works have confirmed that the gold NRs complete their internalization process through an "intermittent rotation" from molecular dynamic simulation and experiment test [69, 70]. These findings inform us that shape may contribute to the high mucus-penetrating ability of particles. The correlation between the shape of NPs and their diffusivity in mucus and the fundamental interaction mechanism, however, is still incomplete, even though this aspect is potentially crucial in the design of NPs-based drug delivery systems. In this section, we will focus on elucidating the mechanism by which the shape of NPs affects their diffusion ability in mucus and propose a new design strategy for improving mucus-penetrating ability of NPs by adjusting their shape.

10.3.1 The Diffusion Behaviors of NPs with Various Shapes in Mucus

To elucidate the effect of shape on the mucus-penetrating property of NPs, three typical mesoporous silica nanoparticles with different shapes but identical surface

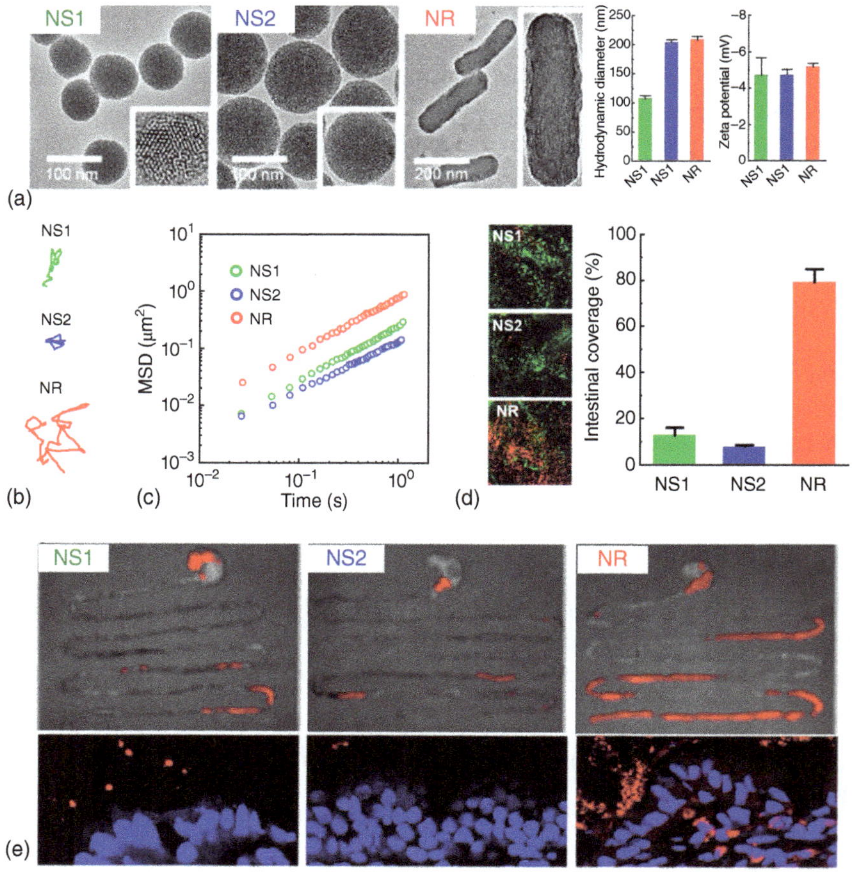

Figure 10.7 Characterization of the transport of NPs with various shapes in rat mucus. (a) TEM images, hydrodynamic diameters, and zeta potentials of the NS1, NS2, and NRs. (b) Representative trajectories of NPs motion in fresh rat intestinal mucus. (c) MSD values as a function of time for NPs in mucus. (d) Particle distribution and penetration in the rat intestinal mucus *ex vivo*. (e) Distribution of NPs in rat intestines. Up: Photographic images of NPs in tissues of the whole small intestine. The images are representative of the average after intragastric administration of 6h. Down: Confocal laser scanning microscopy (CLSM) images of particle distribution in the rat intestine Blue: nuclei of the intestinal villi. Red: NPs. Source: (a,d,e): [71]. Miaorong Yu et al. 2016/Reproduced with permission from American Chemical Society.

chemistries and zeta potentials were fabricated, namely small nanosphere (NS1), large nanosphere (NS2), and nanorods (NRs) (Figure 10.7a). The sizes of NS1, NS2, and NRs were 80 nm, 140 nm, and 80 × 240 nm (AR = 3), confirmed by the TEM images. The hydrodynamic diameters were about 100, 200, and 200 nm for NS1, NS2, and NRs, which keep the shape as the only variable and allow us to directly compare the influence of shape on the diffusivity [71] (Figure 10.7a). The diffusion trajectories of the NPs in fresh intestinal mucus of rats from MPT technology showed that NS1 and NS2 were limited to a small area, suggesting that the mucus network nearly trapped these particles (Figure 10.7b). In contrast, the NRs exhibited a better diffusion pattern and were able to move more freely in a larger area. The MSD based on the moving trajectories of each type of NPs was calculated on a time scale of 1 s (Figure 10.7c), and the MSD of the NRs was approximately 3.3-fold and 5.7-fold higher than that of NS1 and NS2, respectively.

In order to ensure that the observed rapid transport of NRs was not biased by a small fraction of fast-moving outlier particles, we next focused on the rat intestinal loops to investigate whether a better mucus-penetrating ability would lead to a more uniform and deeper distribution of NRs compared with NS1 and NS2 *ex vivo*. In an experiment, the mucin fibers were stained with Alexa Fluor 488-conjugated wheat germ agglutinin (WGA). The particle solutions were then injected into ligated intestinal loops and left to incubate for 30 minutes with gentle agitation. After incubation, the fluorescence images of freshly excised, opened, and flattened intestinal sections were obtained. The results showed that NS1 and NS2 exhibited low mucus coverage and could not penetrate and distribute evenly within the mucus layer. In contrast, the NRs were widely dispersed in the mucus and had a uniform fluorescence intensity in most of the mucosal area (Figure 10.7d). The quantification of the distribution area revealed that approximately 80% of the intestinal surface was covered with NRs, while NS1 and NS2 covered less than 15% (Figure 10.7d). The wider distribution of NRs within the mucus layer of the intestinal loops again provides convincing evidence for the superior mucus-penetrating ability of rod-shaped nanoparticles compared to spherical ones. In addition, a series of *in vivo* tests were conducted to evaluate the transport and distribution of the three types of nanoparticles in the rat gastrointestinal tract after intragastric administration. The results showed that NS1 and NS2 rapidly decreased in amount within the rat gastrointestinal tract, and little fluorescence could be detected after 6 hours (Figure 10.7e). In contrast, a considerable amount (approximately 40%) of the NRs remained in the jejunum and ileum even six hours later and were uniformly distributed close to the intestinal epithelium (Figure 10.7e). These images demonstrate that NRs had a better intestinal dispersion property and were more difficult to wash away by the mucus than NS1 and NS2. Such a prolonged retention time of NRs should benefit the efficient mucus-penetrating property in mucus and sustained drug release and absorption.

Overall, these findings demonstrate the superior mucus-penetrating ability of rod-shaped nanoparticles and highlight their potential applications in drug delivery systems for the gastrointestinal tract.

10.3.2 The Diffusion Mechanisms of NPs with Different Shape in Biological Hydrogels

To further elucidate the mechanism underlying the superior mucus-penetrating ability of rod-shaped nanoparticles, a series of molecular dynamic simulations were used to investigate the diffusion behavior of NPs in mucus. To simplify this problem, a coarse-grained model system composed of a regular cross-linked polymer network, water beads, and NPs is constructed and used to investigate the diffusion behaviors of NPs in the network. In our simulation, two typical NPs with

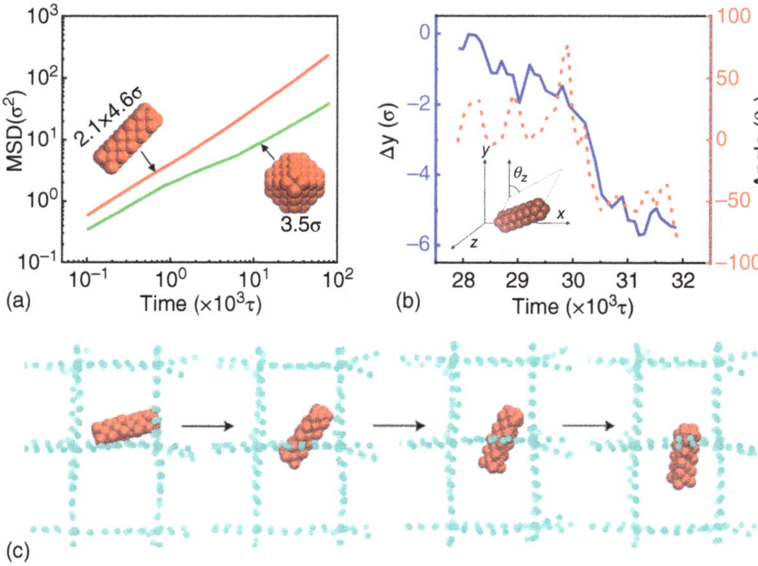

Figure 10.8 The diffusivity behavior of NPs with different shapes in polymer network from molecular simulations. (a) The MSDs for NPs with different shapes diffusion in the polymer network. (b) The displacement variation along the *y* direction and the rotational angle around the *z*-axis (θz) for one typical NR with the simulation time. Insert: Schematic of θz, the angle between the *y*-axis and the projection of the cylinder in an *xy* plane. (c) There is a significant change in the position and orientation of NR at the moment the arrow points to the panel.

the same hydrodynamic diameter, a nanosphere with a diameter of 3.5σ (NS), and nanorods with a size of $2.1 \times 4.6\sigma$ (NR) were studied. The details of the simulation can be found in reference 61.

The simulation results showed that the NR has a faster diffusion velocity and a larger diffusion area than NS, and the calculated MSD value of NR is much higher than that of NS (Figure 10.8a). To explain why NR moves faster than NS, the moving trajectories of both types of particles were analyzed carefully. The results indicated that the translational movement of NR was accompanied by a rotation, particularly when the particle jumped from one unit cell of the network to another (Figure 10.8b,c). This means the rotation of NR around the polymer chains fundamentally contributes to their translational diffusion in mucus. To further verify the conclusion that rotation-facilitated enhanced diffusion contributes to the efficient movement of NRs in mucus, additional simulations were conducted in which the rotation of nanorods was constrained. The results from these simulations confirmed that the rotation of NRs plays a key role in enhancing their diffusion in mucus. When the rotation of the NRs was constrained, their translational diffusion was significantly reduced, indicating that rotation is necessary for facilitating efficient movement through the mucus layer.

Based on these results, we can conclude that the movement of nanorods consists of two parts: rotation around the polymer fibers and translational diffusion. With this mechanism elucidated, we could explain the observed anomalous phenomenon in experiments and simulations: nanorods move faster than nanospheres.

10.3.3 Theoretical Model of Diffusion of Rod-Like Nanoparticles in Polymer Networks

Based on the above findings from both series of experiments and coarse-grained simulations, it can be concluded that the rotation process caused by anisotropic shapes enables rod-like particles to diffuse more effectively in mucus than spherical particles. However, in addition to the difference in morphology, the aspect ratio (AR) of rod-like particles is also an important parameter that affects the interface between particles and physiological media and the biological effects of particles. For example, previous studies have found that gold nanorods' AR significantly affects their intracellular kinetics process [69, 70]. In addition, particle size and the interface interaction between the particle and the medium are also important parameters that may determine the biological effects of particles in physiological media [72]. However, due to the limitations of experimental and simulation technology, it is difficult to fully study the impact of a single parameter under the same conditions as other parameters. Therefore, it is necessary to establish a theoretical model to describe the diffusion behavior of rod-like particles in the network structure. With the help of a theoretical model, on one hand, the physical mechanism that affects the anomalous diffusion of particles in the network structure can be elucidated. On the other hand, using this model, we can systematically study the regulating roles of these parameters on the diffusion ability of particles.

There has been a considerable amount of work studying the anomalous diffusion phenomena of particles in gel networks with different properties, and corresponding theoretical models have been proposed. For example, Xue et al. [73] developed a nonequilibrium thermodynamics model to capture the effect of a tumor matrix, which also has a protein network structure like mucus, on drug transport. Recently, an obstruction-scaling model was developed for the diffusion of spherical solutes in covalently cross-linked hydrogels and later used to describe the diffusion of spherical NPs in mucus to estimate the pore size of human cervicovaginal mucus [74, 75]. Other diffusion models based on the obstruction effects [60], hydrodynamic interactions [76], microscopic statistical mechanical theory [77], and so on have been developed to describe the diffusion of NPs in polymer solutions and gels. However, most of these models focus only on describing the diffusion of spherical particles in the network structure, with little consideration for the interaction between the network and the particles. Therefore, it is necessary to develop a corresponding theoretical model for the motion of rod-like particles in the network, as well as the theoretical elucidation of the impact of parameters such as the AR of particles and the interaction between particles and the network on their diffusion.

10.3.3.1 Nonadhesive Diffusion Model

The diffusivity of the rod-like NPs in the nonadhesive network can be extended to Amsden's obstruction-scaling model [63], which had been used to describe the diffusivity of spherical NPs in polymer gels through the probability of one NP encountering a series of openings between the polymer chains whose spacing is larger than the size of the NPs. Generally, the normalized diffusivity of spherical NPs is

$$\frac{D_g}{D_0} = \int_R^{\infty} g(r)\mathrm{d}r = \mathrm{P}(r > R), \tag{10.23}$$

where D_g is the diffusivity of the NPs in the polymer solution, $D_0 = k_B T/6\pi\mu \cdot R_H$ is the diffusivity of the NPs in water, $g(r)$ is the distribution function of the radii r of the openings between polymer chains, and R is the radius of the spherical NPs. μ is the dynamic viscosity of water, k_B is the Boltzmann constant, and T is temperature. Ogston has showed that the size distribution of spherical spaces in a random network of straight fibers (negligible width) in his work following [78]

$$g(r) = (4\pi\upsilon Lr + 4\pi\upsilon r^2) \cdot \exp\left[-\left(2\pi\upsilon Lr^2 + \frac{4\pi}{3}\upsilon r^3\right)\right], \tag{10.24}$$

where υ is the average number of fibers per unit volume, and L is the half-length of the fibers. The normalized diffusivity of spherical NPs in the network can be obtained easily using this model. In fact, mathematically, Eq. (10.23) is equivalent to the probability of no fibers in contact with the spherical NPs when the NPs are placed in the random network. In addition, the actual number N of fibers in contact with the NP follows a Poisson distribution, i.e.

$$P(N = k) = \frac{\overline{N}^k \exp(-\overline{N})}{k!} \tag{10.25}$$

where $\overline{N}$ is the average number of fibers in contact with the NPs. Following this viewpoint, the normalized diffusivity of the rod-like NPs in the network can be obtained when $k = 0$

$$\frac{D_g}{D_0} = P(N = 0) = \exp(-\overline{N}) \tag{10.26}$$

The average number of fibers in contact with randomly dispersed rod-like NPs in a fiber network can be calculated as follows. As shown in Figure 10.9a, O is the center of the NP, Ω is a spherical surface expanded from point O until it reaches the fibers, and r is the radius of Ω and contact distance [79]. Depending on the contact manner between the NPs and Ω, the fibers can be classified into four kinds: fiber is in tangential contact with the sphere surface, and the contact points are within the NP (l_1); the end point of the fiber contacts the sphere surface, and the contact points are within the NP (l_2); fiber is in tangential contact with the sphere surface, and the contact points are outside the NP (l_3); the end point of the fiber contacts the sphere and the contact points are outside the NP (l_4). The average numbers of the fibers with different contact manners are denoted as N_1, N_2, N_3, and N_4, respectively. And the expected total number N of fibers in contact with the NP is

$$\overline{N} = N_1 + N_2 + N_3 + N_4 \tag{10.27}$$

The details of calculating these contact numbers can be found in refer [71]. After the contact number is obtained, one can calculate the diffusivity of the rod-like NPs

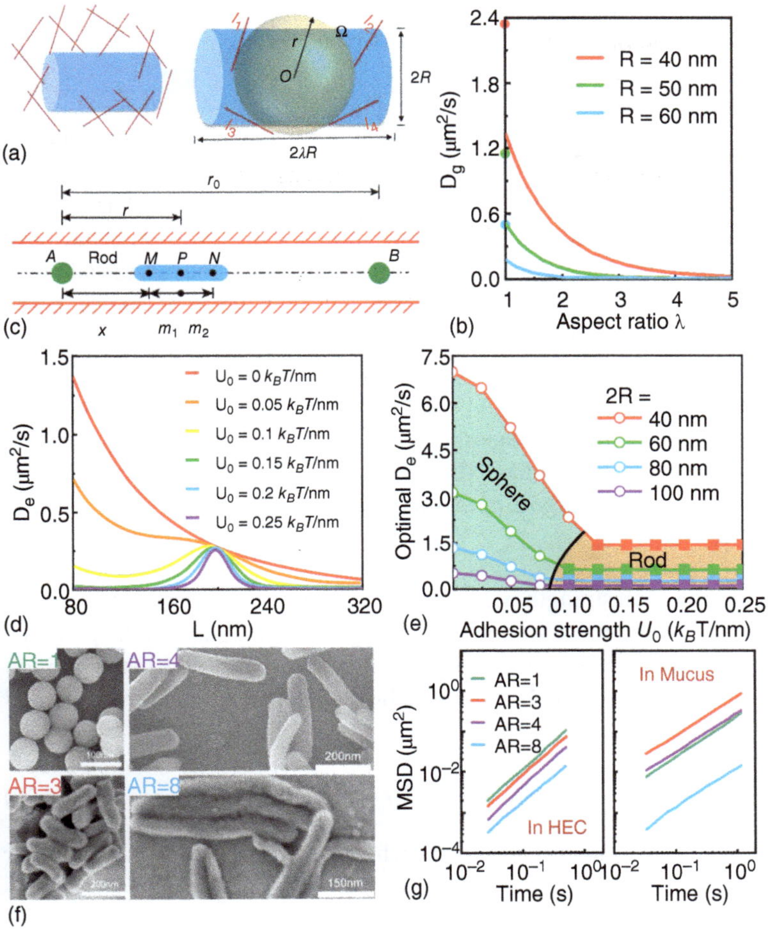

Figure 10.9 The diffusivity of nanorods in polymer network from theoretical prediction. (a) Schematic showing the contact between fibers and a rod. Ω is a spherical surface expanded from point *O* until it reaches the fibers, and *r* is the radius of Ω and referred to as the contact distance. (b) Diffusivity (D_g) of rod-like NPs with the same minor-axis diameter *R* as a function of the aspect ratio λ in a nonadhesive gel from the prediction of the extended obstruction-scaling model. The circles represent the diffusivities of spherical NPs with the same diameter. (c) Schematic showing NP diffusion in a 1D cylindrical nanopore with adhesion regions A and B. (d) Effective diffusivity (D_e) of NPs in 1D nanopores at different adhesion strengths as a function of the NP length, *L*. The diameter of the NPs considered is 80 nm. (e) Optimal effective diffusivity for nanorods with the same minor-axis diameter at different adhesion strengths. (f) SEM images of the nanorod with various aspect ratios. Source: [83]. Jiuling Wang et al. 2018/Reproduced with permission from Elsevier. (g) MSD values as a function of time for NRs in mucus and HEC from multiple-particle tracking technology.

with various AR using Eq. (10.26). Figure 10.9b shows the variation of the diffusivity of nanorods with the same diameter but different lengths. The diffusivity of the rod-like NPs rapidly decreases with the increase of particle length, whatever the diameter of the particle is, especially for the particle with a large diameter. Moreover, the diffusivity of the spherical NPs is much higher than that of the rod-like NPs in the nonadhesive network. This is not surprising as the NPs with rod-like shapes always have a larger hydrodynamic radius than the spherical NPs.

10.3.3.2 Adhesive Diffusion Model

In addition to the steric hindrance effect on the movement of NPs, as mentioned in the above nonadhesive diffusion model, the fibrous of the protein network in mucus also exhibits heterogeneous interactions with NPs, including van der Waals. These electrostatic and hydrophobic interactions can influence the diffusion behavior of NPs in the mucus. According to the adhesive characteristics of mucus, the adhesive property of mucus was assumed to have discrete adhesion zones. Moreover, based on the experimental and molecular simulation results, we can also assume that the NPs undergo hopping diffusion in the network [80, 81], i.e. they will be adsorbed to one adhesion region first (region A) and then detach after a while before being re-adsorbed to another adhesion region (Figure 10.9c). Without loss of generality, the interaction between the rod-like NPs and the adhesion region is described by the Morse potential for the purpose of mathematical manipulation

$$V(x) = U_0(e^{-2x/\rho} - 2e^{-x/\rho}) \tag{10.28}$$

where U_0 is the adhesion strength per unit area, x is the distance between one point on the surface of NP and the adhesion region (Figure 10.9c), and ρ is the potential range. Therefore, the interaction energy $U_A(r)$ between the whole NP in the position r and one adhesion region A can be expressed by the integral of the Morse potential (Figure 10.9c)

$$U_A(r) = \int_0^{r/2} V(|r - m_1|)\mathrm{d}m_1 + \int_0^{r/2} V(|r + m_2|)\mathrm{d}m_2 \tag{10.29}$$

Similar, the interaction energy between the whole NP in the position r and another adhesion region B is

$$U_B(r) = U_A(r_0 - r) \tag{10.30}$$

and the total interaction energy of the NP in a nanopore with two separated adhesive regions (A, B) and spacing of r_0 is

$$U(r) = U_A(r) + U_B(r) \tag{10.31}$$

Finally, from Kramers' theory, the time for an NP to move from A to B is given by the mean first passage time (MFPT) [61]

$$t = \frac{1}{D_g} \cdot \int_0^{r_0} \exp\left(-\frac{U(r)}{k_B T}\right) \cdot \left[\int_r^{r_0} \exp\left(\frac{U(y)}{k_B T}\right) \mathrm{d}y\right] \cdot \mathrm{d}r \tag{10.32}$$

where D_g is the diffusivity of the NPs outside the adhesion region, i.e. the diffusivity of NPs in the nonadhesive network. And the effective diffusivity of the NPs can be obtained through

$$D_e = \frac{r_0^2}{2t}. \tag{10.33}$$

Combining Eq (10.29) with Eq. (10.33), we can calculate the effective diffusivity of the NPs. Further, we can analyze how the length of the rod-like NPs and adhesive strength influence their diffusivity in the adhesive network and further optimize the length of the NP to possess the highest diffusivity. Figure 10.9d showed the effect of the length (L) and adhesive strength. U_0 on the diffusivity of rod-like NP with a given diameter of $2R = 80$ nm. The effective diffusivity decreases with the length of the NPs when the adhesive interaction is weak ($U_0 \leq 0.05k_BT/\text{nm}$), especially the variation of the diffusivity will become consistent with the results from nonadhesive model for $U_0 = 0.0k_BT/\text{nm}$. However, when$U_0 \geq 0.1k_BT/\text{nm}$, the effective diffusivity first increases then decreases with the NP length. This means that there exists an optimal length at which the diffusivity of the nanoparticles is highest. In addition, the optimal diffusivity is completely related to the length of rod-like NP, and further analysis shows that the optimal length is typically close to or slightly smaller than the size of the pores in the porous network regardless of the NP diameter. To systematically understand how the shape of NPs and the adhesive strength of the network affect their diffusion, we conducted a series of calculations. We derived the variation of the optimal diffusivity of those NPs with the adhesive strength and size and shape of NPs (Figure 10.9e). Generally, with increasing adhesion strength, the profiles of the diffusivities decrease first, then tend to become flat, and the NPs with the optimal diffusivity shift from nanospheres to nanorods (with lengths near the pore size of the network). Thus, for NPs diffusing in a porous medium with discrete adhesion regions, nanorods whose length is comparable to the spacing of the adhesion regions can achieve high diffusivity.

Finally, to validate the predicted results of the theoretical model, we prepared a series of mesoporous silica nanoparticles with the same radius but different lengths (Figure 10.9f). These particles were then dispersed in mucus and in hydroxyethyl cellulose (HEC) solution with the pore size of network structure similar to mucus (~200 nm), and the movement of the particle was studied using multi-particle tracking techniques. The results showed that the diffusivity of the particles in nonadhesive HEC networks decreased with their length (Figure 10.9f), which is consistent with the conclusion predicted from the nonadhesive model. However, the diffusivity of particles with lengths near close to the protein mesh size (AR = 3) exhibited the best diffusion ability in adhesive mucus (Figure 10.9f). This conclusion is also consistent with the prediction of the adhesive model. These results all demonstrate the validity of the theoretical model and the high accuracy of the predicted results.

10.3.4 The Effect of the Surface Polyethylene Glycols (PEGs) Distribution on the Diffusivity of Rod-Like NPs

Previous studies have pointed that the surface grafting polyethylene glycols (PEGs) can improve the diffusivity of nanoparticles (NPs) in mucus due to the effective shielding effect of PEG, while the present of the PEG chains severely impede cellular uptake of NPs [32, 40, 82]. The surface-grafted PEG chains separate the contact between NPs and biological medium and thus reduces the adhesive interaction, which alters the surface property of NPs and influences their biological effect in a different medium. Moreover, some works have addressed that the structure of surface-grafted polymer chains can undergo a conformational transition from the "mushroom" regime to the "brush" conformation as the density of the polymer chains increases. In the "mushroom" regime, the polymer chains are relatively short and densely packed, resulting in a compact conformation resembling a mushroom cap. In the "brush" regime, the polymer chains are elongated and extend outward from the surface, resulting in a brush-like appearance. This feature offers a powerful tool for controlling the surface properties and interactions of nanoparticles (NPs) with biological systems, even for single-component polymers. For example, brush-like PEG chains, in particular, have been widely used to create hydrophilic surfaces on NPs that can resist protein adsorption and prolong their circulation time in the bloodstream. By contrast, mushroom-like PEG chains can enable hydrophobic interactions between the NPs and proteins, which can be preferable for specific cellular adhesion. Generally, oral drug vehicles are often required to efficiently pass through the GI mucus and cellular barriers in a short time, but PEGylated NPs can only facilitate one of the two requirements, consequently impairing their drug delivery efficiency. A recent study found that the PEG chains' conformation can change with the degree of curvature, ranging from a "mushroom" conformation on highly curved surfaces to a "brush" conformation on flatter surfaces. In other words, the conformational structure of the grafted PEG chains can be tuned by the curvature of the NPs. Therefore, we can imagine whether two different conformations will be formed simultaneously on the surface of rod-like particles with anisotropy curvature by precise control.

To test this idea, a type of mesoporous silica nanoparticle with a size of 80×220 nm was prepared in the experiment [83]. Then, PEG chains with different density, which can be obtained by adjusting the content (weight ratio of PEG to NPs) of PEG during the grafting reaction process, were grafted onto the surface of the particles. After that, the hydrated PEG corona (HPEGC) around the particles was carefully detected using AFM under fluid conditions in order to verify the density distribution and structure of PEG (Figure 10.10a). The results showed that when the content of PEG was relatively low, a negligible HPEGC formed on the surface of the particles, indicating that only a small amount of PEG could be grafted onto the surface of the particles (2% PEGylation). In addition, PEG chains were uniformly distributed on the surface of the particles, forming a mushroom-like structure. As the content of PEG increases, the HPEGC on the particle surface exhibited an anisotropic structure that was thick at the tips but much thinner around the rod body (4% PEGylation). This means that the PEG distribution on the particle surface is nonuniform, and the coexistence of extended brush and packed mushroom chains in this condition. When the PEG content is further increased to a high value, the HPEGC on the particle surface exhibits an isotropic, thick structure. This indicates that PEG is uniformly distributed on the surface of particles under this condition and forms a brush-like structure due to the high grafting density.

Next, many MPT tests were performed to investigate whether the PEG chain structures could affect the diffusivity of the rod-like NPs in fresh intestinal mucus. The tracking results showed that the NR-4% PEGylation exhibited a sharp growth in diffusivity compared to the NR-2% PEGylation and had an almost similar diffusivity to that of the high-PEG grafted NPs (NR-6% PEGylation), implying

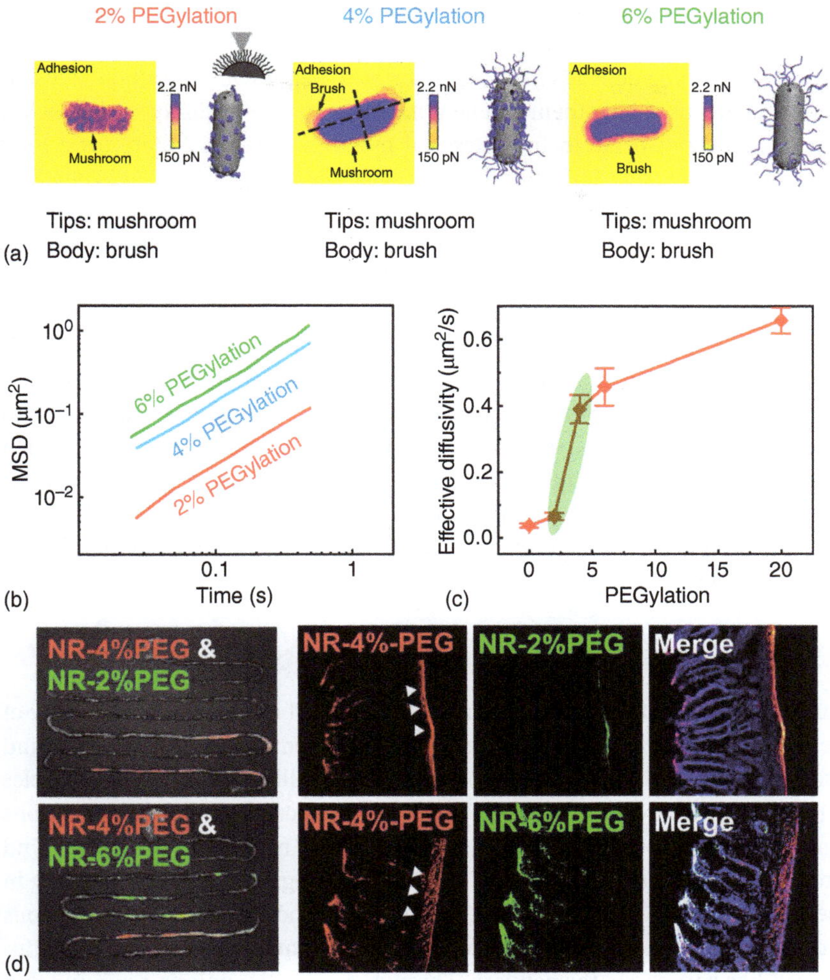

Figure 10.10 The diffusivity of nanorods with various PEG grafting density in the mucus. (a) The HPEGC structure of NPs with various PEGylation was detected using AFM. The right schematic of each structure showed the PEG chain structure. (b) MSD of NPs diffusing in freshly obtained rat intestinal mucus. (c) Effective diffusivities of NPs. The green region indicates a rapid growth in diffusivity. (d) In vivo transportation of NPs in mucosal. The left showed the NP transportation through and retention in rat small intestines after 6H, and the right showed the intestinal uptake examined by CLSM after intragastric (i.g.) administration.

that the structure of HPEGC played a dominant role in controlling the diffusivity of the NPs and the formation of brush-like PEG chains at tips greatly improved diffusivity (Figure 10.10b,c). Given these promising *in vitro* results, we then determined whether the advantageous properties of anisotropic PEG grafting indeed benefited NRs in vivo delivery efficiency. To do this, the isotropic grafted NRs were labeled with a green dye [fluorescein isothiocyanate (FITC)] and mixed with red RITC-labeled anisotropic grafted NRs. After intragastric administration, the NPs were evidently transported along the small intestines at different rates

(Figure 10.10d). The NRs with high-PEG grafting density (NR-6%PEGylation) can better be retained in the small intestines within a long time (six hours) because the high-PEG grafting density effectively reduces the adhesive interaction between particles and mucus, giving them excellent mucus-penetrating ability, which makes them can quickly disperse in the intestinal mucus and remain there for a long time. In contrast, most of NRs with low PEG grafting density (NR-2%PEGylation) can be trapped in the surface of the mucus and removed by peristalsis in a short time. Due to the better mucus diffusion ability, most of the anisotropic grafted NR (NR-4%PEGylation) can also be retained in the small intestines. However, efficient delivery requires not only prolonged retention in mucus but also prompt absorption. Therefore, the intestine was sliced and carefully examined by CLSM, at six hours after administration. The results showed that the villi could internalize a considerable number of anisotropic grafted NR and clearly transported to the basolateral (BL) side, in sharp contrast to the high-PEG grafted NRs, which showed limited fluorescent intensity on the BL side. However, they have superior mucus-penetrating ability, indicating poor uptake efficiency (Figure 10.10d). These data also suggested that anisotropic PEG-grafted nanorods may be a more effective option than other NPs for overcoming in vivo mucosal barriers.

10.3.5 Summary

In this section, we systematically discussed the effect of shape on the diffusivity of particles in mucus. Firstly, through a series of experimental observations, we found that rod-like particles have better mucus-penetrating ability than spherical particles with the same hydrodynamic radius. Then, using molecular dynamics simulations to analyze the motion trajectory particles in the protein network of mucus and find that rod-like particles exhibit a distinct rotation-hopping process when diffusing in the network, providing a microscopic explanation for rod-like particles' anomalous high diffusion rate. Next, we constructed a theoretical model to describe the diffusion of rod-like particles in the network. Through systematic analysis, we found that when the length of the rod-like particle is close to the size of the grid in the network, their diffusion behavior becomes best because the energy barrier required to overcome is the lowest in this condition. Finally, we explored the effect of surface PEG grafting density on the diffusion ability of rod-like particles, and found that nonuniform PEG grafting can simultaneously meet the high mucus-penetrating ability and high cell internalization ability requirements of rod-like carriers for efficient oral delivery. These conclusions can provide theoretical guidance for the rational design of new efficient delivery carriers.

10.4 Conclusion and Outlook

Drug delivery is a complex process that requires nanocarriers (NCs) to overcome various physiological barriers during delivery. In the past, most studies focused only on how the physicochemical properties of the NCs affect their cellular uptake ability and elucidated the corresponding mechanisms, establishing theoretical regulatory models. However, NCs often need to overcome the hindrance of extracellular mucus before reaching target cells. In addition, when designing oral drug carriers, the ability of carriers to penetrate gastrointestinal mucus should also be fully considered. Therefore, it is necessary to elucidate the diffusion process of carriers in mucus and corresponding microscopic mechanisms and theoretical regulatory models should be deeply understood and developed. In this chapter, we systematically analyzed the effects of rigidity and shape on the diffusion ability of particles in mucus by combining experimental observations, multiscale molecular dynamics simulations, and theoretical frameworks. We have successively discovered the "stiffness effect," "shape effect," and "asymmetric surface modification effect" that affect the diffusion ability of carriers, and elucidated the microscopic mechanisms behind these anomalous phenomena and established theoretical models. Finally, based on these conclusions, the key design parameters for two types of carriers that meet the efficient oral delivery capacity were chosen. The key parameter of liposome carriers is Young's modulus, which was required to be close to 50 MPa, while the design parameters for high delivery rod-like silica-based carriers are close to the protein mesh size in mucus in length and nonuniform PEG modification on the surface.

Although we have identified key design parameters for carriers with high delivery efficiency through the above research, carriers designed and prepared with these parameters still exhibit low delivery efficiency in real *in vivo* delivery. The main reason is that the environment of *in vivo* delivery process is much more complex than the studying system we designed in the above works, leading to potentially significant deviations between the results. For example, there are various types of interface interaction forces between the carrier and physiological media in real delivery processes, and they contain numerous weak forces, nonsteady-state, and nonequilibrium state, posing great challenges to experimental design and theoretical simulation. Therefore, in order to make the results more realistic, we believe that future research should improve in the following aspects in terms of content and methodology.

(i) Refine the interaction between carriers and various physiological barriers. The carrier will meet many complex interface interactions at different scales when they enter the body, including specific strong and nonspecific weak interactions. These intertwined interactions determine the biological effects and delivery efficiency of drug carriers in the body. In addition, these interface interactions are closely related to the surface properties of the carrier. A systematic understanding of the mechanisms of these interactions and the relationship between these interactions and the surface properties of the carriers can guide the rational functional modification of carrier surfaces.

(ii) The influence of protein corona on the surface of carriers. As we all know, when the carrier enters the body, it will adsorb various types of proteins to form a protein corona, changing its original surface properties and affecting the interactions between the carrier and various physiological barriers, causing it to lose its original functional surface design. In addition, the protein corona can even change the rigidity, shape, and size of the carriers and further alter their fate, stability, and toxicity. Recent research indicates that the formation of the protein corona is a dynamic and competitive process, which is closely related to the carrier's properties (such as chemical composition, size, surface properties, etc.) and the physiological environment in which it is located (such as ion concentration, pH, temperature, body fluid composition, etc.). This leads the protein corona structure to exhibit a dynamic exchange process, which further has a significant impact on the behavior of liposomes in cells and tissues. Therefore, the protein corona structure endows carriers with new biological characteristics and ultimately affects their biological effects and delivery efficiency. By deeply understanding the formation mechanism of protein corona and establishing corresponding regulatory strategies, it is not only can avoid the loss of the surface functional design of carriers, but also can potentially enhance the surface functionalization of the carriers through the regulation of the composition and structure of the protein corona, providing possibilities for effective transport in various complex environments in the body.

(iii) Expanding the research system from static to dynamic. Currently, most of the *in vitro* experiments and theoretical models regard the physiological barriers as static media. However, under physiological conditions, such barriers are all influenced by complex external loads. For example, the mucous layer on the surface of the gastrointestinal tract is renewed every few minutes, and particles trapped in the mucous layer will be cleared. In addition, the new treatments are often aided by external devices such as magnetism, heat, and ultrasound for assistance. Under these external loads, the protein network and interstitial fluid in the mucus may deform or flow, inevitably affecting the movement of particles within it. At this point, the results from *in vitro* experiments or static theoretical models will deviate from *in vivo* results, making it impossible to provide precise guidance for the design of drug carriers. Therefore, studying the effects of external loads on the movement behaviors of carriers in biological media is urgently needed and of great significance for the design and development of novel drug delivery systems.

References

1 McAuley, J.L., Linden, S.K., Png, C.W. et al. (2007). MUC1 cell surface mucin is a critical element of the mucosal barrier to infection. *The Journal of Clinical Investigation* 117 (8): 2313–2324.

2 Johansson, M.E., Ambort, D., Pelaseyed, T. et al. (2011). Composition and functional role of the mucus layers in the intestine. *Cellular and Molecular Life Sciences* 68: 3635–3641.

3 Bansil, R. and Turner, B.S. (2018). The biology of mucus: composition, synthesis and organization. *Advanced Drug Delivery Reviews* 124: 3–15.

4 Girod, S., Zahm, J., Plotkowski, C. et al. (1992). Role of the physiochemical properties of mucus in the protection of the respiratory epithelium. *The European Respiratory Journal* 5 (4): 477–487.

5 Lai, S.K., Wang, Y.-Y., Wirtz, D., and Hanes, J. (2009). Micro-and macrorheology of mucus. *Advanced Drug Delivery Reviews* 61 (2): 86–100.

6 Liu, J., Leng, P., and Liu, Y. (2021). Oral drug delivery with nanoparticles into the gastrointestinal mucosa. *Fundamental & Clinical Pharmacology* 35 (1): 86–96.

7 Leal, J., Smyth, H.D., and Ghosh, D. (2017). Physicochemical properties of mucus and their impact on transmucosal drug delivery. *International Journal of Pharmaceutics* 532 (1): 555–572.

8 Linden, S., Sutton, P., Karlsson, N. et al. (2008). Mucins in the mucosal barrier to infection. *Mucosal Immunology* 1 (3): 183–197.

9 Zhang, S., Jia, Y., Liu, J. et al. (2023). A viscoelastic alginate-based hydrogel network coordinated with spermidine for periodontal ligament regeneration. *Regenerative Biomaterials* 10: rbad009.

10 Rossi, S., Ferrari, F., Bonferoni, M.C., and Caramella, C. (2001). Characterization of chitosan hydrochloride–mucin rheological interaction: influence of polymer concentration and polymer: mucin weight ratio. *European Journal of Pharmaceutical Sciences* 12 (4): 479–485.

11 King, M. (1987). The role of mucus viscoelasticity in cough clearance. *Biorheology* 24 (6): 589–597.

12 Bansil, R. and Turner, B.S. (2006). Mucin structure, aggregation, physiological functions and biomedical applications. *Current Opinion in Colloid & Interface Science* 11 (2-3): 164–170.

13 France, M.M. and Turner, J.R. (2017). The mucosal barrier at a glance. *Journal of Cell Science* 130 (2): 307–314.

14 Rose, M.C. and Voynow, J.A. (2006). Respiratory tract mucin genes and mucin glycoproteins in health and disease. *Physiological Reviews* 86 (1): 245–278.

15 Thornton, D.J. and Sheehan, J.K. (2004). From mucins to mucus: toward a more coherent understanding of this essential barrier. *Proceedings of the American Thoracic Society* 1 (1): 54–61.

16 Button, B., Cai, L.-H., Ehre, C. et al. (2012). A periciliary brush promotes the lung health by separating the mucus layer from airway epithelia. *Science* 337 (6097): 937–941.

17 Fahy, J.V. and Dickey, B.F. (2010). Airway mucus function and dysfunction. *The New England Journal of Medicine* 363 (23): 2233–2247.

18 Wong, C.Y., Al-Salami, H., and Dass, C.R. (2018). Microparticles, microcapsules and microspheres: a review of recent developments and prospects for oral delivery of insulin. *International Journal of Pharmaceutics* 537 (1-2): 223–244.

19 Patra, J.K., Das, G., Fraceto, L.F. et al. (2018). Nano based drug delivery systems: recent developments and future prospects. *Journal of Nanbiotechnology* 16 (1): 1–33.

20 Barua, S. and Mitragotri, S. (2014). Challenges associated with penetration of nanoparticles across cell and tissue barriers: a review of current status and future prospects. *Nano* 9 (2): 223–243.

21 Sosnik, A., das Neves, J., and Sarmento, B. (2014). Mucoadhesive polymers in the design of nano-drug delivery systems for administration by non-parenteral routes: a review. *Progress in Polymer Science* 39 (12): 2030–2075.

22 Ensign, L.M., Cone, R., and Hanes, J. (2012). Oral drug delivery with polymeric nanoparticles: the gastrointestinal mucus barriers. *Advanced Drug Delivery Reviews* 64 (6): 557–570.
23 Khanvilkar, K., Donovan, M.D., and Flanagan, D.R. (2001). Drug transfer through mucus. *Advanced Drug Delivery Reviews* 48 (2-3): 173–193.
24 Bourganis, V., Karamanidou, T., Samaridou, E. et al. (2015). On the synthesis of mucus permeating nanocarriers. *European Journal of Pharmaceutics and Biopharmaceutics* 97: 239–249.
25 Lai, S.K., Wang, Y.-Y., and Hanes, J. (2009). Mucus-penetrating nanoparticles for drug and gene delivery to mucosal tissues. *Advanced Drug Delivery Reviews* 61 (2): 158–171.
26 Cone, R. (2009). Barrier properties of mucus. *Advanced Drug Delivery Reviews* 61 (2): 75–85.
27 Abdulkarim, M., Agulló, N., Cattoz, B. et al. (2015). Nanoparticle diffusion within intestinal mucus: three-dimensional response analysis dissecting the impact of particle surface charge, size and heterogeneity across polyelectrolyte, pegylated and viral particles. *European Journal of Pharmaceutics and Biopharmaceutics* 97: 230–238.
28 Garcia-Diaz, M., Birch, D., Wan, F., and Nielsen, H.M. (2018). The role of mucus as an invisible cloak to transepithelial drug delivery by nanoparticles. *Advanced Drug Delivery Reviews* 124: 107–124.
29 Blanco, E., Shen, H., and Ferrari, M. (2015). Principles of nanoparticle design for overcoming biological barriers to drug delivery. *Nature Biotechnology* 33 (9): 941–951.
30 Murgia, X., Loretz, B., Hartwig, O. et al. (2018). The role of mucus on drug transport and its potential to affect therapeutic outcomes. *Advanced Drug Delivery Reviews* 124: 82–97.
31 Murgia, X., Pawelzyk, P., Schaefer, U.F. et al. (2016). Size-limited penetration of nanoparticles into porcine respiratory mucus after aerosol deposition. *Biomacromolecules* 17 (4): 1536–1542.
32 Huckaby, J.T. and Lai, S.K. (2018). PEGylation for enhancing nanoparticle diffusion in mucus. *Advanced Drug Delivery Reviews* 124: 125–139.
33 Schuster, B.S., Suk, J.S., Woodworth, G.F., and Hanes, J. (2013). Nanoparticle diffusion in respiratory mucus from humans without lung disease. *Biomaterials* 34 (13): 3439–3446.
34 Schneider, C.S., Xu, Q., Boylan, N.J. et al. (2017). Nanoparticles that do not adhere to mucus provide uniform and long-lasting drug delivery to airways following inhalation. *Science Advances* 3 (4): e1601556.
35 Dawson, M., Wirtz, D., and Hanes, J. (2003). Enhanced viscoelasticity of human cystic fibrotic sputum correlates with increasing microheterogeneity in particle transport. *The Journal of Biological Chemistry* 278 (50): 50393–50401.
36 Zhou, H., Fan, Z., Deng, J. et al. (2016). Hyaluronidase embedded in nanocarrier PEG shell for enhanced tumor penetration and highly efficient antitumor efficacy. *Nano Letters* 16 (5): 3268–3277.

37 Lai, S.K., O'Hanlon, D.E., Harrold, S. et al. (2007). Rapid transport of large polymeric nanoparticles in fresh undiluted human mucus. *Proceedings. National Academy of Sciences. United States of America* 104 (5): 1482–1487.

38 Bandi, S.P., Kumbhar, Y.S., and Venuganti, V.V.K. (2020). Effect of particle size and surface charge of nanoparticles in penetration through intestinal mucus barrier. *Nanoparticle Research* 22: 1–11.

39 Yu, Z., Fan, W., Wang, L. et al. (2019). Effect of surface charges on oral absorption of intact solid lipid nanoparticles. *Molecular Pharmaceutics* 16: 5013–5024.

40 Li, H., Guissi, N., Su, Z. et al. (2016). Effects of surface hydrophilic properties of PEG-based mucus-penetrating nanostructured lipid carriers on oral drug delivery. *RSC Advances* 6: 84164–84176.

41 Liu, Y., Jiang, Z., Hou, X. et al. (2019). Functional lipid polymeric nanoparticles for oral drug delivery: rapid mucus penetration and improved cell entry and cellular transport. *Nanomedicine* 21: 102075.

42 Yamamoto, H., Kuno, Y., Sugimoto, S. et al. (2005). Surface-modified PLGA nanosphere with chitosan improved pulmonary delivery of calcitonin by mucoadhesion and opening of the intercellular tight junctions. *Journal of Controlled Release* 102 (2): 373–381.

43 Shan, W., Zhu, X., Tao, W. et al. (2016). Enhanced oral delivery of protein drugs using zwitterion-functionalized nanoparticles to overcome both the diffusion and absorption barriers. *ACS Applied Materials & Interfaces* 8 (38): 25444–25453.

44 Dai, X., Zhu, Z., Li, Y. et al. (2022). "Shutter" effects enhance protein diffusion in dynamic and rigid molecular networks. *Journal of the American Chemical Society* 144 (41): 19017–19025.

45 Woodley, J. (2001). Bioadhesion: new possibilities for drug administration? *Clinical Pharmacokinetics* 40 (2): 77–84.

46 Zhou, S., Deng, H., Zhang, Y. et al. (2020). Thiolated nanoparticles overcome the mucus barrier and epithelial barrier for oral delivery of insulin. *Molecular Pharmaceutics* 17: 239–250.

47 Truong, N.P., Whittaker, M.R., Mak, C.W., and Davis, T.P. (2015). The importance of nanoparticle shape in cancer drug delivery. *Expert Opinion on Drug Delivery* 12: 129–142.

48 Hui, Y., Yi, X., Hou, F. et al. (2019). Role of nanoparticle mechanical properties in cancer drug delivery. *ACS Nano* 13 (7): 7410–7424.

49 Dai, X., Zhang, X., Gao, L. et al. (2022). Topology mediates transport of nanoparticles in macromolecular networks. *Nature Communications* 13 (1): 4094.

50 Xu, Z., Dai, X., Bu, X. et al. (2021). Enhanced heterogeneous diffusion of nanoparticles in semiflexible networks. *ACS Nano* 15 (3): 4608–4616.

51 He, X., Yang, Y., Han, Y. et al. (2023). Extracellular matrix physical properties govern the diffusion of nanoparticles in tumor microenvironment. *Proceedings of the National Academy of Sciences of the United States of America* 120 (1): e2209260120.

52 Guo, R., Mao, J., and Yan, L.-T. (2013). Unique dynamical approach of fully wrapping dendrimer-like soft nanoparticles by lipid bilayer membrane. *ACS Nano* 7 (12): 10646–10653.

53 Bao, C., Li, Z., Liang, S. et al. (2021). Microneedle patch delivery of capsaicin-containing α-lactalbumin nanomicelles to adipocytes achieves potent anti-obesity effects. *Advanced Functional Materials* 20 (31): 2011130.1–2011130.11.

54 Liu, B., Liu, B., Wang, R. et al. (2021). α-Lactalbumin self-assembled nanoparticles with various morphologies, stiffnesses, and sizes as Pickering stabilizers for oil-in-water emulsions and delivery of curcumin. *Journal of Agricultural and Food Chemistry* 69: 2485–2492.

55 Yi, X., Shi, X., and Gao, H. (2011). Cellular uptake of elastic nanoparticles. *Physical Review Letters* 107 (9): 098101.

56 Sun, J., Zhang, L., Wang, J. et al. (2015). Tunable rigidity of (polymeric core)-(lipid shell) nanoparticles for regulated cellular uptake. *Advanced Materials* 27 (8): 1402–1407.

57 Behzadi, S., Serpooshan, V., Wei, T. et al. (2017). Cellular uptake of nanoparticles: journey inside the cell. *Chemical Society Reviews* 46: 4218–4244.

58 Chen, Y., Ju, L., Rushdi, M. et al. (2017). Receptor-mediated cell mechanosensing. *Molecular Biology of the Cell* 28 (23): 3134–3155.

59 Yu, M.R., Xu, L., Tian, F.L. et al. (2018). Rapid transport of deformation-tuned nanoparticles across biological hydrogels and cellular barriers. *Nature Communications* 9 (1): 2607.

60 Johansson, L., Elvingson, C., and Loefroth, J.E. (1991). Diffusion and interaction in gels and solutions. 3. Theoretical results on the obstruction effect. *Macromolecules* 24 (22): 6024–6029.

61 Hänggi, P., Talkner, P., and Borkovec, M. (1990). Reaction-rate theory: fifty years after Kramers. *Reviews of Modern Physics* 62 (2): 251.

62 Helfrich, W. (1973). Elastic properties of lipid bilayers: theory and possible experiments. *Zeitschrift für Naturforschung. Section C* 28 (11-12): 693–703.

63 Amsden, B. (1999). An obstruction-scaling model for diffusion in homogeneous hydrogels. *Macromolecules* 32 (3): 874–879.

64 Yu, S., Tian, F., and Shi, X. (2022). Diffusion of deformable nanoparticles in adhesive polymeric gels. *Journal of the Mechanics and Physics of Solids* 167: 105002.

65 Mazaheri, M., Ehrig, J., Shkarin, A. et al. (2020). Ultrahigh-speed imaging of rotational diffusion on a lipid bilayer. *Nano Letters* 20 (10): 7213–7219.

66 Xiao, L., Wei, L., Liu, C. et al. (2012). Unsynchronized translational and rotational diffusion of nanocargo on a living cell membrane. *Angewandte Chemie, International Edition* 51 (17): 4181–4184.

67 Liu, Y., Zheng, X., Guan, D. et al. (2022). Heterogeneous nanostructures cause anomalous diffusion in lipid monolayers. *ACS Nano* 16 (10): 16054–16066.

68 Chen, P., Yue, H., Zhai, X. et al. (2019). Transport of a graphene nanosheet sandwiched inside cell membranes. *Science Advances* 5 (6): eaaw3192.

69 Shi, X., von dem Bussche, A., Hurt, R.H. et al. (2011). Cell entry of one-dimensional nanomaterials occurs by tip recognition and rotation. *Nature Nanotechnology* 6 (11): 714–719.

70 Wang, Y.F., Zhang, Q., Tian, F. et al. (2022). Spatiotemporal tracing of the cellular internalization process of rod-shaped nanostructures. *ACS Nano* 16: 4059.

71 Yu, M., Wang, J., Yang, Y. et al. (2016). Rotation-facilitated rapid transport of nanorods in mucosal tissues. *Nano Letters* 16 (11): 7176–7182.

72 Shi, X. and Tian, F. (2019). Multiscale modeling and simulation of nano-carriers delivery through biological barriers—a review. *Advanced Theory and Simulations* 2 (1): 1800105.

73 Xue, S.L., Bo, L., Feng, X.Q., and Gao, H. (2016). Biochemomechanical poroelastic theory of avascular tumor growth. *Journal of the Mechanics and Physics of Solids* 94: 409–432.

74 Amsden, B. (1998). Solute diffusion within hydrogels. Mechanisms and models. *Macromolecules* 31 (23): 8382–8395.

75 Hadjiev, N.A. and Amsden, B.G. (2015). An assessment of the ability of the obstruction-scaling model to estimate solute diffusion coefficients in hydrogels. *Journal of Controlled Release* 199: 10–16.

76 Altenberger, A.R., Tirrell, M., and Dahler, J.S. (1986). Hydrodynamic screening and particle dynamics in porous media, semidilute polymer solutions and polymer gels. *The Journal of Chemical Physics* 84 (9): 5122–5130.

77 Zhang, R. and Schweizer, K.S. (2016). Statistical mechanical theory of penetrant diffusion in polymer melts and glasses. *Macromolecules* 49 (15): 5727–5739.

78 Ogston, A. (1958). The spaces in a uniform random suspension of fibres. *Transactions of the Faraday Society* 54: 1754–1757.

79 Wang, J., Yang, Y., Yu, M. et al. (2018). Diffusion of rod-like nanoparticles in non-adhesive and adhesive porous polymeric gels. *Journal of the Mechanics and Physics of Solids* 112: 431–457.

80 Xue, C., Huang, Y., Zheng, X., and Hu, G. (2022). Hopping behavior mediates the anomalous confined diffusion of nanoparticles in porous hydrogels. *Journal of Physical Chemistry Letters* 13 (45): 10612–10620.

81 Xue, C., Shi, X., Tian, Y. et al. (2020). Diffusion of nanoparticles with activated hopping in crowded polymer solutions. *Nano Letters* 20 (5): 3895–3904.

82 Wang, Y.Y., Lai, S.K., Suk, J.S. et al. (2008). Addressing the PEG mucoadhesivity paradox to engineer nanoparticles that "slip" through the human mucus barrier. *Angewandte Chemie, International Edition* 120 (50): 9872–9875.

83 Yang, Y., Tian, F., Nie, D. et al. (2020). Rapid transport of germ-mimetic nanoparticles with dual conformational polyethylene glycol chains in biological tissues. *Science Advances* 6 (6): eaay9937.

11

Physical Attributes of Nanoparticle Transport in Macromolecular Networks: Flexibility, Topology, and Entropy

Xiaobin Dai, Xuanyu Zhang, Lijuan Gao, Yuming Wang, and Li-Tang Yan

Tsinghua University, State Key Laboratory of Chemical Engineering, Department of Chemical Engineering, No 30, Shuangqing Road, Haidian District, Beijing 100084, P. R. China

11.1 Introduction

Macromolecular networks are among the most widely used polymeric materials, with applications [1] in, e.g. cosmetics [2], adhesives [3], rubbers [4], medical devices [5], soft actuators [6], hydrogels [7], and electronic materials [8]. Biological macromolecular networks, which are composed of hydrated biomacromolecules, are universal structural bases of diverse biological soft matter systems, such as mucus [9], extracellular matrix (ECM) [10], biofilms [11], and the nuclear pore [12]. As a consequence, diffusion transport of nanoparticles in confined environments of macromolecular networks is a fundamental problem underlying many important physical processes and biological responses [9]. For example, the mechanical, viscoelastic, and functional properties of the filled elastomers strongly depend on the diffusional dynamics of the nanoparticles embedded in the matrix of polymer networks, because the diffusivity of these nanoparticles not only affect their aggregation process but also the dynamical properties of the surrounding networks [13]. In biological systems, the ECM is an obstacle to drug delivery, because its mesh prevents large nanoparticles from penetrating deep into tissue or tumors; mucus similarly controls the permeability of nanoparticles, but can also limit diffusion of small molecules such as antibiotics. Therefore, a thorough understanding of the nanoparticle diffusion behaviors in macromolecular networks is significant for the design and fabrication of new materials with targeted properties as well as for the elucidation of physical origin behind many biological phenomena.

A macromolecular network usually consists of hierarchical structures, i.e. from the small to the large scales, strands, loop, cell, and the whole network, which possess diverse intrinsic physical properties (Figure 11.1). For the strands, the typical intrinsic physical property is the bending rigidity of the macromolecular chains. One of the ways to quantify the bending rigidity of the polymer strands embedded in a network is by persistence length (l_p), being essentially the length over which they appear straight in the presence of Brownian forces. For example, $l_p \sim 10^{-2}$ μm for

Dynamics and Transport in Macromolecular Networks: Theory, Modeling, and Experiments, First Edition. Edited by Li-Tang Yan.

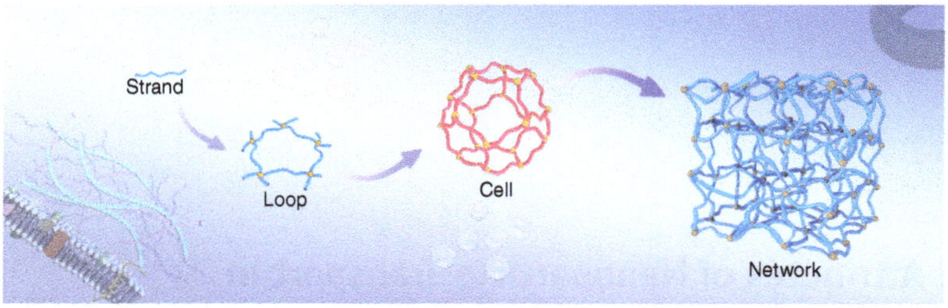

Figure 11.1 The schematic diagram shows the typical hierarchical structures of a macromolecular network.

the flexible polymers, while $l_p \sim 10^{-1}$ μm and $l_p \sim 10^0$–10^1 μm for stiffer strands of DNA [14] and cytoskeleton [15, 16], respectively. For the loops, the typical intrinsic physical property is the loop order which describes the number of strands involved in a loop, such as, for example, a primary loop with only one strand, and the secondary and ternary loops with two and three strands. For the network cells, one of the most important intrinsic physical properties is the topological structure, which can be characterized by functionality k and genus g. The cell topology indicates the spatial organization of the loops involved in the network cell. These physical attributes can significantly modify the free energy landscape experienced by a nanoparticle, which, in turn, determines the diffusion dynamics of the nanoparticle. Thus, the intrinsic physical property of the networks belongs to one key factor dictating the diffusion dynamics of nanoparticles in various macromolecular networks.

The current chapter highlights the advances in investigating the physical attributes of nanoparticle transport in macromolecular networks, focusing on the theoretical and simulation aspects of two typical intrinsic physical properties, that is, strand flexibility [17] and cell topology [18]. The entropic effects arisen in these both properties are explored, leading to mechanistic interpretation regarding the physics behind the phenomena. In particular, we will firstly introduce the enhanced heterogeneous diffusion of nanoparticles in the macromolecular networks with semiflexible loops. Next, we will move on to how the cell topology impacts the transport of nanoparticles in macromolecular networks. Finally, the challenges and perspectives of this field will be presented.

11.2 Effects of the Chain Flexibility of Strands

The motion of a guest particle in the macromolecular networks can be mightily hampered by its surrounding network strands. From a physical point of view, the main differences between a variety of biological or synthetic polymer strands are their intrinsic physical properties, for example, the network chain flexibility [9, 15, 17]. Thus, understanding the bending rigidity effects from the network strands is essential for revealing the fundamental physics of nanoparticle dynamics in various macromolecular networks, which has not been established. One particular aspect of the issue that is of great relevance concerns the nanoparticle transport in semiflexible networks, where the bending rigidity of a strand is large enough so that the bending energy, that tends to a straight conformation, can just outcome the entropic tendency of the strand to crumple up into a Gaussian or random coil [19–21]. As a distinct class of soft condensed matter with striking properties, semiflexible cross-linked networks are especially essential in biology, as they form a major structural component of living cells and tissue, e.g. intracellular scaffolds known as the cytoskeleton and ECMs of collagen [22, 23]. As a consequence, exploring the transport of nanoparticles in semiflexible networks is important in biology and relevant biomedical applications [24, 25]. In this section, we introduce the results by combining numerical simulations and theoretical analysis to present a fundamental study of the transport of nanoparticles confined in cross-linked macromolecular networks with various bending rigidities.

The transport of a nanoparticle in the cross-linked macromolecular networks is simulated by using dissipative particle dynamics (DPD), which has been widely used to model the structural and dynamic properties of macromolecular networks [26, 27]. The network consists of 3456 strands and 1963 cross-links (Figure 11.2a) with connectivity $f = 4$ (Figure 11.2b). Besides, networks with different cross-link connectivities, such as the hexa-functional network (Figure 11.2c) and the octa-functional network (Figure 11.2d), can make significance to the transport of a nanoparticle. The network strands are modeled as a sequence of N_s beads connected by harmonic bonds with an interaction potential given by $U_{\text{bond}} = K_b[(r - b)/r_c]^2$, where $K_b = 64k_BT/r_c{}^2$ and is large enough to prevent bond crossing [28]; $b = 0.5r_c$, with the length unit r_c, Boltzmann constant k_B, and temperature T. We also employ the bending potential along a network strand of the form $U_{\text{angle}} = K_a[1 + \cos(\theta)]$; K_a controls the chain stiffness and θ is the angle between two successive bonds. Using $k_BT = 1$, the length of bond is $l_b \approx 0.59r_c$, the contour length of the strands is $L_s = (N_s - 1)l_b$, and the persistence length is $l_p \approx K_a l_b$ [29]. The nanoparticle, which has weak interaction with the networks, is fabricated by the beads arranged on an fcc lattice with a lattice constant of $0.35r_c$, ensuring that the nanoparticle is not penetrated by other beads [30]; all beads constituting a sphere nanoparticle move as a rigid body. For all systems, the diameter of the nanoparticle d is fixed at $d = 6.4r_c$, while a variety of values of l_p and L_c (L_c is the contour length of a loop consisting of six strands [Figure 11.2a]) are set to evaluate their effects on the nanoparticle motion. This results changes confinement parameter ξ, defined as the ratio between d and the network mesh size a_x, ranging from about 1.01 to 3.26 as illustrated in Figure 11.2a. The total simulation time is $t = 5 \times 10^5\tau$, where τ is the time unit of the simulation. The simulation box is $40r_c \times 40r_c \times 40r_c$ in size and with a wall at the bottom but periodic boundary condition in other directions.

11.2.1 Dynamical Heterogeneity of a Semiflexible Network

We perform simulations of the diffusion dynamics of a nanoparticle in a typically semiflexible macromolecular network. Here, both the persistence length and the loop length of the strands are set as the similar values, that is, $l_p = 23.6r_c$ and $L_c = 28.3r_c$. As illustrated by the spatial trajectory in Figure 11.2e (middle), the

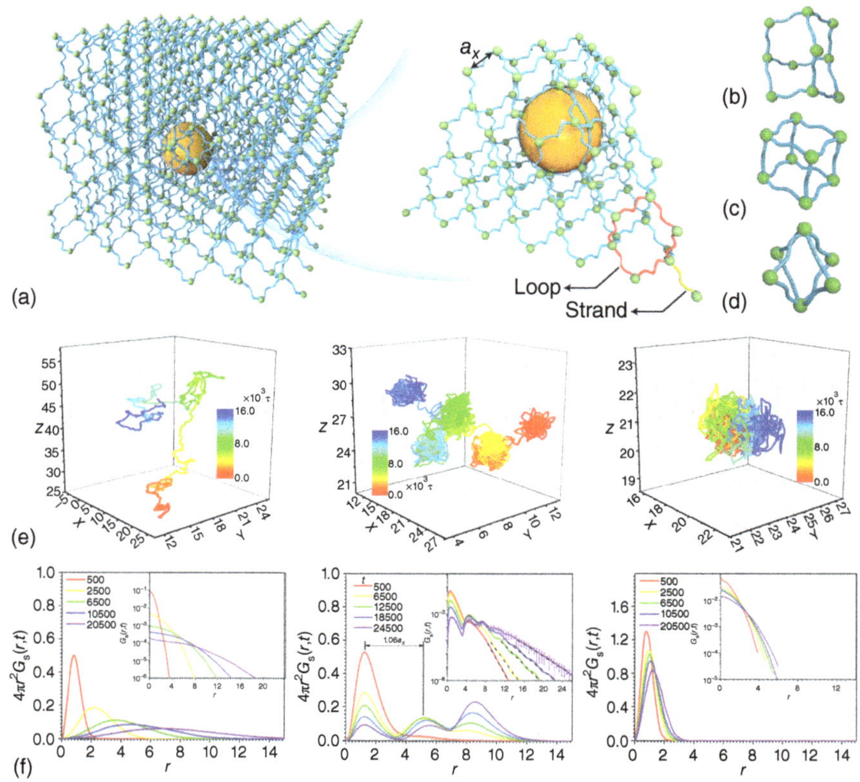

Figure 11.2 (a) Schematic of a nanoparticle in a tetrafunctional network with strands and loops, as highlighted by red and yellow, respectively. (b–d) Schematic diagrams of elementary network cages with the cross-link connectivity: (b) $f = 4$, (c) $f = 6$, and (d) $f = 8$. (e) Representative trajectories of nanoparticle diffusion dynamics in the macromolecular network at $l_p = 23.6r_c$ with $L_c = 35.4r_c$ (left), $L_c = 28.3r_c$ (middle), and $L_c = 17.7r_c$ (right), respectively. (f) The displacement probability distribution, $4\pi r^2 G_s(r,t)$, of nanoparticle diffusion dynamics in the macromolecular networks at $l_p = 23.6r_c$ with $L_c = 35.4r_c$ (left), $L_c = 28.3r_c$ (middle), and $L_c = 17.7r_c$ (right), respectively. Inset: $G_s(r,t)$ is plotted logarithmically against linear displacement. Dashed lines are fits to exponential functions. Source: Reprinted with permission from Ref. [17]; © 2021, American Chemical Society.

nanoparticle undergoes constrained motion punctuated by large-scale jumps. The time scale of these jumps is very short compared to the residence time within the network cells. This demonstrates that nanoparticle rapidly and randomly jumps between neighboring network cells wherein it is constrained, taking non-Gaussian and hopping diffusion as confirmed by several peaks in the self-part of the van Hove correlation function [31], $G_s(r,t)$, and the exponential distribution of nanoparticle displacements (Figure 11.2f, middle).

11.2.2 Nonmonotonic Feature

To assess the detailed range for the emergence of this dynamic heterogeneity, we systematically investigate the diffusion behaviors of the nanoparticle in the

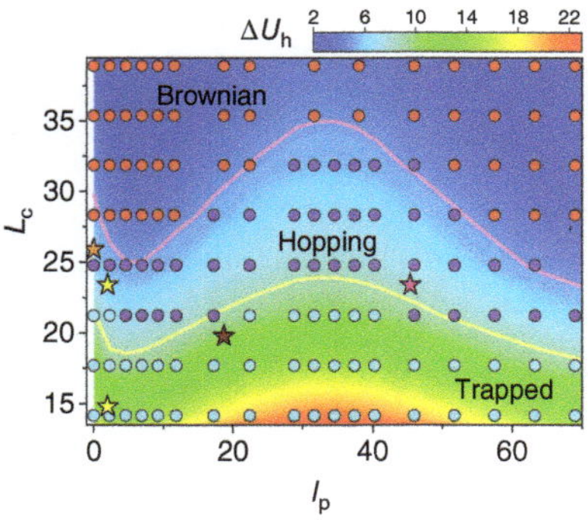

Figure 11.3 Diagram of nanoparticle dynamics interrelating to L_c and l_p. The contour map: energy barrier for the nanoparticle hopping ΔU_h, which is derived from the network deformation energy, in the L_c–l_p plane. The pink and yellow lines: theoretically upper and lower bounds of the hopping region. The nanoparticle diameter d is $6.4r_c$ for all simulations. Representative experimental results: Polyacrylamide (Ref. [32], orange star), Hyaluronan (Ref. [33], yellow star), Fd-virus (Ref. [34], wine star), and Actin and Microtubules (Ref. [35], pink star). Source: Reprinted with permission from Ref. [17]; © 2021, American Chemical Society.

macromolecular networks as a function of L_c and l_p, allowing us to construct a diagram of the diffusion dynamics corresponding to these both parameters (Figure 11.3). Here the circle points in Figure 11.3 are all obtained from the DPD simulations. In this diagram, the strand stiffness is tuned to yield l_p from 0 to about $70r_c$, covering flexible, semiflexible, and rod-like strands for the present range of L_c, i.e. $14r_c < L_c < 39r_c$. The shaded regions in the diagram approximately discriminate three typical regimes: in the bottommost regime, trapped state is preferred (Figure 11.2f, right); in the upmost regime, the nanoparticle favors Brownian dynamics (Figure 11.2f, left); within the regime in between, the nanoparticle diffusion is dominated by hopping between neighboring network cells. Interestingly, the heterogeneous diffusion exhibits a nonmonotonic dependence on l_p: the networks with semiflexible strands, where l_p close to L_c, possess a larger area of hopping dynamics (Figure 11.3). Besides, for different types of networks by changing the cross-link connectivity from the current 4 to 6 and 8, the same simulation is also performed, and similar regimes can be identified from the dynamics diagrams (see Figure 11.2e,f), indicating the universality of the rule.

To delineate the correlation between l_p and the dynamic heterogeneity in more detail, we analyze the mean square displacement (MSD) $\langle \Delta r^2(t) \rangle$ and the three-dimensional (3D) non-Gaussian parameter $\alpha(t) \equiv 3/5\langle \Delta \mathbf{r}^4(t) \rangle / \langle \Delta \mathbf{r}^2(t) \rangle^2 - 1$, where $\mathbf{r}(t)$ is the position vector of nanoparticle at time t. Figure 11.4a depicts $\langle \Delta r^2(t) \rangle$ and $\alpha(t)$ for different l_p at $L_c = 28.3r_c \cdot \langle \Delta r^2(t) \rangle$ scales as t^2 at short time scales, demonstrating that the nanoparticle in all the networks takes ballistic motions. Above $l_p = 59r_c$ and below $l_p = 11.8r_c$, the nanoparticle evolves directly from the ballistic regime to the Fickian regime with $\langle \Delta r^2(t) \rangle \sim t$, behaving as Brownian dynamics. However, between these both values of l_p, there is an intermediate regime with the subdiffusive behavior and the nanoparticle diffuses via hopping

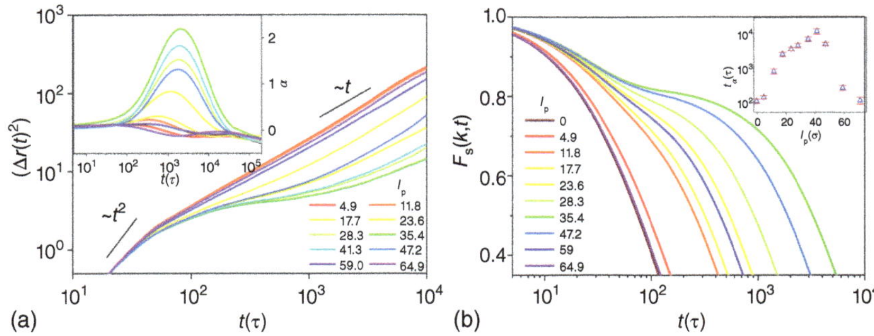

Figure 11.4 (a) $\langle\Delta r^2(t)\rangle$ and α (inset) for different l_p at $L_c = 28.3r_c$. (b) $F_s(k,t)$ for different l_p at $L_c = 28.3r_c$, where $k = 2r_c^{-1}$. Inset: t_α as a function of l_p. Source: Reprinted with permission from Ref. [17]; © 2021, American Chemical Society.

motions and, i.e. $\langle\Delta r^2(t)\rangle \sim t^\eta$, emerges, where the exponent $\eta < 1$ and depends on l_p, and $\alpha(t)$ demonstrates clearly that dynamics shows stronger heterogeneity.

Interestingly, even within the region of hopping diffusion, the intermediate regime and $\alpha(t)$ show nonmonotonic dependence on l_p: with the increase of l_p, both the range of intermediate regime and the peak of $\alpha(t)$ firstly increase at small l_p and then they crossover to decrease at large l_p. Particularly, the values of α seem to be smaller than those in the glass-forming liquids, which can be fundamentally ascribed to the physical nature of the dynamic heterogeneity in these different media [36, 37]. The maximum of α occurs at $l_p = 35.4r_c$, close to L_c, indicating that polymer networks with semiflexible strands are more inclined to the enhanced heterogeneous diffusion. The nonmonotonic feature and the enhanced heterogeneity can also be accurately captured by the self-part of the intermediate scattering function [38, 39] $F_s(k,t) = \langle\exp\{i\mathbf{k}\cdot[\mathbf{r}(t)-\mathbf{r}(0)]\}\rangle$ in Figure 11.4b, where the height of the shoulder indicates, the relaxation time, t_α, from $F_s(k,t)$ as $F_s(k = 2, t = t_\alpha) = e^{-1}$ (inset of Figure 11.4b).

The probability of a hopping motion increases with time interval and becomes remarkable enough at the residence time scale t_c, for a successful escape of nanoparticle from a network cell. This escape is activated by a large fluctuation and relaxation of network strands. To further pinpoint the influence of strand rigidity on the activated event, we present the relation of t_c, determined based on continuous time random walk (CTRW) formalism [40], to l_p, as shown in Figure 11.5b. Combining with Figure 11.3, one can find that within the region of hopping diffusion, the residence time become longer; moreover, t_c exhibits a nonmonotonic dependence on l_p and the peaks occur at the points where l_p approximates to L_c, demonstrating that semiflexible strands favor extended time to relax and thereby exert stronger constraints for slowing down the nanoparticle motion, thereby causing the enhanced dynamic heterogeneity. This can also be verified from the plots of t_c against confinement parameter ξ for a set of l_p (Figure 11.5c). Figure 11.5c clarifies that semiflexible strand at $l_p = 35.4r_c$ results in much longer residence time than the very stiff ($l_p = 59r_c$) and completely flexible ($l_p = 0r_c$) strands. In light of the

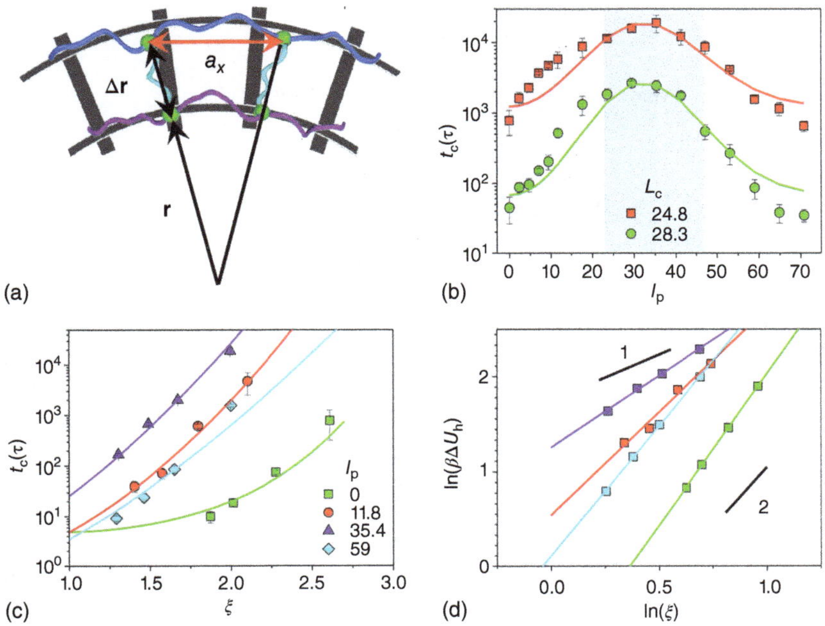

Figure 11.5 (a) Diagram scheming the railway track model of worm-like chains. (b) t_c as a function of l_p for different L_c. (c) t_c as a function of ξ for various l_p. (d) ΔU_h as a function of ξ in the log–log scale. The solid lines in (b), (c), and (d) are predicted results based on the theoretical analysis. Source: Reprinted with permission from Ref. [17]; © 2021, American Chemical Society.

CTRW formalism, the relation between t_c and the activation energy of nanoparticle hopping, ΔU_h, takes the form [41]:

$$\frac{1}{t_c/\tau} \cong e^{-\beta \Delta U_h} \tag{11.1}$$

where $\beta = 1/k_B T$. ΔU_h measures the magnitude of free energy barrier for a nanoparticle escaping from a network cell. It mainly arises from the deformation energy of the network due to the inserted nanoparticle. To give an expression of ΔU_h, we developed an analytical model based on railway-track model (Figure 11.5a), and more details are explained in Ref. [17]. From Eq. (11.1), we thus obtain ΔU_h and present the ξ dependence of ΔU_h for various l_p in the inset of Figure 11.5c. As shown in Figure 11.5d, for the flexible strands, ΔU_h scales as a power law for ξ, $\beta \Delta U_h \sim \xi^{\nu}$, with the exponent $\nu \approx 2$ which agrees well with previous scaling analysis of Cai et al. [42] However, our results demonstrate that, for the semiflexible strands with l_p close to L_c ($l_p = 35.4 r_c$), ΔU_h turns to an asymptotically linear dependence on ξ.

11.2.3 Validation by MC Simulations and Experimental Data

Our results demonstrate that the bending-rigidity-controlled hopping diffusion and heterogeneous dynamics can be originally attributed to the competition between the bending energy and the conformation entropy, which is regulated by the strand

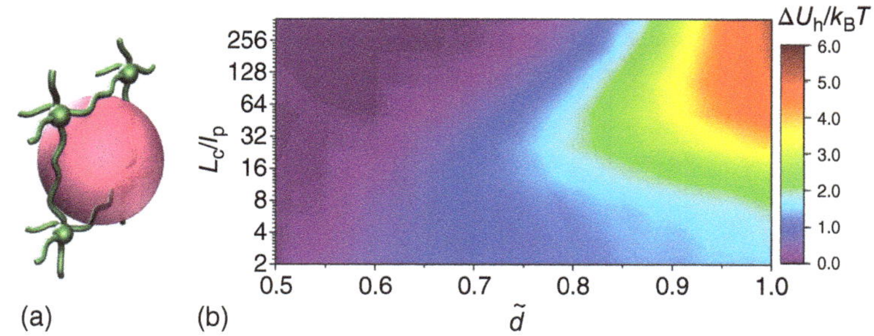

Figure 11.6 (a) Schematic of MC simulation for the translocation of a nanoparticle of normalized size, $\tilde{d} = d/a_x$, across a network loop with contour length L_c and persistence length l_p. (b) The free energy barriers during a translation event, ΔU_h, as a function of L_c/l_p and $\tilde{d}$. Source: Reprinted with permission from Ref. [17]; © 2021, American Chemical Society.

rigidity and consequently results in the nonmonotonic feature. To further verify such a competition between entropic and energetic effects, we perform additional Monte Carlo (MC) simulations of the translocation of a nanoparticle across a single-network loop with four fixed cross-links, as schemed in Figure 11.6a. The nanoparticle is originally placed above the loop center, and then it is moved to translocate through the loop center along z-axis. The strand flexibility can be characterized by L_c/l_p, and a larger value of L_c/l_p indicates a higher flexibility of the strands. One can just identify a nonmonotonic dependence of the free energy barrier on the strand rigidity (Figure 11.6b), and the energy barriers of loops with semi-flexible strands become stronger upon the insertion of a nanoparticle, reproducing aforementioned results and corroborating the generality of the nonmonotonic behavior.

11.3 Effects of Network Topology

Over the past few decades, with the development of modern organic synthesis, the manipulation of the topological structures in macromolecular networks has advanced significantly, leading to well-regulated and even programmable network topologies [43–45]. For the transport of nanoparticles confined in macromolecular networks, previous theories and models based on the free energy barrier of the local network loop cannot fully express the topology of the macromolecular network [42, 46, 47]. To explain the physical origin of topological effects, a comprehensive understanding of the free energy landscape tailored by network topology, as well as its impact on the dynamics of nanoparticles, are necessary. However, owing to the complex topologies and microenvironments, detecting the free energy landscape experienced by a nanoparticle in the macromolecular network becomes very difficult using experimental techniques. Therefore, the theoretical approach may be a good option to explicitly quantify the local free energy landscape of the nanoparticle, predict its anomalous dynamic behaviors, and reveal the topological effects of network on it. In this section, we introduce the studies through developing a theoretical framework to provide a rigorous analysis of the relation of the free energy landscape and diffusive dynamics to the topological structure for a particle in network cells of permanently cross-linked macromolecular networks.

11.3.1 Analytical Model for Free Energy Landscape

First, we develop a theoretical framework to provide a rigorous analysis of the relationship between the topological structure for a particle in network cells of permanently cross-linked macromolecular networks and the free energy landscape and diffusive dynamics in order to understand the nature of topological effects on particle transport dynamics. Full details can be found in Ref. [18]. Briefly, in this model, only a hard spherical nanoparticle of radius R in a network was considered, cross-linked by Gaussian chains without dangling ends. In this model, to encode the topology of the network, we suppose (i) the set of cross-links $k = \{\mathbf{r}_i\}_{i=1}^{M}$, with M cross-links between the efficiently bridged Gaussian chains, (ii) the collection of linker connections marked as (i, j), and (iii) the continue curve path of linked strands $\mathbf{r}_{ij}(s)$ with contour parameter $s \in [0, 1]$. For a Gaussian chain of N bonds of Kuhn length b, the contour length $L = Nb$, and the average mesh size, $a_x = N^{1/2}b = 1$, is the unit length of the system. For simplicity, we assume that the cross-linked chains of the network dissolve in the theta-solvents, where excluded volume interactions of cross-linked chain are screened. Based on these assumptions, in the canonical ensemble, the partition function $Z(\mathbf{r}_{np})$ takes the form [48, 49]:

$$ZZ(\mathbf{r}_{np}) = \prod_{k} \int d\mathbf{r}_k \prod_{(i,j)} \int D\mathbf{r}_{ij}\delta(\mathbf{r}_{ij}(0) - \mathbf{r}_i)\delta(\mathbf{r}_{ij}(1) - \mathbf{r}_j) \times \exp\left[-\beta \sum_{(i,j)} H(\mathbf{r}_{ij} - \mathbf{r}_{np})\right] \tag{11.2}$$

$\mathbf{r}_k$ is the position vector of the cross-link point k, $\mathbf{r}_{ij}(s)$ is the path vector of the strand with its start $\mathbf{r}_{ij}(0) = \mathbf{r}_i$ and end $\mathbf{r}_{ij}(1) = \mathbf{r}_j$, $\mathbf{r}_{np}$ is the position vector of the particle, and δ is the delta function. Coupling the excluded volume effect of a hard sphere with radius R, the modified Hamiltonian of the strand between cross-link pair (i, j) is given by:

$$H(\mathbf{r}_{ij}, \mathbf{r}_{np}) = \frac{3k_\mathrm{B}T}{2Nb^2}\int_0^1 ds \left\| \frac{\partial \mathbf{r}_{ij}}{\partial s} \right\|^2 \Phi\left(\left|\mathbf{r}_{ij} - \mathbf{r}_{np}\right| - R\right) \tag{11.3}$$

where $s \in [0, 1]$ is the contour variable, N is the number of bonds in a strand, b is the Kuhn length, and $\Phi(x)$ is the Heaviside step function. Hence, the Helmholtz free energy of the particle-network system can be given as $F(\mathbf{r}_{np}) = -k_\mathrm{B}T\ln Z(\mathbf{r}_{np})$, which describes the free energy landscape as a function of the position of the nanoparticle in the network $\mathbf{r}_{np}$.

11.3.2 Network Topology and Free Energy Landscape

To quantitatively examine the effect of network topology on the free energy landscape of the particle-network system, we develop a new approach to determine the

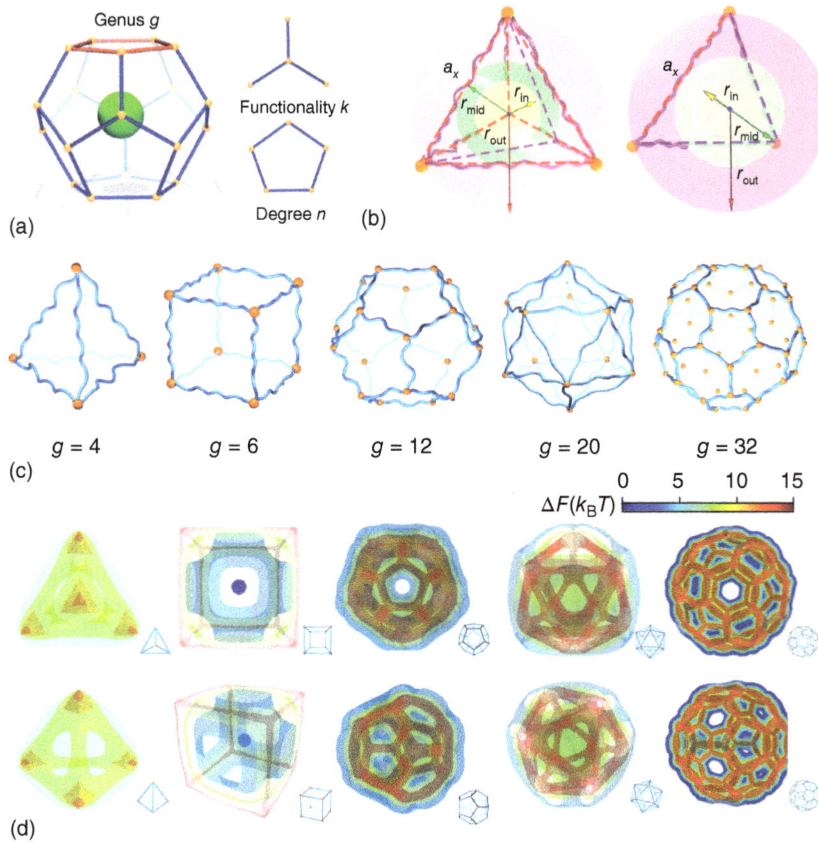

Figure 11.7 Detailed overview of network topology and the free energy landscape. (a) Schematic representation of a particle in a macromolecular network with topological parameters: genus g, degree n, and functionality k. (b) Schematic diagram of the scaling parameters of a network cell at $g = 4$, where arrows shown in yellow, green, and red denote, respectively, inradius r_{in}, midradius r_{mid}, and circumradius r_{out} of the cell from different views. (c) Schematic representation of macromolecular network cells with different topologies. (d) Isosurfaces of the free energy change ΔF of a particle in networks with different topologies marked at the right bottom, where the scaled diameter $d/a_x = 1.4$. The color bar on the top right corner encodes the value of ΔF. Source: Reprinted with permission from Ref. [18]; © 2022, Springer Nature.

topological parameters, consisting the genus g, the functionality k, and the degree n, of the network (Figure 11.7a). In this study, the mesh size a_x, defined as the root-mean-square end-to-end distance of the strand, is the same for every strand in a network cell [42]. Hence, a series of Platonic or Archimedean polyhedra are modeled as the representative network cells, in which the vertices and edges of polyhedra are employed as the strands and cross-links, as shown in Figure 11.7b,c. Owing that these polyhedra have central symmetry and are homeomorphic to spheres, the topologies of corresponding network cells can be simply determined by genus g, proposed by topological properties of these polyhedral cells [50]. Figure 11.7c shows the representative network cells with increasing g. As all Platonic or Archimedean polyhedra have an inscribed sphere tangent to the faces, a midsphere tangent to the edges, and a circumscribed sphere through the vertices (Figure 11.7b), the normalized radii of these spheres, i.e. r_{in}/a_x, r_{mid}/a_x and r_{out}/a_x, allow scaling parameters to characterize the cell topologies and can be given based on g and degree n:

$$\begin{aligned}\frac{r_{in}}{a_x} &= \frac{1}{2}\cot\left(\frac{\pi}{n}\right)\tan\left(\frac{\theta}{2}\right)\\ \frac{r_{mid}}{a_x} &= \frac{1}{2}\cot\left(\frac{\pi}{n}\right)\sec\left(\frac{\theta}{2}\right)\\ \frac{r_{out}}{a_x} &= \frac{1}{2}\cot\left(\frac{\pi}{n}\right)\sqrt{\sec^2\left(\frac{\theta}{2}\right)+\tan^2\left(\frac{\pi}{n}\right)}\end{aligned} \tag{11.4}$$

where θ represents the dihedral angle between any two faces.

Figure 11.7d shows the isosurfaces of the Helmholtz free energy ΔF for a nanoparticle, with diameter $d/a_x = 1.4$, in some representative network cells corresponding to Figure 11.7c. Even if the nanoparticle and the network mesh are of the same size, the free energy landscape experienced by the nanoparticle can vary significantly in different topologies of network cells. Moreover, the anisotropic free energy landscape has a strong dependence on g, leading to a variety of free energy barriers for nanoparticle diffusion across network cells with different topologies, in contrast to the permanent energy barrier of the local network loop. However, no matter how the cell topology changes, the free energy reaches its global minimum when the nanoparticle locate at the center of each cell, and a local free energy minimum can be identified when the nanoparticle locate the center of each face. Connecting the global and local free energy minima, a transition path for a nanoparticle traversed from a network cell to its neighbor is found, which has been proved to be a minimum energy path (MEP) [51].

11.3.3 Topology-Dictated Scaling Regimes of Free Energy Change

In order to depict the free energy encountered by a nanoparticle within a network cell, we investigate the MEP for nanoparticles of different sizes when $g = 6$. In pursuit of this objective, we select one MEP as the z-axis, aligning its origin at the center of the cell (refer to the inset in Figure 11.8a). Figure 11.8a presents the free energy change, $\Delta F(z) = F(z) - F(0)$, with the zero-point free energy $F(0)$, upon increasing z from the origin to the face center for various d in the log–log scale. To validate our theoretical findings, we additionally conduct MC simulations, and we compare the MC results to the theoretical outcomes for a selection of particle sizes in Figure 11.8b, with a standard error estimated to be within $0.3k_BT$. A good agreement between the simulated and theoretical results suggests that the theoretical model accurately captures the variations in free energy encountered by the nanoparticle confined within network cells.

An analysis of the $\Delta F \sim z$ profiles depicted in Figure 11.8a,b reveals an intriguing finding: the path dependencies of $\Delta F(z)$ can be categorized into four distinct regimes based on the particle size, which are roughly delineated by shaded regions. In the case of a large nanoparticle, $\Delta F(z)$ demonstrates a power law relationship

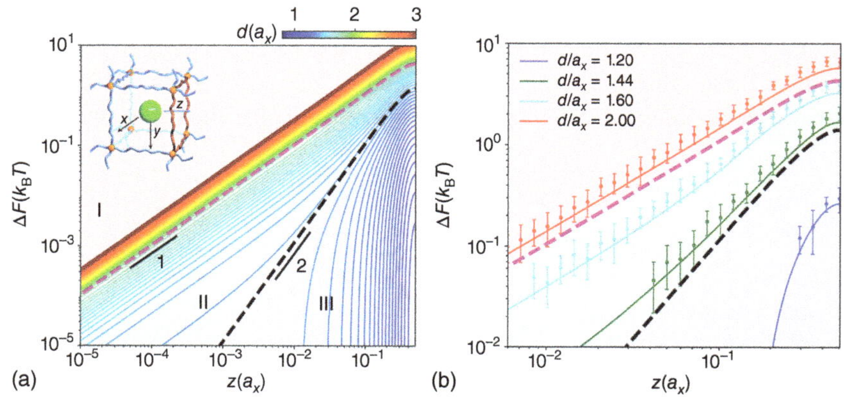

Figure 11.8 Novel scaling regimes of free energy change. (a) Dependence of $\Delta F(z)$ on the position z for various d in the log–log scale, obtained from the theoretical model of a particle-network system with $g = 6$. The upper left inset shows the axes of the system, where z-axis is along the minimum energy path (MEP). The dashed lines represent the theoretical boundaries separating different regimes. (b) The comparison between MC (points) and theoretical results (lines) for some representative particle sizes. The error bar indicates the standard error. Source: Reprinted with permission from Ref. [18]; © 2022, Springer Nature.

with z, specifically $\Delta F \sim z^{\eta}$, where η represents the scaling exponent and η equals 1 within this particular regime. As the nanoparticle size decreases, in the subsequent regime, η undergoes a crossover from 1 to 2. The phenomenon of crossover becomes more apparent as d increases. When d is further reduced, the profiles become discontinuous, wherein $\Delta F(z)$ initially remains at zero but then abruptly jumps to a significantly higher value. This sudden jump reminds us of the hindrance effect of diffusion within a polymer gel network, as observed in experimental studies [52]. When d reaches a sufficiently small value, the influence of the network cell becomes negligible, resulting in $\Delta F(z)$ remaining at zero throughout the entire path in the final regime. To visually present these four scaling regimes in a clear manner, we systematically compute the profiles of $\Delta F(z) \sim z$ for nanoparticles within network cells of varying g (Figure 11.9a), thereby consolidating the universal nature of these regimes. Prior investigations that primarily investigated the elastic deformation of local network loops have also hinted at the existence of Regime I [42, 47]. In contrast, our theoretical methodology enables a thorough exploration of the MEP, thereby uncovering the presence of Regimes II and III. Notably, upon comparing the distributions of these regimes for varying values of g, it becomes evident that larger g values correspond to smaller areas occupied by Regimes II and III. This observation emphasizes the connection between scaling regimes and cell topologies.

Expanding our analysis, we investigate the profiles of free energy at specific values of r_{out}, r_{mid}, and r_{in}. As illustrated schematically in Figures 11.7b and 11.9b, r_{out} can be interpreted as the critical particle size that triggers deformation of the vertices (cross-links) within the network cells. By setting $R = r_{\text{out}}$, we establish the boundary between Regimes I and II, represented by the magenta dashed curve. Similarly, r_{mid} can be regarded as the critical size that induces deformation of the edges (strands).

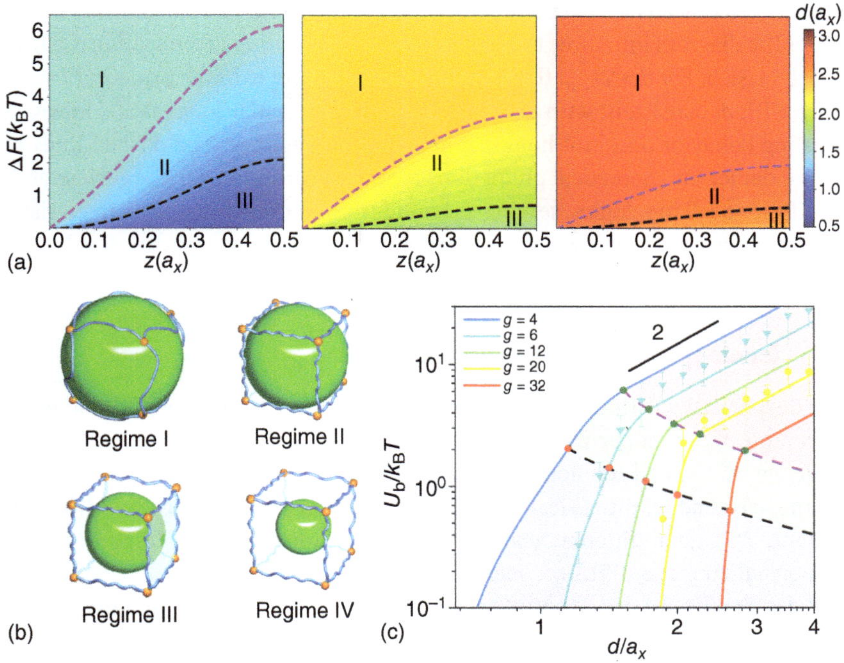

Figure 11.9 Topology-dictated scaling regimes of free energy change. (a) Heat map of $\Delta F \sim z$ for various d in network cells at $g = 4$ (left), 20 (middle), and 32 (right). The color bar indicates the values of the particle diameter d. The boundaries of Regimes I and II and Regimes II and III are represented by purple and black dashed lines, respectively. Specifically, the hidden Regime IV in each plot gives $\Delta F(z) = 0$. (b) Schematic of spherical particles with various sizes in a network cell for different regimes: Regime I: $R > r_{out}$, Regime II: $r_{mid} < R < r_{out}$, Regime III: $r_{in} < R < r_{mid}$, and Regime IV: $R < r_{in}$. (c) Free energy barrier of the particle between neighboring cells U_b against d for various cell topologies. MC results are also plotted for topologies of $g = 6$ (blue), and 20 (yellow). Source: Reprinted with permission from Ref. [18]; © 2022, Springer Nature.

When $R = r_{mid}$, the boundary between Regimes II and III can be defined by the black dashed curve. Consequently, the boundaries that separate these distinct regimes can be determined based on r_{in}, r_{mid}, and r_{out}, as depicted in Figure 11.9b. As expressed in Eq. (11.4), an increase in g leads to a decrease in the aspheric parameters r_{mid}/r_{in} and r_{out}/r_{in}, gradually approaching 1.0. This trend signifies a transition from anisotropic to isotropic behavior within the network cell. Consequently, the boundaries between $R = r_{out}$ and r_{mid}, as well as between $R = r_{mid}$ and r_{in}, converge, resulting in a contraction of Regimes II and III. In particular, in the case of a network cell with a very large g, where r_{in}, r_{mid}, and r_{out} are approximately equal, these boundaries are expected to overlap, causing Regimes II and III to vanish. This reverts the system to previous findings that focused on the isotropic deformation of a circular loop. This trend is also evident in the free energy barrier encountered by a nanoparticle during its migration from one cell to a neighboring cell, that is, U_b. In Figure 11.9c, we show the plots of U_b against d for various g, and fitting U_b at $d = 2r_{out}$ and

$2r_{mid}$ in each plot leads to two boundaries (magenta and black dashed lines) which separates the diagram into three typical regions corresponding to the scaling regimes in Figure 11.8a,b. For $d > 2r_{out}$, the hopping energy barrier exhibits a quadratic relationship with d, consistent with previous research on circular loops [42]. However, this scaling behavior does not hold true in the other regimes. The dashed lines in the figure depict the theoretical predictions for $U_b(2r_{out})$ and $U_b(2r_{mid})$ corresponding to different cell topologies, serving as boundaries that separate the characteristic regimes. Notably, as g increases, these two boundaries of the regimes approach each other asymptotically, aligning with the aforementioned findings.

11.3.4 Topology-Mediated Dynamical Regimes

To gain deeper insights into the underlying mechanism of the network cell topology, we delve into the nanoparticle dynamics within these regimes and examine their dependence on the cell topology. Utilizing the previously constructed free energy landscape, we conduct numerical simulations to obtain the nanoparticle's diffusion coefficient, D, under different combinations of d and g values. (Figure 11.10). Upon normalizing the diffusion coefficient, D, by its value at $2r_{in}$ and plotting it against d normalized by $2r_{in}$, an intriguing observation emerges. The nanoparticle diffusivities at various g values collapse onto a single master curve, exhibiting a power law slope of -3 for $d \leq 2r_{in}$. This finding indicates that the diffusivity is solely dictated by the local Rouse dynamics of the strand chains. (Figure 11.10a). Nevertheless, for nanoparticles of intermediate size ($2r_{in} < d < 2r_{out}$), the diffusion coefficient, D, undergoes a pronounced decrease. When $d \geq 2r_{out}$, the presence of a larger nanoparticle triggers a radial expansion of the cell vertices (as depicted in Figure 11.9b, Regime I). This radial dilation leads to an almost isotropic deformation of the cell, resembling the deformation observed in a circular loop. Consequently,

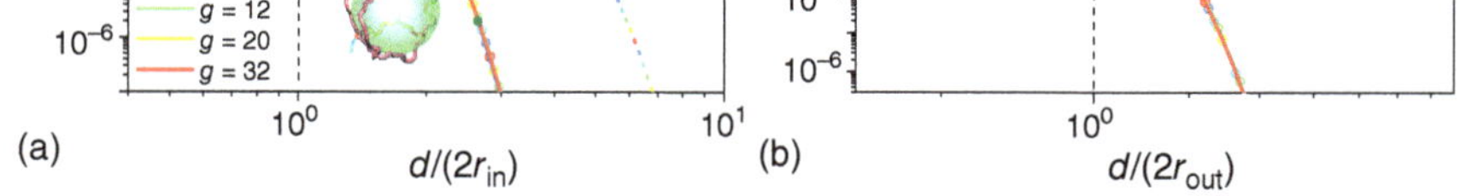

Figure 11.10 Dynamical regimes mediated by network topologies. (a) $D/D(2r_{in})$ against $d/2r_{in}$ for various network cell topologies in the log–log scale. Inset: schematic diagrams of the deformation based on a network cell (left) and the deformation based on a network loop (right). (b) $D/D(2r_{out})$ against $d/2r_{out}$ for various network cell topologies in the log–log and the log–linear (inset) scales. The scaling parameters r_{in} (purple), r_{mid} (green), and r_{out} (blue) are presented on each plot. Power law and exponential dependences on the ratio between d and $2r_{in}$ (dashed line) or $2r_{out}$ (circular scatter) are depicted. Source: Reprinted with permission from Ref. [18]; © 2022, Springer Nature.

the diffusivity transitions back to an exponential dependence, which is supported by the circular data points in Figure 11.10a,b, as well as the collapsed curves of $D/D(2r_{\text{out}}) \sim d/(2r_{\text{out}})$ in Figure 11.10b. Let us recall the previous investigations that concentrated on the localized deformation of a circular loop. It was found that the diffusion of small nanoparticles, normalized by the size of a_x, followed the dynamical regime of $D \sim (d/a_x)^{-3}$, while large nanoparticles exhibited the regime of $D \sim \exp(-d^2/a_x{}^2)$. In contrast, the appearance of the intermediate regime and the shift in the switching points from a_x to r_{in} and r_{out} for different regimes, as illustrated in Figure 11.10, emphasize the influence of cell topology on the dynamics of nanoparticle diffusion. For $d > 2r_{\text{in}}$, once the nanoparticle enters the regime illustrated in Figure 11.9b as Regime II, it begins to interact with the free energy landscape present on the faces of the network cell. This interaction activates the cell topology effect, leading to a departure from the exponential dependence observed in the diffusion dynamics. This deviation is represented by the colored dashed curves in Figure 11.10a.

11.4 Summary and Outlook

We introduce the advances in investigating the physical attributes of nanoparticle transport in macromolecular networks, focusing on the theoretical and simulation aspects of two typical intrinsic physical properties, that is, strand flexibility and cell topology. In particular, the enhanced heterogeneous diffusion of nanoparticles in the macromolecular networks with semiflexible loops is described firstly. Then, we introduce how the cell topology impacts the transport of nanoparticles in macromolecular networks. The results might bring insight into the fundamental physics of substance transport in the confined environments of networks constituted by widely biological or synthetic polymers, and could lead to advantageous applications in biological and biomedical aspects and beyond.

Although considerable advances have already been made in understanding how the physical attributes of macromolecular networks affect the nanoparticle transport in them, the theoretical aspect of this field is still in its infancy. At present, it is clear that some physical attributes, such as the chain flexibility of the strands and the topology of network cell, profoundly impact the dynamical mechanisms of the nanoparticle transport. Many other physical properties including the size and shape of nanoparticles will also significantly impact their dynamical behaviors. Another key factor mediating the nanoparticle transport is the deficiency of the networks which can evidently modify the molecular architectures of the macromolecular networks. Clarifying the deficiency effect in the nanoparticle transport is essential for understanding the nanoparticle transport in a real network and may lead to new physics regarding the nanoparticle transport the environments of networks. The formulated theoretical approaches can serve as a foundation for further exploration of physical effects on the dynamic behavior in various networks, synthetic or biological. We believe that this work will certainly stimulate new efforts into the aforementioned promising topics of interest to physicists and materials scientists.

Acknowledgments

The authors thank Lijun Dai, Duo Xu, Haixiao Wan, Xueqing Jin, Wenlong Chen, and Wenjie Wei for helpful discussions. This work is supported by the National Natural Science Foundation of China under grant nos. 22025302 and 21873053.

References

1 Gu, Y., Zhao, J., and Johnson, J.A. (2019). A (macro) molecular-level understanding of polymer network topology. *Trends in Chemistry* 1 (3): 318–334.
2 Peppas, N.A., Hilt, J.Z., Khademhosseini, A., and Langer, R. (2006). Hydrogels in biology and medicine: from molecular principles to bionanotechnology. *Advanced Materials* 18 (11): 1345–1360.
3 Yang, J., Bai, R., Chen, B., and Suo, Z. (2020). Hydrogel adhesion: a supramolecular synergy of chemistry, topology, and mechanics. *Advanced Functional Materials* 30 (2): 1901693.
4 Ramarad, S., Khalid, M., Ratnam, C.T. et al. (2015). Waste tire rubber in polymer blends: a review on the evolution, properties and future. *Progress in Materials Science* 72: 100–140.
5 Lee, K.Y. and Mooney, D.J. (2001). Hydrogels for tissue engineering. *Chemical Reviews* 101 (7): 1869–1880.
6 Ionov, L. (2014). Hydrogel-based actuators: possibilities and limitations. *Materials Today* 17 (10): 494–503.
7 Liu, X., Liu, J., Lin, S., and Zhao, X. (2020). Hydrogel machines. *Materials Today* 36: 102–124.
8 Yang, C. and Suo, Z. (2018). Hydrogel ionotronics. *Nature Reviews Materials* 3 (6): 125–142.
9 Lieleg, O. and Ribbeck, K. (2011). Biological hydrogels as selective diffusion barriers. *Trends in Cell Biology* 21 (461): 543–551.
10 Chaudhuri, O., Cooper-White, J., Janmey, P.A. et al. (2020). Effects of extracellular matrix viscoelasticity on cellular behaviour. *Nature* 584 (7822): 535–546.
11 Molloy, S. (2013). Biofilm microanatomy. *Nature Reviews Microbiology* 11 (5): 301–301.
12 Schmidt, H.B. and Görlich, D. (2016). Transport selectivity of nuclear pores, phase separation, and membraneless organelles. *Trends in Biochemical Sciences* 41 (1): 46–61.
13 Danielsen, S.P.O., Beech, H.K., Wang, S. et al. (2021). Molecular characterization of polymer networks. *Chemical Reviews* 121 (8): 5042–5092.
14 Bustamante, C., Marko, J.F., Siggia, E.D., and Smith, S. (1994). Entropic elasticity of λ-phage DNA. *Science* 265 (5178): 1599–1600.
15 Broedersz, C.P. and MacKintosh, F.C. (2014). Modeling semiflexible polymer networks. *Reviews of Modern Physics* 86 (3): 995.
16 Chaikin, P.M. and Lubensky, T.C. (1995). *Principles of Condensed Matter Physics*. Cambridge: Cambridge University Press.

17 Xu, Z., Dai, X., Bu, X. et al. (2021). Enhanced heterogeneous diffusion of nanoparticles in semiflexible networks. *ACS Nano* 15 (3): 4608–4616.

18 Dai, X., Zhang, X., Gao, L. et al. (2022). Topology mediates transport of nanoparticles in macromolecular networks. *Nature Communications* 13 (1): 4094.

19 Lin, Y.-C., Yao, N.Y., Broedersz, C.P. et al. (2010). Origins of elasticity in intermediate filament networks. *Physical Review Letters* 104: 058101.

20 Yao, N.Y., Broedersz, C.P., Depken, M. et al. (2013). Stress-enhanced gelation: a dynamic nonlinearity of elasticity. *Physical Review Letters* 110: 018103.

21 Zhu, G., Huang, Z., Xu, Z., and Yan, L.-T. (2018). Tailoring interfacial nanoparticle organization through entropy. *Accounts of Chemical Research* 51: 900–909.

22 Fletcher, D.A. and Mullins, R.D. (2010). Cell mechanics and the cytoskeleton. *Nature* 463: 485–492.

23 Lieleg, O., Claessens, M.M.A.E., Luan, Y., and Bausch, A.R. (2008). Transient binding and dissipation in cross-linked actin networks. *Physical Review Letters* 101: 108101.

24 Theocharis, A.D., Skandalis, S.S., Gialeli, C., and Karamanos, N.K. (2016). Extracellular matrix structure. *Advanced Drug Delivery Reviews* 97: 4–27.

25 Wälde, S. and Kehlenbach, R.H. (2010). The part and the whole: functions of nucleoporins in nucleocytoplasmic transport. *Trends in Cell Biology* 20: 461–469.

26 Yong, X., Kuksenok, O., Matyjaszewski, K., and Balazs, A.C. (2013). Harnessing interfacially-active nanorods to regenerate severed polymer gels. *Nano Letters* 13: 6269–6274.

27 Raos, G. and Casalegno, M. (2011). Nonequilibrium simulations of filled polymer networks: searching for the origins of reinforcement and nonlinearity. *The Journal of Chemical Physics* 134: 054902.

28 Nikunen, P., Vattulainen, I., and Karttunen, M. (2007). Reptational dynamics in dissipative particle dynamics simulations of polymer melts. *Physical Review E* 75: 036713.

29 Nikoubashman, A., Vega, D.A., Binder, K., and Milchev, A. (2017). Semiflexible polymers in spherical confinement: bipolar orientational order versus tennis ball states. *Physical Review Letters* 118: 217803.

30 Yan, L.-T., Popp, N., Ghosh, S.-K., and Böker, A. (2010). Self-assembly of Janus nanoparticles in diblock copolymers. *ACS Nano* 4: 913–920.

31 Hansen, J. and McDonald, I. (2006). *Theory of Simple Liquids*. London: Elsevier Academic Press.

32 Poling-Skutvik, R., Krishnamoorti, R., and Conrad, J.C. (2015). Size-dependent dynamics of nanoparticles in unentangled polyelectrolyte solutions. *ACS Macro Letters* 4: 1169–1173.

33 Burla, F., Sentjabrskaja, T., Pletikapic, G. et al. (2020). Particle diffusion in extracellular hydrogels. *Soft Matter* 16: 1366–1376.

34 Kang, K., Gapinski, J., Lettinga, M.P. et al. (2005). Diffusion of spheres in crowded suspensions of rods. *The Journal of Chemical Physics* 12: 044905.

35 Wulstein, D.M., Regan, K.E., Garamella, J. et al. (2019). Topology-dependent anomalous dynamics of ring and linear DNA are sensitive to cytoskeleton crosslinking. *Science Advances* 5: eaay5912.

36 Li, B., Lou, K., Kob, W., and Granick, S. (2020). Anatomy of cage formation in a two-dimensional glass-forming liquid. *Nature* 587: 225–229.

37 Song, S., Park, S.J., Kim, M. et al. (2019). Transport dynamics of complex fluids. *Proceedings of the National Academy of Sciences of the United States of America* 116: 12733–12742.

38 Debenedetti, P.G. and Stillinger, F.H. (2001). Supercooled liquids and the glass transition. *Nature* 410: 259–267.

39 Michel, E., Filali, M., Aznar, R. et al. (2000). Percolation in a model transient network: rheology and dynamic light scattering. *Langmuir* 16: 8702–8711.

40 Montroll, E.W. and Weiss, G.H. (1965). Random walks on lattices. II. *Journal of Mathematical Physics* 6 (2): 167–181.

41 Laidler, K.J. (1969). *Theories of Chemical Reaction Rates*. New York: McGraw-Hill.

42 Cai, L.-H., Panyukov, S., and Rubinstein, M. (2015). Hopping diffusion of nanoparticles in polymer matrices. *Macromolecules* 48: 847–862.

43 Darmanis, S., Sloan, S.A., Zhang, Y. et al. (2015). A survey of human brain transcriptome diversity at the single cell level. *Proceedings of the National Academy of Sciences* 112 (23): 7285–7290.

44 Furukawa, H., Cordova, K.E., O'Keeffe, M., and Yaghi, O.M. (2013). The chemistry and applications of metal-organic frameworks. *Science* 341 (6149): 1230444.

45 DeForest, C.A. and Anseth, K.S. (2011). Cytocompatible click-based hydrogels with dynamically tunable properties through orthogonal photoconjugation and photocleavage reactions. *Nature Chemistry* 3 (12): 925–931.

46 Bailey, E.J. and Winey, K.I. (2020). Dynamics of polymer segments, polymer chains, and nanoparticles in polymer nanocomposite melts: a review. *Progress in Polymer Science* 105: 101242.

47 Dell, Z.E. and Schweizer, K.S. (2014). Theory of localization and activated hopping of nanoparticles in cross-linked networks and entangled polymer melts. *Macromolecules* 47 (1): 405–414.

48 Deam, R.T. and Edwards, S.F. (1976). The theory of rubber elasticity. *Philosophical Transactions of the Royal Society of London. Series A, Mathematical and Physical Sciences* 280 (1296): 317–353.

49 Schmid, F. (2013). Self-consistent field approach for cross-linked copolymer materials. *Physical Review Letters* 111 (2): 028303.

50 Torquato, S. and Jiao, Y. (2009). Dense packings of the Platonic and Archimedean solids. *Nature* 460 (7257): 876–879.

51 Ting, C.L., Appelö, D., and Wang, Z.G. (2011). Minimum energy path to membrane pore formation and rupture. *Physical Review Letters* 106 (16): 168101.

52 Amsden, B. (1999). An obstruction-scaling model for diffusion in homogeneous hydrogels. *Macromolecules* 32 (3): 874–879.

Index

Dynamics and Transport in Macromolecular Networks: Theory, Modeling, and Experiments, First Edition. Edited by Li-Tang Yan.

r

s

t

u

v

www.ingramcontent.com/pod-product-compliance
Ingram Content Group UK Ltd.
Pitfield, Milton Keynes, MK11 3LW, UK
UKHW051035091026
15624UKWH00011B/439

9 783527 35098